SYNGRESS®

Business Continuity & Disaster Recovery

for IT Professionals

Susan Snedaker, MCSE, MCT

KEY	SERIAL NUMBER
001	HJIRTCV764
002	PO9873D5FG
003	829KM8NJH2
004	QX34TYP89Z
005	CVPLQ6WQ23
006	VBP965T5T5
007	HJJJ863WD3E
008	2987GVTWMK
009	629MP5SDJT
010	IMWQ295T6T

PUBLISHED BY
Syngress Publishing, Inc.
Elsevier, Inc.
30 Corporate Drive
Burlington, MA 01803

Business Continuity and Disaster Recovery Planning for IT Professionals

Printed in the United States of America
  3 4 5 6 7 8 9 0

ISBN 13: 978-1-59749-172-3

Publisher: Amorette Pedersen   Cover Designer: Michael Kavish
Project Manager: Anne McGee   Page Layout and Art: Patricia Lupien
Copy Editor: Adrienne Rebello   Indexer: Richard Carlson

For information on rights, translations, and bulk sales, contact Matt Pedersen, Commercial Sales Director and Rights, at Syngress Publishing; email m.pedersen@elsevier.com.

Acknowledgments

Thanks first to Syngress for publishing this book and for the efforts of Amy Pedersen and all the others behind the scenes in helping shepherd this project to completion. Thanks to my friend and colleague, and a brilliant attorney, Deanna Conn, for contributing a piece on legal aspects of data security within the context of business continuity and disaster recovery (BC/DR) planning. Thanks also go out to my good friend and colleague, Patty Hoenig, whose expertise in crisis communications will help you understand and address the challenges of communicating effectively during and after a crisis. Additional thanks to my friend and colleague Nels Hoenig for reviewing sections of the material and providing invaluable feedback along the way. Thanks to my colleague Debbie Earnest who shared her expertise and experience (bumps and bruises) garnered from working in the field of BC/DR. Her advice and comments helped me in crafting this book from start to finish. Her contribution offers time-tested techniques for overcoming some of the common challenges to BC/DR planning. Last but not least, many thanks to the voice of reason, my cohort and draft reviewer, Lisa Mainz, for reading through early versions of chapters and helping me find the clearest, most direct path through the material.

Finally, thanks to my long-time friend, Shirley, who passed away just after I started writing this book. She had recently reminded me of this oft-quoted phrase: *Hope for the best, plan for the worst. Nothing will ever surprise you.*

About the Author

Susan Snedaker, Principal Consultant and founder of Virtual Team Consulting, LLC has over 20 years experience working in IT in both technical and executive positions including with Microsoft, Honeywell, and Logical Solutions. Her experience in executive roles at both Keane and Apta Software provided extensive strategic and operational experience in managing hardware, software and other IT projects involving both small and large teams. As a consultant, she and her team work with companies of all sizes to improve operations, which often entails auditing IT functions and building stronger project management skills, both in the IT department and company-wide. She has developed customized project management training for a number of clients and has taught project management in a variety of settings. Ms. Snedaker holds a Masters degree in Business Administration (MBA) and a Bachelors degree in Management. She is a Microsoft Certified Systems Engineer (MCSE), a Microsoft Certified Trainer (MCT), and has a certificate in Advanced Project Management from Stanford University. She recently completed an Executive program in International Management at Thunderbird University's Garvin School of International Management.

Contents

Introduction

Let's start with the obvious. Business continuity and disaster (BC/DR) planning is not the most uplifting topic.

Few people want to spend their day (or even an hour) in BC/DR planning, but like brushing or flossing your teeth, the payoff far exceeds the investment of time and effort. In today's environment, where technology reaches into every corner of almost every organization, BC/DR planning has become imperative. Unfortunately, it falls very low on a long list of IT priorities. By the time BC/DR hits the radar screen, most IT staff are already over-utilized and overwhelmed.

The one statistic that should remind you of the importance of creating and maintaining a BC/DR plan is this: In a study of companies that experienced a major data loss without having a solid BC/DR plan in place, 43% never reopened, 51% closed within two years, and only 6% survived long-term. (see Chapter 1 for more on this). Let's repeat that: 6% survive long-term. If you're reading this foreword, it means you've at least thought about this. The good news is that you can dramatically improve your odds of your company surviving a major disaster by creating a BC/DR plan. The purpose of this book is to provide a framework within which you can develop an effective BC/DR plan for your company. It's targeted at small and medium-sized businesses, though it can easily be used in larger companies.

We'll cover the important elements of BC/DR, point you to additional resources and provide some real-world advice that you can put to use immediately. The book is intended to be scalable to fit your needs. If you've avoided creating a BC/DR plan because your organization tends toward "just-in-time planning" (also known as "seat of your pants planning"), this book will help you by giving you the bottom line and the minimum requirements, whenever possible. If you and your organization are very detail-oriented and you were looking for the right framework to use, this book will help you develop one suited to the unique needs of your company.

This book adheres to industry standards and best practices, but it will not prepare you for a formal certification in business continuity or disaster recovery planning. It's not exhaustive on any of the BC/DR topics, either. If you're looking for an extremely detailed, comprehensive, exhaustive (and exhausting) look at BC/DR, look elsewhere. If you're looking for a fast, effective framework that you can actually use, this book is for you. It is a roll-up-your-sleeves-and-get-your-hands-dirty kind of book meant to be used to quickly step you through the process of creating an effective BC/DR plan with the least effort.

If you've created a plan in the past, you can use this book to make sure your plan is comprehensive and up-to-date. If you don't yet have a plan, you can use this to get going. In either case, you will find this book to be a great resource that you pick up time and again or that you hand over, dog-eared and highlighted, to your successor (as you move up the career ladder).

My commitment to you – to keep the gloom and doom to a minimum and to make the book as interesting and useful as possible. We've delayed the inevitable this long, so let's go ahead and take the plunge.

Chapter 1

Business Continuity and Disaster Recovery Overview

Solutions in this chapter:

- **Business Continuity and Disaster Recovery Defined**
- **Components of Business**
- **The Cost of Planning versus the Cost of Failure**
- **Types of Disasters to Consider**
- **Business Continuity and Disaster Recovery Planning Basics**

☑ **Summary**

☑ **Solutions Fast Track**

☑ **Frequently Asked Questions**

Introduction

Powerful Earthquake Triggers Tsunami in Pacific. Hurricane Katrina Makes Landfall in the Gulf Coast. Avalanche Buries Highway in Denver. Tornado Touches Down in Georgia. These headlines not only have caught the attention of people around the world, they have had a significant effect on IT professionals as well. As technology continues to become more integral to corporate operations at every level of the organization, the job of IT has expanded to become almost all-encompassing. These days, it's difficult to find corners of a company that technology does not touch. As a result, the need to plan for potential disruptions to technology services has increased exponentially. Business continuity and disaster recovery (BC/DR) plans were certainly put to the test by many financial firms after the terrorist attacks in the United States on September 11, 2001; but even six years later, there are many firms that still do not have any type of business continuity or disaster recovery plan in place. It seems insane not to have such a plan in place, but statistics show that many companies don't even have solid data backup plans in place. Given the enormous cost of failure, why are many companies behind the curve? The answers are surprisingly simple. Lack of time and resources. Lack of a sense of urgency. Lack of a process for developing and maintaining a plan. This book will help you overcome some of those challenges.

A study released by Harris Interactive, Inc. in September 2006 indicated that 39% of CIOs who participated in the survey lacked confidence in their disaster readiness. There's good news and bad news here. The bad news, clearly, is the fairly high lack of confidence in disaster plans in firms with revenues of $500M or more annually. The good news is that only 24% of CIOs in 2004 felt their disaster plans were inadequate. Although the *increase* in lack of confidence may appear to be a negative, it also highlights the increasing awareness of the need for comprehensive disaster readiness and a more complete understanding of what that entails. Back in 2000, some companies might have thought a "good" disaster readiness plan was having off-site backups. After the terror attacks, bombings, anthrax incidents, hurricanes, and floods that hit the United States (and other major incidents worldwide) since that time, most IT professionals now understand that off-site backups is just a small part of an overall strategy for disaster recovery.

In today's environment, no company can afford to ignore the need for BC/DR planning, regardless of the company size, revenues, or number of staff. The statistics on the failure rate of companies after a disaster are alarming (discussed later in this chapter) and alone should serve as a wake up call for IT professionals and corporate executives. Granted, the cost of planning must be proportionate to the cost of failure, which we'll address throughout this book.

Let's face it—very few of us want to spend the day thinking about all the horrible things that can happen in the world and to our company. It's not a cheery subject, one that most of us would rather avoid—which also helps explain the glaring lack of BC/DR plans in many small and medium companies (and a share of large companies as well). Stockholders of publicly held companies are increasingly demanding well thought-out BC/DR plans internally as well as from key vendors, but in the absence of this pressure, many companies expend their time and resources elsewhere. Business continuity and disaster recovery planning projects have to compete with other urgent projects for IT dollars. Unless you can create a clear, coherent, and compelling business case for BC/DR, you may find strong executive resistance at worst or apathy at best.

You may wonder why you should have to champion this cause and push for some sort of budget or authorization to create such a plan. The truth is that you shouldn't, but since a disaster will probably have a disproportionately high impact on the IT department, it's very much in your own self-interest to try to get the OK to move forward with a planning project.

In this chapter, we'll look at some of the impediments to BC/DR planning as well as some of the compelling reasons why spending time, money, and staff hours on this is well worth the expenditure. We'll provide you with specific, actionable data you can use to convince your company's executive or management team to allocate time and resources to this project. We'll also look at the different types of disasters that need to be addressed—they're not all obvious at first glance. Finally, we'll provide a framework for the rest of the book and for your BC/DR planning.

Business Continuity and Disaster Recovery Defined

Before we go too far, let's take a moment to define *business continuity* and *disaster recovery*. These two labels often are used interchangeably, and though there are overlapping elements, they are not one and the same. *Business continuity planning* (BCP) is a methodology used to create and validate a plan for maintaining continuous business operations before, during, and after disasters and disruptive events. In the late 1990s, BCP came to the forefront as businesses tried to assess the likelihood of business systems failure on or after January 1, 2000 (the now infamous "Y2K" issue). BCP has to do with managing the operational elements that allow a business to function normally in order to generate revenues. It is often a concept that is used in evaluating various technology strategies. For example, some companies cannot tolerate any downtime. These include financial institutions, credit card processing companies and perhaps some high volume online retailers. They may decide that the cost for fully redundant systems is a worthwhile investment because the cost of downtime for even five or ten minutes could cost millions of dollars. These companies require their businesses run continuously, and their overall operational plans reflect this priority. Business continuity has to do with keeping the company running, regardless of the potential risk, threat, or cause of an outage.

Continuous availability is a subset of business continuity. It's also known as a zero-downtime requirement, and is extremely expensive to plan and implement. For some companies, it may be well worth the investment because the cost of downtime outweighs the cost of implementing continuous availability measures. Other companies have a greater tolerance for business disruption. A brick-and-mortar retailer, for example, doesn't necessarily care if the systems are down overnight or during nonbusiness hours. Although it may be an inconvenience, a retailer might also be able to tolerate critical system outages during business hours. Granted, every business that relies on technology wants to avoid having to conduct business without that technology. Every business that relies on technology will be inconvenienced and disrupted to some degree to have to conduct business without that technology. The key driver for business continuity planning is how much of a disruption to your business is tolerable and what are you able and willing to spend to avoid disruption. If money were no issue, every business using technology would probably elect to implement fully redundant, zero-downtime systems. But money *is* an issue. A small retailer or even a small online company can ill afford to spend a million dollars on fully redundant systems when their revenue stream for the year is $5 to $10 million. The cost of a business disruption for a company of that size might be $25,000 or even $100,000 and would not justify a million dollar investment. On the other hand, a million dollar investment in fully redundant systems for a company doing $5 billion annually might be worth it, especially if the cost of a single disruption would cost more than $1 million. As previously mentioned, your BC/DR plan must be appropriate to your organization's size, budget, and other constraints. In later chapters, we'll look at how to assess the cost of disruption to your operations so you can determine the optimal mitigation strategies.

Disaster recovery is part of business continuity, and deals with the immediate impact of an event. Recovering from a server outage, security breach, or hurricane all fall into this category. Disaster recovery usually has several discreet steps in the planning stages, though those steps blur quickly during implementation because the situation during a crisis is almost never exactly to plan. Disaster recovery involves stopping the effects of the disaster as quickly as possible and addressing the immediate aftermath. This might include shutting down systems that have been breached, evaluating which systems are impacted by a flood or earthquake, and determining the best way to proceed. At some point during disaster recovery, business continuity activities begin to overlap, as shown in Figure 1.1. Where to set up temporary systems, how to procure replacement systems or parts, how to set up security in a new location—all are questions that relate both to disaster recovery and business continuity, but which are primarily focused on continuing business operations. Figure 1.1 shows the cycle of planning, implementation, and assessment that is part of the ongoing BC/DR maintenance cycle. We'll discuss this in more detail later, but it's important to understand how the various elements fit together at the outset.

Figure 1.1 Business Continuity and Disaster Recovery Planning, Implementation, and Revision Cycle

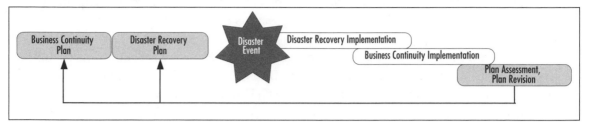

Components of Business

There are many ways to break down the elements of business, but for the purposes of BC/DR planning, we'll use three simple categories: people, process, and technology. As an IT professional, you understand the importance of the interplay among these three elements. Technology is implemented by people using specific processes. The better defined the processes are, the more reliable the results (typically). Technology is only as good as the people who designed and implemented it and the processes developed to utilize it. As we discuss BC/DR planning throughout this book, we'll come back to these three elements repeatedly. When planning for BC/DR, then, we have to look at the people, processes, and technology of the BC/DR planning itself as well as the people, processes, and technology of the plan's implementation (responding to an emergency or disaster). Let's look at each of the three elements in this light. Figure 1.2 depicts the relative relationship of people, process, and technology in most companies. Infrastructure is part of the technology component, but is listed separately for clarity.

Figure 1.2 How People, Process, and Technology Interact

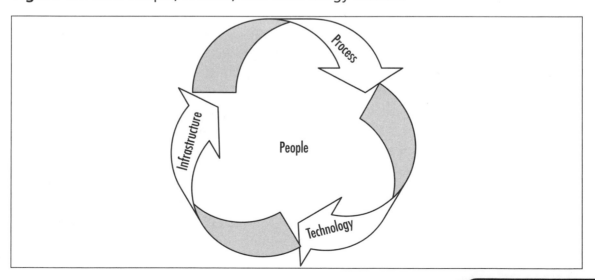

People in BC/DR Planning

Clearly, people are the ones who do the actual planning and implementation of a business continuity and disaster plan, but there are many aspects to the people element that often are overlooked during the planning process. In this section, we'll look at a few of the commonly missed elements. However, as you read through this, keep your own organization in mind. Every company is different and therefore every BC/DR planning process will have to be different. A small retail outlet's IT planning for BC/DR will be very different from a call center or a manufacturing facility. There is no "one size fits all" approach, so although we can point out the major elements, you'll need to fill in the specifics for your company.

Let's begin with one very interesting fact. According to a recent IBM white paper, 80% of all data loss is human-caused. That's the *people* part of the equation. People are responsible for designing, implementing, and monitoring processes intended to safeguard data. However, people make mistakes every single day. As one National Transportation Safety Board official put it when interviewed about a plane crash, there are multiple layers of systems in place to ensure the plane doesn't crash, but sometimes a series of bad choices or errors leads to a critical event. The same is true with your IT infrastructure. Hopefully, there are multiple layers of processes, procedures, and cross-checks in place to prevent human-caused disasters, but sometimes these fail. If 80% of data loss is attributable to human error, that leaves 20% of data loss attributable to other causes such as equipment malfunction, natural disasters, and terrorism (which is in the same general category of "human-caused" but at a different level altogether).

We'll discuss the specific steps needed to form your BC/DR plan later in this chapter and in subsequent chapters. Now, though, let's look at some general guidelines. Your BC/DR plan requires people from across your organization in order to be effective. As an IT professional, you may know who has which laptop and how applications are secured across the network, but you very likely have no idea how things run, on a day-to-day basis, in other parts of the company. You may not know what data, what processes, what parts of the technology puzzle are critical to various departments. You certainly will not know critical dates, key milestones, or other information that people in other departments know. To create a plan without input from across the company almost guarantees the plan will fail—if not during the planning stage then certainly in the implementation stage. Getting key people in the company to participate in the planning helps you develop a more robust plan and, just as important, helps you identify the key people needed to implement the plan, should that become necessary.

Another key aspect to people in BC/DR planning is that it's critical to remember that if a disaster hits your company, people will have a wide variety of responses. Some people, especially those with emergency preparedness training, will rise to the occasion and start taking effective action through leadership roles. Others will be completely overwhelmed and unable to act effectively (or at all). Understanding this is important when creating your

BC/DR plan because it will not be "business as usual" when an emergency hits. Emotional and physical stress may reduce effectiveness of even the most prepared individuals, so working with this assumption will help ensure a successful plan and more importantly, a successful outcome when the plan needs to be implemented.

As an IT professional, it may be that you do not have primary responsibility for your company's BC/DR planning. That said, you may be the only person in the company that recognizes the need for this type of planning. Therefore, you may have to champion the cause and rally resources to get the planning going. If you're a senior manager in a small or medium-sized firm, you may, in fact, be the go-to resource for both the planning and implementation of a BC/DR plan. Regardless of your role, we will discuss the broader implications of BC/DR throughout so that you can either include them yourself or ensure that others in the organization are including them. Our objective is to help you create an effective BC/DR plan for IT, but that cannot be accomplished in a vacuum. It will need to be integrated across the organization in order to be effective when it counts—when things go wrong.

Process in BC/DR Planning

Process in BC/DR planning also has two phases: the planning phase and the implementation phase. The processes your company uses to run the day-to-day business are key to the long-term success of the business. These processes were developed (and hopefully documented) in order to manage the recurring business tasks. Things outside the normal recurring tasks typically are handled as exceptions until they recur often enough to create a new process, and the cycle continues. If your business is suddenly hit by a disaster—fire, flood, earthquake, chemical spill—your processes are immediately interrupted. How quickly you recover from this and either re-implement or re-engineer your processes to get the business up and running again relies on the processes delineated in your BC/DR plan. By developing a process for handling various types of emergencies and disasters, you can rely on these when people are stressed and business is interrupted. Trying to develop effective processes in the face of an emergency is usually not at all successful. Having simple, well-tested processes to rely on when disaster strikes is often the difference between eventual recovery and business failure.

As you'll see later in this book, the processes used by the company in day-to-day operations need to be evaluated and prioritized. What processes are critical to the ability of the company to conduct operations? What processes can be put on hold during an emergency? Circumstances surrounding the emergency certainly come into play—time of year, where you are in various business cycles, and so on. When looking at your payroll process during an emergency, for example, you'll also need to understand the normal timing of these processes within the company. A power outage right *after* payroll is processed may be far less critical than a power outage just before payroll is processed. As we look at processes within the company, we'll keep these kinds of timing issues in mind. However, this is another justification for having a wide array of interests represented during the BC/DR planning phases, so you can evaluate these aspects and factor them in appropriately. Let's look at an example from the Human Resources department. In Figure 1.3, you can see a portion of a simple flowchart that HR could construct to assist both IT and HR in the aftermath of a disaster.

Figure 1.3 Sample Human Resources Process Flowchart

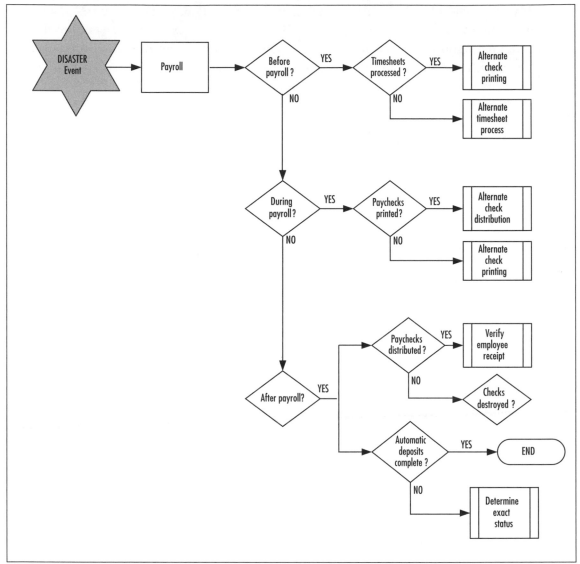

As you can see in Figure 1.3, there are defined steps in your company's payroll process. These steps become the framework for a decision flowchart to help HR staff determine what steps need to be taken in the aftermath of a significant event with regard to payroll processing. The first step is to determine the exact status of payroll—did the disaster hit before, during, or after payroll? Then, depending on the status, what would be the appropriate steps to take and how can these steps be taken if key systems are down? Although you might think that payroll should be the least of your company's concerns in the immediate aftermath of a disaster, your company's employees will think otherwise. They may need to seek alternate accommodations such as staying in a nearby hotel or they may need to purchase food, medical supplies, or transportation. They may be relying on that very paycheck in order to provide them adequate funds to pay rent or eat that week. Without addressing payroll needs, your company will be unnecessarily increasing the stress levels for all employees, even those who may not be dependent on receiving those funds immediately. Perhaps more importantly, this issue might not matter on the first day or two after an event, but what happens if your company's building was destroyed in a fire and it will be weeks before you resume normal operations?

This procedure clearly helps HR understand the current process they use and what processes may be needed in the event of a minor, major, or catastrophic event. It might also help them see ways to improve processes in their current day-to-day operations since few of us ever take the time to map out key processes. You don't need to use flowcharts, though they do provide a good visual, but you do need to find some standardized method of evaluating processes and creating contingency plans. We'll discuss this later in the book in more detail.

Technology in BC/DR Planning

Technology is clearly the piece of the puzzle that you, as an IT professional, will be most familiar with. As you participate in your company's BC/DR planning (or head it up, as previously mentioned), you will be in the best position to understand what happens with various technology components during different types of disasters. Part of the reason for BC/DR planning is to look at your use of technology and understand which elements are vulnerable to which types of disasters. A power outage, for example, impacts all the technology in a building. Suppose you have battery backup or generators for lights and certain computers but no power for air conditioning in Miami in July? Timing and circumstance come into play and working closely with your facilities person, for example, will help you look at the plan in a more holistic (and realistic) manner than you might on your own.

As we look at BC/DR planning, we'll also look at various vulnerabilities of different technologies and discuss, in broad strokes, strategies, tools, and techniques that might be helpful to mitigate or avoid some of these risks. We won't delve into specific technology solutions as those are ever-evolving, but we will look at common methods used today and what needs to be considered as you look at your unique circumstances. In some cases, your BC/DR planning may yield information you can use to make the business case for why the firm should authorize the purchase of a particular technology or service. For example, if you've been trying to get funding approved for collocation services to speed up user access to critical business data across a wide geographic area, you can use the results of your BC/DR planning to add to the business case. Clearly, collocation can be part of a solid business operations management strategy and can also be an integral part of a business continuity and disaster recovery plan. When you can add strength to your business case, you're more likely to find executive support for funding.

As an IT professional, you will need to work closely with members of other departments to understand the technology needs in an emergency—not only what technology is needed to get the business back up and running (business continuity) but also what is needed to manage the crisis. These are two distinct (but probably overlapping) concerns that should be assessed and addressed by your plan.

Looking Ahead...

Business Continuity and Disaster Recovery Planning Resources

There are numerous organizations, worldwide, that focus on business continuity planning and/or disaster recovery planning. Many of these organizations provide training, methodologies, and certification tracks. For anyone interested in becoming a focused specialist in one of these areas, you would do well to investigate these various organizations. If you're involved with business continuity or disaster recovery planning and want to stay current on the latest trends from the field, be sure to bookmark a few of these sites. We've listed just a few here, but a quick Internet search will yield more resources. Please keep in mind, as with any URL listed in this book, Web sites and URLs can change.

> The Business Continuity Institute (UK): www.thebci.org/
>
> DRI International (US): /www.drii.org/DRII/index.htm
>
> GlobalContinuity.com (South Africa): www.globalcontinuity.com/
>
> Department of Homeland Security Business Readiness (US): www.ready.gov/business/index.html
>
> Disaster Recovery Journal (US): www.drj.com/

The Cost of Planning versus the Cost of Failure

Companies typically look at their "top" line and their "bottom" line. Top line is revenue, and many publicly held companies chase after top-line growth, meaning they want to aggressively increase revenues. This often means they are grabbing a larger share of the market or are pushing the market to expand. It does not, however, account for the cost of doing so. If you pick up another $100 worth of business but it costs you $125 to do so, you may have top-line growth, but your bottom line (profitability) will suffer. In some cases, this makes sense in the short term—you can capture market share that becomes profitable at some later point in time. Other companies look just for bottom-line growth—revenues minus expenses (and other things) equals profit—so if a company's revenues minus expenses is greater than last year's, it means that the company has generated a larger profit (generally speaking). However, if your company is losing market share and lays off three-quarters of the workforce and closes four locations, things are not going well, even if you end up with short-term bottom line growth. Therefore, most companies look for a balance between top and bottom line growth.

You might be wondering what all this has to do with BC/DR planning, so let's connect the dots. The cost of planning might be significant in terms of staff time, resources, and the like, and might impact your bottom line (depending on many factors). If your company is concerned only with top line growth, they may not be overly concerned with the cost of a BC/DR project plan. You may also find that key customers desire or demand that your company have such a plan, so you might argue that creating this plan could contribute to top-line growth. If you're able to capture a new customer because you have a BC/DR plan, that's clearly going to help your case. On the other hand, if you work for a company strictly concerned with bottom-line growth, you may have a much bigger challenge. You can certainly see if having such a plan would improve operational efficiencies or land you a new client. Short of that, you might have to point out the potential hit to the bottom line if you experienced a disaster without a BC/DR plan in place. However, you can be sure that failure to mitigate the impact of a disaster will absolutely impact both your top and bottom lines, and will likely put your company's very existence in peril. Therefore, when you compare the cost of planning to the cost of failure, there is only one approach that makes business sense—and that is to plan to the extent it makes financial sense to do so.

Disasters can result in enormous losses—financial, investor confidence, and corporate image. It can also lead to serious legal issues, especially when more and more private data is being captured, stored, and transmitted across the very public Internet. These losses and legal challenges can have a small, short-term impact but more often than not, they have a significant, long-term impact, and in some cases imperil the existence of the company. For more information on the legal implications of disasters and data security, be sure to read the case study that follows this chapter.

In companies that do have some sort of disaster plan in place, it more than likely resides in or originates from the IT department. IT staff have long understood the business implications of the outage of even one server (phones ringing off the hook is one measure of the importance of even a single server or business application). However, it's also clear that IT equipment—routers, servers, switches, hubs, firewalls, and more—are just part of the overall business equation. Certainly, without these technology components in place, business as usual will be limited at best. However, without also considering the way in which your company earns income and the way in which it conducts its business, all the IT planning in the world won't protect a company if a disaster strikes. A holistic approach to the business is needed in order for any business continuity and disaster recovery planning to be realistic and effective. This involves every key area of your business and the various stakeholders that represent those business units. It won't help if you can keep your Web site's e-commerce functions up and running if your warehouse operations have come to a screeching halt.

Most IT departments have some minor disaster recovery procedures in place. If your firm performs backups of critical data on servers, you have basic disaster recovery capabilities, assuming those backups are taken off-site or are stored (or performed) remotely. Though you might think this is quite obvious, you would probably be surprised to know how many companies (and IT professionals themselves) either fail to make backups or fail to store them in a safe location. However, most small, medium, and certainly most large companies at least have a reasonable data backup solution in place. This, in and of itself, is a good start but does not constitute a BC/DR plan. For example, if your area was flooded and you were unable to enter your building, could the company continue operations? If this is one location out of many, perhaps. If this is your only location, perhaps not. It depends, of course, on the nature of your business. If you have a warehouse full of product that is also underwater, you might have contracted with your suppliers to direct ship to customers in the event of a disaster. Did you also develop a plan for how customers would place orders or how you would track and invoice those orders? Clearly, the technological component is a critical link in the chain, but it's not the only link. Throughout the remainder of the book, we'll look not only at the IT components but the other non-IT elements that need to be in place as you develop your BC/DR plan so that you don't overlook any crucial aspects of the business.

Disaster planning is about recovering after an event, but business continuity planning is not just about recovering from outages of key technical components, it is a way of looking at and managing business. BC planning is about looking ahead and seeing what could potentially disrupt your company's operations and then finding ways to mitigate or avoid those events. It really is a coordinated and integrated approach that spans the entire company and all its operations. As in any other area of life, one or two poor decisions can usually be corrected or overcome, but when things get stressful it's highly likely that a string of poor decisions could literally spell disaster for your company. The point of BC/DR planning is to help avoid those pitfalls that can be avoided and to provide a sane, rational, well thought-out approach to managing the disaster when an event does occur. If the number of poor decisions can be held to a minimum, there is a stronger likelihood that you will avoid compounding the problem and perhaps even be able to come out of it quickly and in relatively good shape.

Human nature is a funny thing. When we're young, we think nothing bad can ever happen to us. When we're older, we may we think we can play the odds. As trite as it may sound, failing to plan is planning to fail. In an *Entrepreneur* magazine article in 2003, author Dan Tynan included a quote that sums up the situation with small businesses. "Small companies often spend more time planning their company picnics than for an event that could put them out of business," explains Katherine Heaviside, principal of Epoch 5, a Huntington, New York, public relations firm that specializes in crisis communications (www.entrepreneur.com/magazine/entrepreneur/2003/april/60242.html).

In a study of companies that experienced a major data loss without having a solid BC/DR plan in place, 43% never reopen, 51% will close within two years, and only 6% will survive long-term. (Cummings, Haag & McCubbrey, 2005.). Let's repeat that: only 6% will survive long-term. That's a 94% mortality rate for companies that experience a major data loss. In August 2002, the American Management Association released a study indicating that more than half of the surveyed companies had no disaster recovery or crisis management plan in place. Another report from Gartner, Inc. indicated that less than 10% of small and medium businesses had disaster plans, and that 40% of companies that experience a disaster without a disaster recovery plan will go out of business within five years. Looking specifically at fires, the most common disasters businesses experience, it is estimated that 44% of companies whose premises experience a significant fire do not recover at all, primarily because they have no BC/DR plans in place. The World Trade Center bombing in Manhattan in 1993 resulted in 150 out of the 350 businesses located in the center going out of business—that's about a 42% failure rate. Contrast that with many of the financial firms who had well-developed and tested BC/DR plans that were located in the Twin Towers on September 11, 2001—a majority of them were back up and running within days.

An October 2005 survey by the Advertising Council found that 92% of businesses say it is important to plan for emergencies; 88% agreed having some sort of emergency plan would make sense; 39% said they actually had a plan. What's interesting is that 12% of companies did not think having an emergency plan would make sense. Although the question was not posed to these companies, it would certainly be interesting to understand why these companies feel a plan is not needed. Studies point to the broadly-held but incorrect notion that the time and expense of creating a plan will far outweigh any upside return. However, as you've learned, that's just not true. Many BC/DR planning activities can be accomplished relatively quickly and with little or no funding. If the 12% of companies that feel no need to plan for emergency understood there was an almost even chance the company would go out of business if it did experience an emergency, their thinking might shift.

Small businesses, those most likely to avoid, delay, or short-cut business continuity planning and disaster recovery planning are most susceptible to the long-term impact of emergencies and disasters. Yet, these same small companies are the economic engine of many economies around the world. In the United States, small businesses account for 99% of all employers (that's right, large companies employ 1%), 75% of all new jobs, and 97% of all U.S. exporters. It can therefore be argued that BC/DR planning is extremely important for companies of all sizes, even small companies.

Regardless of the size of your company, the odds are high that if your company experiences any sort of disaster—natural or man-made—it has 40% to 50% chance of going out of business as a result. Certainly, the strength of the company, the industry, and other factors come into play when looking at long-term survival of companies hit by disasters, but it's clear that if your company doesn't have a business continuity and disaster recovery plan, it is essentially taking a 50% chance on failing. Without a well-conceived business continuity and disaster recovery plan, that's an enormous gamble to take. It impacts not just the corporate entity itself, but the lives of all the employees and suppliers as well. The ripple effect can be massive and it will impact you, your staff, their families, and the rest of the community.

There are many people who will counter with the argument that a company could spent a lot of money on planning and *never* have to deal with a disastrous event. True. Many people drive their entire lives and never have a single auto accident, but they probably all have auto insurance. Clearly, the question is one of balance. If your company does $50M in annual revenue, a cost of $1M for BC/DR planning is very little to pay for that type of insurance. If your company does $1.25M annually, you probably don't need to (and can't) spend $1M on BC/DR planning. Obviously, the cost of planning must be balanced with the cost of doing nothing and the risk of going out of business. Like auto insurance, you certainly hope you'll never need to use it, but you don't want to get caught without it either. Ultimately, it's less expensive to expend an appropriate and proportionate amount of time and resources to create and maintain the plan than to face even one disaster without a plan. As we proceed through this book, we'll take this into account. For example, if your company is in the Gulf States region of the United States, you need to have an emergency plan in place in the event a hurricane hits the area, as has happened repeatedly in the past few years. On the other hand, if your firm is located in the desert Southwest of the United States, you don't need to plan for hurricanes but you will have to plan for power outages and lightning strikes. Even though this is obvious, it bears repeating because you don't need to over-engineer your BC/DR plan. You will need to evaluate the potential impact to your company of various types of events and then create a plan for just those events most likely to occur and most likely to have a critical impact on operations. When you do this, you use your planning time effectively, and the cost of planning will certainly be far lower than creating an all-encompassing plan or the cost of facing a disaster empty-handed.

While we're on this topic, let's take a moment to look at how the cost of planning (investment) and the cost of failure (loss) impact the people, processes, and technology of a company. The impact, though not immediately apparent, is significant and worth exploring briefly.

WARNING

A bad plan or incomplete plan is often worse than no plan at all. An ill-conceived or incomplete plan may lead people to mistakenly assume that emergency and contingency plans are in place when, in fact, they are not. A false sense of security can lead to an even bigger problem than the disaster event itself precipitates. Remember, if a disaster strikes your area, emergency personnel will be going to hospitals, nursing homes, day care centers, and schools to help. Your business will be pretty low on the list of priorities, so you need to be prepared to take matters into your own hands. If employees falsely believe the company is prepared for disaster, you're facing a whole host of problems. A poorly conceived plan may also lead to significant financial penalties and legal liabilities since it might be argued you had the opportunity to plan and failed to do so. We'll explore some of the potential legal issues later in this book.

People

Spending time and resources to plan for emergency responses, from an organizational perspective, is an excellent investment for many reasons. One that might not be immediately evident is that when employees understand that the company has contingency plans in place, they tend to feel that the company is organized, positioned for success, and concerned for their safety. It provides an opportunity for the company to demonstrate its commitment to its employees' well-being, which can help retain key employees. Companies that run in a perpetual ad hoc manner are often more at risk of losing key employees for this same reason. Will a solid BC/DR plan keep employees happy? Of course not. It does contribute to an overall environment that fosters respect and concern for employee well-being.

In addition, a crisis that is well-managed by the company is less likely to cause key employees to seek employment elsewhere. A well-managed event also keeps employees calm and focused so business can get back to usual as quickly as possible. A well-managed crisis can also enhance a company's reputation, leaving it stronger than it was before the incident. One example of excellent crisis management (not IT-related) was when the Extra Strength Tylenol pain product was contaminated with cyanide in 1982. The company quickly asked retailers to pull *all* of its products from store shelves until it could understand the nature and extent of the "attack." The year prior to the incident, Tylenol had about 35% of the billion dollar analgesic market, or about $350 million in annual sales. Immediately afterward, its market share was 0%. However, within four years, the company has regained almost all its former market share (98% of precontamination sales revenues). Although this example is outside the domain of IT professionals, it points to the opportunity a company has to manage an emergency. It gets one shot to get it right and its future reputation rides on the decisions made during the crisis. Today, the "Tylenol incident" as it is sometimes referred to, is held up as an excellent example of how a company can and should respond to a crisis.

The effect of stress on people during an emergency cannot be overemphasized. Having a well-thought-out and well-rehearsed BC/DR plan will reduce that stress considerably. In turn, people will be able to function again and return to their jobs more quickly. Thus, the very act of planning how to take care of the people in your organization during an emergency can quickly impact the company's ability to return to normal operations—and revenue generation. BC/DR planning, then, directly impacts the top and bottom line and the cost of planning will quickly offset the cost of an unmanaged event.

Process

BC/DR planning can provide an opportunity for a company to evaluate and improve its business processes. As your project team (we'll discuss the team later in the book) evaluates business processes as it relates to BC/DR, it might discover new ways to streamline operations. For example, in planning for a major disruption due to a natural disaster, your team might uncover new methods while determining "bare minimum" requirements. If a process takes 20 steps and four departments now, you might find that the pared down approach discussed in a post-disaster scenario would actually work well all the time. When you're forced to look at everything from the ground up (which is what happens when you're dealing with a disaster), you discover that you don't need all the bells and whistles. This can sometimes translate into streamlined processes that can be incorporated into the day-to-day operations.

In addition, documenting critical business processes can truly mean the difference between life and death for the corporate entity. If you are unable to resume some sort of operations in a reasonable time frame after a disaster, your company is not likely to survive. The cost, then, may be the ultimate corporate cost—failure to exist. This is not only unfortunate for the corporate shareholders (whether publicly or privately held), but it impacts the lives of all the company's employees and their families and takes a toll on the community as well. The ripple effect is enormous and should not be quickly discounted.

Technology

Scrambling to deal with technology issues once a disaster has hit is guaranteed to cost your firm more than if you have a solid plan in place. For example, if you need temporary computing facilities, it's less costly to have a contingency contract in place beforehand than to desperately call various facilities looking for assistance while the smoke clears. Not only will you be in a better frame of mind emotionally in the planning phase (vs. the reaction phase after a disaster), you'll be in a much stronger position to negotiate the details of a contingency contract.

In addition, if the disaster impacts other companies, it might also create a competitive situation that drives the price for technology components up. Again, being able to calmly negotiate and procure commitments for emergency services beforehand almost always generates lower costs when those contracts are activated by an emergency. Finally, it is customary for most companies to provide service to contract holders before they provide service to noncontract holders. If you're currently a customer, you're going to get service before the person who just called in today looking for assistance. So, prenegotiating anticipated emergency services can generate lower costs and a higher ROI on your BC/DR planning process.

Common Challenges...

Dealing with Optimists and Pessimists

When developing your BC/DR plan, you have to find some balance between the optimists and the pessimists. The optimists will dismiss many potential risks and dangers and will often minimize the potential impact of events. On the other hand, pessimists believe every possible danger is likely to occur and to have a much larger impact than it likely would should it occur. Part of your job is to try to remain balanced and realistic, especially when it comes to developing mitigation strategies, which we'll discuss later in this book. Additionally, many BC/DR planners place a disproportionate amount of time and attention on major catastrophes. As you'll see throughout this book, we first look at the most common scenarios and then turn our attention to major events. The thinking is this: If you spend time to prepare for the common, smaller events, you can then perform a second round of planning for major catastrophes, or create two different planning teams. If you're ready for the next Category 5 hurricane but you fail to have a solid plan in place for a workplace fire (the most common business emergency), you'll be doing yourself, your employees, your company, and your community a disservice. So, in the end, you will need to balance the need for disaster planning with the financial and organizational constraints of your company and focus on the smaller, more likely events first. This can best be accomplished by listening to both the optimists and the pessimists and finding acceptable middle ground.

Types of Disasters to Consider

So far, we've spent time talking about *why* it's important to plan for disasters. Now, let's turn our attention to the types of disasters that might occur. The reason for this is that there may be a few you don't think of immediately (or at all) that might potentially impact your company. Although this list is extensive, it is certainly not exhaustive. Throughout this book, we'll continually give examples of a variety of disasters because we want to make sure you cover all your bases and think through all potential threats to your company. You and your BC/DR planning team should be sure to look at your company's specific location(s), your industry, and your operations to determine exactly what types of disasters and events could have a significant impact on you. This list should be a good starting point and might also spark ideas about other elements that could be essential to include in your company-specific plan. Not only is it important to review the entire list and be sure you've covered your bases, you also have to start with the more likely events and move outward from there. As mentioned, fire is the most common business emergency companies face. So, if you don't have an established fire plan, you're really a sitting duck. As you'll see in Chapter 3, risk assessment should be holistic and broad in scope, but it should also then narrow down your focus to those risks that are most likely to occur and that will have the biggest impact on your company's operations.

As an IT professional, your job may be limited to dealing with just the technology aspects of the BC/DR plan, but you need to be aware of all the various threats because your company will be relying on you to understand and address the potential impact of threats on the company's technological operations. Technology is so pervasive in most organizations these days that IT will be one of the key drivers in both the planning phase and the implementation/recovery phase. Therefore, it's critical that you and your IT team be well-versed in all aspects of BC/DR planning.

Threats or hazards come in three basic categories:

- Natural hazards
- Human-caused hazards
- Accidents and technological hazards

Clearly, natural hazards are ones that can sometimes be anticipated and the effects mitigated; other times, they come without warning and must be responded to. Human-caused hazards also can sometimes be anticipated and other times come as a surprise. Finally, accidents can happen and accidents span the range from minor to major to catastrophic. Included in this category are what often are termed "technological threats" because they involve the failure of buildings or infrastructure technology. Let's look at these categories in more detail.

Natural Hazards

Natural hazards are the types of disaster events that usually come immediately to mind when we mention business continuity and disaster recovery planning. In the past several years, the news has been full of headlines about natural disasters. Clearly, the tsunami that hit Indonesia December 26, 2005 was a catastrophic event. The hurricanes that hit the Gulf Coast of the United States, most notably Hurricane Katrina, also caused catastrophic failures. Keep in mind that natural hazards can be minor, major, or catastrophic in nature, so your planning should account for these three potential threat levels. In addition, natural hazards can be anticipated or unanticipated. The tsunami that hit Indonesia was anticipated to the degree that scientists around the world measured the power of the undersea earthquake. It was unanticipated in that few knew that a massive wall of water would hit various shorelines from Thailand to Kenya within a matter of hours. Hurricane Katrina was, for all intents and purposes, anticipated, but grossly underestimated by the public and private sector. Regardless, the impact in both cases was total devastation. There are numerous cases of companies and organizations that had disaster plans that went into limited effect because, prior to these events, no one could realistically imagine that level of devastation. In the United States, the Federal Emergency Management Agency (FEMA) was widely criticized for failing to effectively implement and manage a disaster plan in New Orleans in the aftermath of Hurricane Katrina. The news was rife with heart-rending stories of human suffering and death due to massive confusion and disorganization. Individual companies were fairly powerless in this situation, but companies that had plans in place for evacuation, remote data backup, storage, and such undoubtedly got back up on their feet faster and more effectively than those that did not have such plans in place.

Cold Weather Related Hazards

- Avalanche
- Severe snow
- Ice storm, hail storm
- Severe or prolonged wind

Warm Weather Related Hazards

- Severe or prolonged rain
- Heavy rain and/or flooding
- Floods
    - Flash flood
    - River flood
    - Urban flood
- Drought (can impact urban, rural, and agricultural areas)

- Fire
    - Forest fire
    - Wild fire—urban, rural, agricultural
    - Urban fire
- Tropical storms
- Hurricanes, cyclones, typhoons (name depends on location of event)
- Tornado
- Wind storm

Geological Hazards

- Earthquake
- Tsunami
- Volcanic eruption
    - Volcanic ash
    - Lava flow
    - Mudflow (called a *lahar*)
- Landslide (often caused by severe or prolonged rain)
- Land shifting (*subsidence* and *uplift*) caused by changes to the water table, man-made elements (tunnels, underground building), geological faulting, extraction of natural gas, and so on

Although this list does not contain every possible variation, it should give you a good starting point for determining which hazards are applicable to your geographic locations. Remember that BC/DR planning should involve people across your organization and that includes various locations. If you have offices in London, Mumbai, Perth, and New York, each location should be represented in order to effectively review potential hazards based on the geography of the area, the key business functions at each location, and other factors we'll discuss in subsequent chapters.

Human-Caused Hazards

Human-caused hazards, also known as *anthropogenic* hazards, are a bit more diverse in their nature. Some of the items on the list may surprise you. Since most are intentional, we will just list them without categorization.

- Terrorism
    - Bombs
    - Armed attacks
    - Hazardous material release (biohazard, radioactive)
    - Cyber attack
    - Biological attack (air, water, food)
    - Transportation attack (airports, water ports, railways)
    - Infrastructure attack (airports, government buildings, military bases, utilities, water supply)
    - Kidnapping (nonterrorist)
- Bomb
    - Bomb threat
    - Explosive device found
    - Bomb explosion
- Explosion
- Fire
    - Arson
    - Accidental
- Cyber attack
    - Threat or boasting
    - Minor intrusion
    - Major intrusion
    - Total outage
    - Broader network infrastructure impaired (Internet, backbone, etc.)
- Civil disorder, rioting, unrest
- Protests
    - Broad political protests
    - Targeted protests (specifically targeting your company, for example)
- Product tampering
- Radioactive contamination

- Embezzlement, larceny, theft

- Kidnapping

- Extortion

- Subsidence (shifting of land due to natural or man-made changes causing building or infrastructure failure)

Accidents and Technological Hazards

Accidents and technological hazards often are related to man-made hazards but differ only in that they are usually unintentional. If intentional, they fall under the category of man-made hazards. Regardless of the category in which we place them, they are issues that can be over-looked in the planning process.

- Transportation accidents and failures
    - Highway collapse or major accident
    - Airport collapse, air collision, or accident
    - Rail collapse or accident
    - Water accident, port closure
    - Pipeline collapse or accident
- Infrastructure accidents and failures
    - Electricity—power outage, brown-outs, rolling outages, failure of infrastructure
    - Gas—outage, explosion, evacuation, collapse of system
    - Water—outage, contamination, shortage, collapse of system
    - Sewer—stoppage, backflow, contamination, collapse of system
- Information system infrastructure
    - Internet infrastructure outage
    - Communication infrastructure outage (undersea cables, satellites, etc.)
    - Major service provider outage (Internet, communications, etc.)
    - Systems failures
- Power grid or substation failure
- Nuclear power facility incident
- Dam failure

- Hazardous material incident

  - Local, stationary source

  - Nonlocal or in-transit source (e.g., truck hauling radioactive or chemical waste crashes)

- Building collapse (various causes)

The list of disasters here is enough to make you want to hide your head under a pillow for a while and wait for the images to fade. Unfortunately, these are all incidents that can and have occurred, and the best way to deal with these kinds of unimaginable uncertainties is to imagine them and develop a methodical plan for handling them. To be sure, if one of these more major events occurs and you have to deal with it, it's unlikely you'll follow your plan to the letter. It's impossible to imagine everything you'll be experiencing and have to deal with until you're in the middle of it. Having a solid plan in place that's been tested and practiced will reduce the stress of the situation and increase the likelihood that you've anticipated the major issues you'll need to address. In dire circumstances, that can mean the difference between surviving or not, between recovering or not.

Common Challenges...

Corporatewide Participation

Although your specific role in the company may not bear responsibility for business continuity and disaster planning, you may need to lead the charge. As an IT professional, you understand the immediate implications of a power outage or a cyber attack or even a building evacuation on your business. If you're leading the BC/DR planning, you'll need to educate yourself to the larger business issues for two reasons. First, you'll need to understand the broader business issues involved with business continuity and disaster recovery, not just the IT issues. Second and perhaps more important, you'll need to gain executive support for your BC/DR planning initiative. Executive support is key to success for any type of project and this is no exception. If the folks "upstairs" don't support the project, you'll have a hard time gaining the authority, funding, staffing, or resources needed to create a successful BC/DR plan. Going through the motions without creating a workable plan is almost worse than having no plan at all—it may provide a false sense of security to your organization. If or when disaster strikes, your plan has to work, it can't just be words in a document. Gaining executive support, a topic we'll discuss in the next chapter, is key to success, as is participation across the organization.

Electronic Data Threats

Although this area falls under human threats and accidents, we're going to take a moment to delineate in more details some of the threats facing electronic data. Some of these areas are well known to you and some might serve simply as reminders of elements to include in your company-specific BC/DR plan. Attacks to computer systems, networks, and electronic data occur every single day. The question is not if, but when a network will be attacked—internally or externally, intentionally or accidentally. Recall the IBM study mentioned earlier, in which it was estimated that 80% of all data loss is human-caused.

Personal Privacy

Personal privacy is an area that increasingly has been in the spotlight, and with good reason. More and more personal data is stored in electronic format on servers and hard drives around the world. From students' financial aid records to employees' social security numbers, to patients' private health data, to consumers' credit card numbers, the electronic world is afloat with very private data. Clearly, when this data falls into the wrong hands, either intentionally or inadvertently, bad things can and do happen. Companies in recent years that have failed to safeguard this type of data have faced sometimes daunting charges, including financial ruin, huge and expensive lawsuits, and in some cases, increased regulation from governmental or regulatory agencies. This, in turn, increases the cost of doing business and makes the company that much less competitive in the marketplace. It can throw a company into a death spiral—personal data is compromised, the company receives negative publicity, customers stop doing business with them, legal (or other recovery) expenses skyrocket, and so on until the firm has no choice but to cease operations. Granted, this is the most dire situation but it's also not outside the realm of real possibility for companies that routinely deal with sensitive data. As you develop your BC/DR plan, you'll need to pay special attention to the types of data your company deals with and how those types of data need to be managed, particularly in avoiding or recovering from an incident. The case study that follows this chapter addresses this topic in detail.

Privacy Standards and Legislation

Privacy standards are constantly evolving, so any very specific information provided here will be dated before long. However, it's worth spending just a bit of time reviewing some of the major privacy standards and legislation in the United States, if for no other reason than to remind you to take a look around your firm and determine what data might need special or additional protection and handling in the event of a disaster or security incident. This list is not exhaustive, but should serve as a good starting point to get you thinking along these lines.

Gramm-Leach-Bliley Act (GLBA)

The Gramm-Leach-Bliley Act (GLBA) was enacted in late 1999 and was intended to enhance competition in the financial services industry. Among other provisions, the GLBA requires financial institutions (and others that operate in the financial services industry) to create and implement policies to protect private information from foreseeable threats in both security and data integrity. GLBA also has provisions requiring financial institutions to develop written security plans detailing how the company plans to protect clients' nonpublic personal information. The plan must include, at minimum, four elements:

- At least one employee assigned to manage the safeguards
- A thorough risk management plan for each department that handles nonpublic information
- A plan to develop, implement, test, manage, and monitor data security
- A process for updating and changing the plan as methods of collecting, storing, transmitting, and managing data change

There are many provisions of the GLBA that you must comply with if you work within the financial industry, and chances are good that if these regulations apply to you, you're already well aware of them. However, if for any reason you believe these regulations may apply to your firm, you should consult with appropriate legal and financial counsel to determine your responsibilities. For additional (general) information on GLBA, you can check out Wikipedia at http://en.wikipedia.org/wiki/Gramm-Leach-Bliley_Act#Privacy; you can also get more formal information on the U.S. government Web site at http://www.ftc.gov/privacy/privacyinitiatives/glbact.html.

Health Insurance Portability and Accountability Act (HIPAA)

Anyone working in the health care industry in the United States is well aware of the Health Insurance Portability and Accountability Act (HIPAA). This regulation creates requirements for health care providers (and others in the health care industry) to protect personal data, especially health care information. It requires companies to create policies and procedures to ensure this data is kept confidential and is shared only with authorized parties. In its original form, it was intended to establish national standards for electronic health care transactions and to ensure the security and privacy of health data. For general information on HIPAA, you can visit Wikipedia at http://en.wikipedia.org/wiki/HIPAA or visit the U.S. government's Health and Human Services (HHS) Web site at this URL: www.hhs.gov/ocr/hipaa/.

Common Challenges...

If Disaster Strikes, Will You Still Be Compliant?

It was a major effort for many organizations to become (and remain) compliant when privacy and security regulations went into effect, but can these same companies maintain compliance in the face of disaster? HIPAA and other regulations require that companies plan for foreseeable threats, but it's important to think this all the way through to the practical, day-to-day operations. If, for example, your company is forced to resume operations from alternate locations or with employees working from home, how will your privacy and security plans hold up? Even though you may not be able to address every potential pitfall, you must address privacy and security through your BC/DR plan to the same extent you plan alternative operations. If patient information will be taken by phone and written on paper while computer systems are being reconfigured and communications are being reestablished, for example, how will you protect privacy? Remember that HIPAA and other regulations require you maintain appropriate levels of security at *all times*, not just when everything is humming along just fine.

Although you and your organization may be well-versed in the requirements of regulations that apply to your industry or business sector, also be sure that you include these topic headers in your BC/DR plan so you can remain as compliant as possible in the event of a security incident or natural disaster. At the end of the day, you need to be sure you're addressing the threats most reasonable people could foresee. As an IT professional, you've probably become very familiar with the requirements that apply to your company and your industry, so be sure to bring this expertise to the table when creating your BC/DR plan and consult with outside experts as needed. The fines and regulatory issues that can result from noncompliance could, in themselves, be a disaster for your firm.

Social Engineering

You're well aware of the potential for social engineering when it comes to protecting network data and maintaining security. However, social engineering takes on a whole new meaning when disaster strikes. The unfortunate truth of the matter is that people don't always "rise to the occasion" when a disaster hits. After Hurricane Katrina, it was widely reported that many people who had no legitimate claim were collecting government payments of various kinds. The information systems and methods of validating legitimate "clients" were clearly impaired. If disaster strikes your firm, what kind of data would someone want to get their hands on and what ruses could they use to gather that information? Remember, during any type of incident—whether a major snow storm, a massive earthquake, or anything in between—people are more emotional and vulnerable. Your employees are only human and may be far more vulnerable to social engineering techniques than they might otherwise be. Your BC/DR plan should certainly take this into account in order to help mitigate the impact on security and data integrity as well as on your company's confidential data and operations as a whole. During your planning phase, focus on the kinds of data that might be of value to outsiders, then focus on the most common or likely social engineering scenarios. If you attempt to build in these safeguards at a time when everyone is relatively sane and rational, there's a much better chance that employees will be able to resist social engineering tactics during a disaster.

Fraud and Theft

It's well known within IT circles that most corporate fraud and theft is committed by company insiders. This means that your first priority during a disaster will be to mitigate opportunities for internal fraud and theft, then look for ways to safeguard against potential external sources. There are numerous types of fraud and theft related to computer and IT technology and unfortunately they are constantly evolving. Whatever threats are delineated in this book will undoubtedly be surpassed by new variations and new schemes within days or weeks. No company can remain completely immune to these threats, but every company can put safeguards in place to prevent the most blatant acts. Next, we've listed a few categories of fraud and theft that you can use as a jumping off point in your planning to assess your company's specific and unique vulnerabilities.

General Business Fraud

This is the catch-all category, and is included first because it is the most common area. Each company has a unique risk profile in this area. It would be well worth your time to look into this for both emergency and non-emergency situations. Unfortunately, without monitoring, any area of business is vulnerable to fraud or theft and there are usually signals along the way that are missed or disregarded for a variety of reasons. Managers become busy and miss signs; analysts make incorrect assumptions or ask the very perpetrators of fraud for additional information. The list goes on. Now, take this and multiply it by a factor of 100, which represents the chaos and stress of an emergency or disaster event. People, processes, and technology are all in disarray during and after an event. Often, people are focusing on short-term safety and survival. Only emergency processes are implemented and normal safeguards may be suspended or simply not viable. Technology is rudimentary, if available at all. This leaves the company wide open to potential fraud and theft during an emergency.

Since every business is unique, there is no one failsafe method of preventing, spotting, or stopping internal fraud or theft. However, there are some general guidelines you can use for both normal and emergency operations.

1. If it seems odd, it is.

2. Evaluate your assumptions.

3. If it can't be tracked, monitored, and accounted for, evaluate the activity.

4. If it falls outside standard operating procedures, scrutinize it.

5. Spot inspections and random detailed reviews on normal activities can reveal problems.

Most of the time, a manager or colleagues can tell when someone is behaving in an odd manner. Even during times of stress and emergency, people typically behave with some consistency. If someone is behaving in an odd manner, investigate it. If a transaction or activity seems odd, investigate it. Turning your back on potential problems can lead to much bigger problems down the road, including legal problems and termination of business operations.

Sometimes an odd activity or behavior, especially during a stressful event or disaster, is written off as being caused by the emergency—and sometimes that's the case. However, rather than assume that is the case, look for evidence that supports or contradicts your assumptions. For example, don't assume someone has authority to remove files, equipment, or assets from the building—look at your disaster recovery plan and determine who actually *is* authorized for these activities. An unauthorized employee could be trying to capitalize on a bad situation, or an outsider could spot an opportunity to turn a quick profit.

Whether in normal operations or in an emergency, your goal should be to ensure that there are processes in place to monitor and track the movement of people, technology, and business assets to the greatest extent possible (within a reasonable limit). Creating a lot of bureaucratic processes causes people to circumvent the system, so be sure that whatever safeguards you have in place are lean and streamlined to the greatest extent possible. If someone is attempting to circumvent operational processes, it should raise a flag. In an emergency or disaster, there may well be exceptions that must be made quickly to ensure the safety of the people and assets of the business. However, if you have emergency processes delineated in your BC/DR plan and you have tested and rehearsed these, the exceptions should be fewer and more reasonable. This can help you and other employees spot unusual activities or behavior, even during a disaster. Often, when a potential thief is directly questioned about an activity, it will cease. Though this is not always the case, it can help thwart these crimes of convenience.

Spot inspections and random data reviews can also be helpful, both during normal operations and emergencies. Smart thieves are good at flying "below the radar" and will work hard to keep up normal appearances; you won't always be alerted to problems by seeing dramatic anomalies. Therefore, it can be helpful to do spot inspections and data reviews. Though you may be limited in your role as an IT member (and not a corporate executive), you can certainly keep an eye out for these issues within your department. For example, perhaps you have an inventory of low-cost spare parts and you see these are being used with a very slight increase in frequency, despite the fact that your company hasn't added any new equipment. At first, you might assume that as the equipment ages, it's very likely to need replacement parts. You could easily dismiss this. However, you could also ask the people who are signing out these parts where the trouble ticket is or where the report is, or which desktop computer this was installed in. Talk with the user whose desktop computer was allegedly worked on—was a technician working on his or her computer? If so, did he or she request it? What was the reported problem? Even though you may have all this data in a trouble ticket, randomly verifying some of these tickets might help you spot a problem. You might discover the parts that were allegedly used are easy to slip into someone's pocket for resale online. Spot checks and random analysis can help prevent this type of fraud and theft. During an emergency, it might be less obvious, but simply keeping your eyes open and asking questions can help. Having easy-to-use processes in place for emergencies can help limit the movement of people, data, and company assets to legitimate purposes.

Remember, your company's data may be of particular interest to employees or competitors, and during times of crisis, standard safeguards may fall by the wayside. What can you do, within the confines of the IT department, to ensure data safety and integrity during an emergency? How can you prevent theft and fraud? What are the likely scenarios to arise and what processes can you use to mitigate or avoid these issues?

Looking Ahead...

IT, Security, Disasters... and the Law

One of the emerging trends in IT and IT security is the increased demand that companies secure private data such as social security numbers, credit card numbers, home addresses and phone numbers, financial data, medical data, and more. As the amount of electronic data collected and stored increases, so too does the risk to individuals. Recent headlines are rife with examples of personal data being lost, stolen, hacked, or modified. Companies can no longer say "we did our best" without proving that their best was at least up to current industry standards. Looking ahead, companies can expect three major trends to impact how they manage IT security. These standards will apply during normal business operations and emergencies—companies won't be able to easily blame breaches and theft on emergencies that were foreseeable and manageable, as is the case with many of the disaster events listed earlier in this chapter. These three key trends, which you should monitor for your IT organization, are:

■ The continuing expansion of the requirement to provide IT (and data) security

Continued

- The emergence of a standard definition of "reasonable security"
- The imposition of the duty to warn

Consumers and regulators alike are raising their expectations regarding IT security, and companies are both legally and ethically bound to make serious, effective efforts to safeguard private data. Emergency and disaster conditions may soften those requirements just a bit, but don't assume your company will be able to hide behind a disaster or event if data is lost, stolen, mishandled, or inappropriately disclosed. If your firm deals with data that is sensitive, confidential, or private in nature, consult with your firm's legal counsel to understand fully the legal and regulatory requirements your firm will be subject to during a crisis, emergency, or disaster. As you've noticed, we've continually emphasized the legal aspects of BC/DR because of the increasing regulation of electronic data. The case study that follows this chapter provides an excellent example of the need for due diligence in handling electronic data regardless of whether you're facing normal operational challenges or a major disaster.

Managing Access

It doesn't get more basic than this: Managing access to data is the first and most important element of managing data security and integrity. It also has to be an integral part of your BC/DR plan. In the aftermath of a major event, it sometimes becomes a "free for all," with everyone (and no one) in charge. Who should have access to data and systems in an emergency? If you allow only one person that access, what if that one person is on vacation halfway across the world when disaster strikes, or worse, injured during the event and unable to perform his or her duties? Clearly, access that is too restrictive could significantly delay the ability of the company to get back up and running after a disaster or event. The opposite is also a danger—giving half the company the authority to make changes, request access, or manage data could also put the company at risk by creating lack of accountability and certainly a tangled trail to unwind should something go wrong.

In a disaster or emergency, physical access is often one of the areas most likely to be impacted. How can you manage physical access to a building that has one whole side blown off or a building that is structurally unsound? You may not be able to do anything about it when faced with the situation, but if you try to imagine how you would manage it and include that in your BC/DR plan, you may have a chance to reduce or avoid this issue later on.

Beyond physical access are the electronic methods of access control that are well known to IT professionals. Again, you need to take a look at your access control methods in light of various disaster/event scenarios. What are the critical systems, who should have access to them in an emergency, and what (if any) failsafe systems can you put in place so that the right people have access to the data when needed? As you'll see in our BC/DR planning later in this book, one of the most effective methods of planning for these kinds of situations is to develop likely scenarios and work your way through all aspects to ensure your plan is thorough. A good plan provides a solid framework without being either too restrictive or too vague. It's a delicate balance to achieve but it can be done using a logical, consistent project management framework, which we'll discuss in the next chapter and throughout the remainder of this book.

Business Continuity and Disaster Recovery Planning Basics

Your role as an IT professional is unique in BC/DR because on one hand, you are not necessarily responsible for the company's comprehensive BC/DR planning; but on the other hand, technology is so integral to most corporate operations, IT can't be completely separated out as a stand-alone issue. As a result, we will continually address BC/DR in a holistic manner and allow you to determine the most appropriate role for your IT group within your company.

The elements that should be included in your plan will extend beyond the walls of the IT department, so you'll need to form a project team with expertise in several areas. Figure 1.4 shows some of the areas that might be included, depending on the type of products and services your company creates.

Figure 1.4 Elements of a Business Continuity and Disaster Recovery Plan

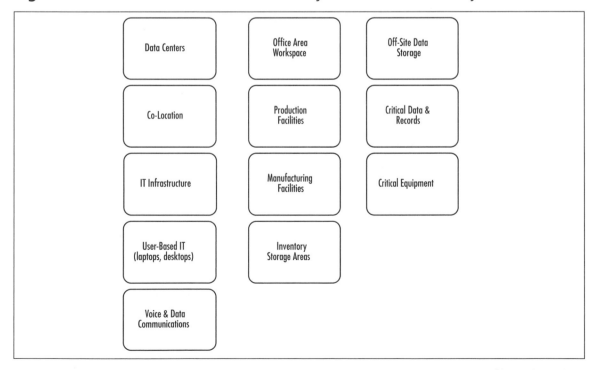

You're no doubt familiar with the concept of *reliable system design* and *single point of failure* when it comes to designing, implementing, managing, and repairing the IT infrastructure for your company. Briefly, these concepts relate to building in redundancies and safeguards so that if one key component fails, the entire company doesn't come to a screeching halt. You probably also understand that having two servers or routers in the same rack leaves your network vulnerable—the single point of failure could be as simple as someone tripping and spilling a large cup of coffee on the rack itself. You might conscientiously make backups, verify the backups and store them securely, but leave them on-site. The single point of failure could be as minor as something falling on the rack holding your tape backups or as major as a serious fire in the server room or building. The reason for discussing this concept at this juncture is that as you look at your business continuity and disaster recovery options, you need to assess your risks with regard to reliable systems and single points of failure. For example, you may want to evaluate your availability solutions as part of an overall business strategy to reduce operational risks, minimize the occurrence and cost of downtime, and maximize data and IT service availability. These availability solutions will also likely impact your compliance with a variety of regulations by providing protection and reliability of information resources as well. Additionally, these solutions will impact your BC/DR risk assessment and planning. If these solutions are not currently in place, this BC/DR planning process may help you build the business case for implementing some of these technologies. If they are currently in place, you can look at them with a fresh perspective to determine how they contribute to an overall business continuity strategy. We'll discuss this in more detail in Chapter 3.

With that, let's look at contingency planning basics: the basic steps to be taken to create a solid BC/DR plan for your company. The basic steps in any BC/DR plan, shown in Figure 1.5, include:

- Project Initiation

- Risk Assessment

- Business Impact Analysis

- Mitigation Strategy Development

- Plan Development

- Training, Testing, Auditing

- Plan Maintenance

Figure 1.5 Business Continuity and Disaster Planning Steps

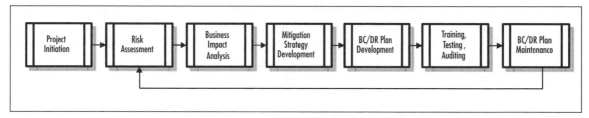

Those of you familiar with project management methodologies will notice the similarity in the BC/DR planning process to PM processes, and with good reason. Creating a BC/DR plan can (and should) be approached as a discrete project that has a defined start, middle, and end. As with many other IT projects, once the BC/DR plan is completed, it must be maintained so that it stays current with changes in the company, its technology, and the broader business landscape. We'll discuss each of the sections here briefly to provide an overview, and we'll delve more deeply into each of these areas in subsequent chapters.

Project Initiation

Project initiation is one of the most important elements in BC/DR planning, because without full organizational support, the plan will be incomplete. As an IT professional, there may be limits to what you can do to create an organizationwide functional BC/DR plan. For example, you may know how to set permissions for a particular business application, but do you really know how users interact with it and what would be required to get the business back up and running with regard to that particular business function? If the application server is destroyed and you have data backups, do you also have a backup server? Do you have a way to allow users to connect to the application securely? Where are users located? How will business resume? Can it resume without that application in the near-term or not? You will not likely be able to answer these questions. It requires the input and assessment from subject matter experts in other departments and divisions. Therefore, getting executive and companywide support for the BC/DR planning process is absolutely key to its success. We'll discuss this in more detail in the next chapter.

Risk Assessment

Risk assessment is the process of sitting down with key members of your company and looking at the potential risks your company faces. These risks run from ordinary to extraordinary—from a fire or minor flood in a server room to a catastrophic loss such as an earthquake or major hurricane and everything in between. Again, as an IT professional, you can certainly lend your expertise to this process by helping define the likely impact to technology components in various types of disasters or events, but you can't do it alone. For example, it's likely that your transportation manager understands the potential business impact of bad weather around the country, not just in your local area. Your marketing manager might best understand the potential business risk of a contaminated product or a Web site breach. Some of these areas may fall into pure business continuity planning and may be more suitable for others in your organization. However, in almost all companies, IT expertise must be included in the business continuity and disaster recovery risk assessment process. In Chapter 3, we'll discuss risk assessment in depth.

Business Impact Analysis

In a sense, this is where "the rubber meets the road." Once you've delineated your risks, you need to turn your attention to the potential impact of these various risks. This is one area that, as an IT professional, you clearly need input from your company's experts. As mentioned earlier, you might understand the technical aspects of an application server going down, but what is the actual business impact, and can that be tolerated? For example, you might determine that your ERP application cannot be down. Period. E-mail and your Web server, however, can go down, even though both events would be disruptive. Once you understand these parameters, you can develop an IT-based strategy to meet the requirements that result from this analysis. We'll look at business impact analysis and how IT interacts with this process in Chapter 4.

Mitigation Strategy Development

If you're part of a small company, your mitigation strategy might be quite simple. Keep several copies of backups off-site and keep several copies of key information such as employee list, phone numbers, emergency service phone numbers, key suppliers, and customers in a binder off-site in a secure but accessible location. That might be the extent to which you choose to mitigate your risks. However, for most companies, the process is a bit more complex. For each identified risk that has a significant business impact, you need to look at your options. How can the risk and impact be tolerated, reduced, avoided, or transferred? We'll discuss mitigation strategies in Chapter 5.

Plan Development

After you've gone through the analysis steps, you'll be ready to develop your plan. As with other types of IT project plans, you'll want to outline the methodology you're going to follow so that you improve your chance of success and reduce your chances for errors and gaps. This includes standard processes like developing business and technical requirements, defining scope, budget, timeline, and quality metrics, and so forth. We'll discuss these elements in Chapter 6 and we'll use standard IT PM methodologies to help you create a solid plan, regardless of the size of your company.

Training, Testing, Auditing

Once the plan has been developed, people need to be trained on how to implement it. In many cases, scenario-based case studies can be a good first step (though this may be part of the plan development stage as well). Running through appropriate drills, exercises and simulations can be of great help, especially for disasters or events that rank high on the list of "likely to occur." In Chapter 7, we'll discuss emergency preparations. Then, in Chapter 8, we'll look at some of the ways you can train, test, and audit your plan so that you can develop a process that closely tracks with your company and the way it operates.

Plan Maintenance

Finally, plan maintenance is the last step in the BC/DR planning process, and in many companies, it is "last and least." Without a plan to maintain your plan, it will become just another project document on a file server or sitting in a binder on a shelf. If it doesn't get maintained, updated, and revalidated from time to time, you'll find that the plan may be rendered useless if a disaster does strike. Maintenance doesn't have to be an enormous task, but it is one that must be done. Most importantly, there must be an organizational commitment to do so and someone within the company to own it. We'll look at this in Chapter 9.

Summary

Business continuity and disaster recovery are not new concepts to business, but the act of consciously assessing and planning for potential problems certainly has been underscored by disastrous events in the past decade including earthquakes, tsunamis, hurricanes, typhoons, and terrorist attacks. Companies need to plan for potential disasters that will impact their ability to continue operations and earn income. Without a plan to recover from any disaster or event, no matter how large or small, many companies fail. The statistics speak for themselves. The odds are between 40% and 50% that a company will fail after a fire or significant data loss, and that only 6% of companies survive long-term after a major incident.

When developing a business continuity and disaster recovery plan, you need to look at the three core components of business: people, process, and technology. When you take a holistic view of the company and its operations through the lens of these three elements, you're more likely to understand the best approach to your own unique BC/DR planning process. People, process, and technology must be considered in an integrated and holistic manner since they are closely tied together.

Through your BC/DR planning process, you may find additional information you can use to support the purchase or implementation of a particular technology or service. If that technology or service will not only help day-to-day operations but will also fit nicely into a BC/DR strategy, you have effectively doubled the usefulness of the technology and reduced the perceived cost. Building the business case can help tilt the budget decisions in your favor. In addition, through reviewing business processes for your BC/DR plan, you may discover new or improved ways to run daily operations, which will add to the perceived value (or reduce the perceived cost) of your BC/DR activities.

In some companies, even a little downtime can be devastating, but in the majority of companies, some downtime can be tolerated, though it will still be disruptive. You and your planning team will need to thoroughly assess the company's tolerance for downtime and disruption in order to develop an effective plan. We'll discuss this in later chapters in more detail. It's important to keep the business failure statistics in mind as you make your plans. Fire is the most common business emergency, and 40% to 50% of businesses fail after a major fire. Without a well-defined BC/DR plan, your company is putting the welfare of its employees, stakeholders (or shareholders), suppliers, vendors, and the community in which it operates at risk. Many BC/DR planning activities and remedies cost little or nothing to implement, so doing nothing is not an acceptable "no cost" option. It is simply imprudent and irresponsible for companies of all sizes to fail to plan.

Disasters impact different types of companies in assorted ways. A flooded retail location has a different set of challenges than a flooded nursing home. The impact of a biohazard incident is far different if it occurs at or near a day care center than near a remote manufacturing facility. As an IT professional, you and your team should fully understand the potential risks to your company in your location(s). As we move through this book, you'll learn how to prioritize and address a variety of risks and threats your company might face.

Also as we progress through this book, we'll rely upon standard project management tools to help develop the BC/DR plan. For those of you with formal PM training and skills, these steps will be familiar. For those of you less familiar with these methods (or for those of you who hate a lot of "process"), don't panic. We'll review the necessary steps and provide plenty of guidance along the way. We'll avoid getting bogged down with the very fine, detailed aspects of project management. Instead, we'll simply use it as our framework to help guide us through the process of risk assessment; business impact analysis; mitigation strategy development; plan development; training, testing and auditing; and plan maintenance.

Solutions Fast Track

Business Continuity and Disaster Recovery Defined

- ☑ Business continuity focuses on a company's ability to continue operations regardless of the nature of potential disruption.

- ☑ Business continuity planning is a formalized methodology that can be studied and for which practitioners can be certified.

- ☑ Disaster recovery planning is typically a subset of business continuity planning because it deals with stopping the effects of the disaster or event.

- ☑ Once the effects of the disaster or event have been addressed, business continuity activities typically begin.

Components of Business

- ☑ Businesses are comprised of people, process, and technology and related infrastructure. Each of these must be addressed in business continuity and disaster recovery planning.

- ☑ People are responsible for creating and implementing BC/DR plans. They are susceptible to the effects of a disaster including being overwhelmed with a variety of emotions, being physically injured (or killed), and being unable to perform their duties due to these and other influences that occur during a disaster.

- ☑ Processes are used in businesses to maintain an orderly and consistent flow of business operations.

- ☑ Business processes must be evaluated during BC/DR planning in order to determine which are the critical business processes and how they should be implemented in the face of a disaster or event.

☑ In some cases, you may find ways to streamline everyday business processes as a result of your BC/DR planning activities.

☑ Technology is implemented through people and processes. Therefore, an integrated approach to emergency planning for technology is needed that considers people, process, and technology.

☑ Understanding how technology is used in day-to-day operations is important for BC/DR planning.

☑ As an IT professional, you may understand how to implement technology but you will need to collaborate with others in your organization to understand the broader business impact of technologies on operations so you can effectively plan for emergencies.

The Cost of Planning versus the Cost of Failure

☑ Fire is the most common emergency (disaster) companies face. 40% to 50% of companies that experience a major fire go out of business because most do not have BC/DR plans in place.

☑ Despite the high likelihood that a company will go out of business after a disaster, more than 90% of small businesses lack a disaster recovery plan.

☑ Even though many companies say they understand the need for a disaster recovery plan, very few actually make it a priority.

☑ There may be substantial financial and legal implications for failing to plan and for failing to take reasonable precautions. This can add to a company's burdens after a disaster strikes.

Types of Disasters to Consider

☑ Disasters fall into three general categories: natural hazards, human-caused hazards, and accidental/technical hazards.

☑ Natural hazards include weather problems in both hot and cold climates as well as geological hazards such as earthquakes, tsunamis, volcanic eruption, and land shifting.

☑ Human-caused hazards can be accidental or intentional. Some intentional human-caused hazards fall under the category of terrorism, some are less severe, and may be "simply" criminal or unethical.

☑ Human-caused hazards include cyber attacks, rioting, protests, product tampering, bombs, explosions, and terrorism, to name a few.

☑ Accidents and technological hazards include such issues as transportation accidents and failures, infrastructure failures, and hazardous materials accidents, to name a few.

☑ The list of potential disasters or incidents is not exhaustive. You will need to evaluate your company's business operations, geographical locations, and other factors to assess your risks.

☑ According to an IBM study, 80% of all data loss is human-caused.

☑ Threats to electronic data include both intentional and inadvertent problems such as inappropriate handling, disclosure, or distribution of private data.

☑ Personal privacy, fraud and theft, and managing access are three key areas to be addressed in a BC/DR plan. Regulations regarding the handling of electronic data are increasing due to the number of high profile problems that have occurred recently.

Business Continuity and Disaster Recovery Planning Basics

☑ Using standard project management methodologies will help you throughout your BC/DR plan development process. It will help reduce errors and avoid potential gaps in your planning activities.

☑ The basic steps of BC/DR planning are project initiation, risk assessment, business impact analysis; mitigation strategy development; plan development; testing, training and auditing; and plan maintenance. Each is discussed in subsequent chapters.

Frequently Asked Questions

The following Frequently Asked Questions, answered by the authors of this book, are designed to both measure your understanding of the concepts presented in this chapter and to assist you with real-life implementation of these concepts. To have your questions about this chapter answered by the author, browse to **www. syngress.com/solutions** and click on the **"Ask the Author"** form.

Q: I work for a company that does not have any business continuity plan in place. Every time I bring up the need for one, people roll their eyes. Any suggestions on how to convince management a plan is needed?

A: Yes. Whenever you run into resistance trying to get your company to do something you believe is in the best interest of the company, the employees, the stockholders, the vendors, the suppliers, and the community, you need to clearly and succinctly spell out the business case. Unfortunately, some IT departments have gotten a bad reputation, either for always planning and never executing or for spending lots of money with little to show for it. Although we all know these reputations are often inaccurate, you may have to overcome that sort of attitude in your firm. If possible, find out your company's annual revenues. If that information is confidential, ask for an "order of magnitude"—a range such as $1M to $25M; $26M to $100M, and so on. Then estimate, as best you can, the cost of an outage from the most likely to occur risks (we'll go through this later in the book)—a flood, a winter ice storm, power outage, and the like. Finally, estimate the cost of putting a BC/DR plan in place—add in staff hours at some average hourly equivalent, add in supplies, extra equipment, and such. Granted, at this point, all you have is an educated guess that will probably end up being inaccurate should you get the green light to actually develop a real plan, but it should help prove your case. Almost without exception, it will cost less to plan for an outage than it will cost you to deal with the aftermath of an outage if you are unprepared. Another approach is to simply (and dramatically, if necessary) announce the failure rate of companies that experience a disastrous event. Do this in your next staff meeting, in your weekly status report, via e-mail, PowerPoint, or meeting with executives (make sure it is an appropriate venue for such a discussion). Sometimes pure, simple statistics can be a great wake-up call for reluctant executives.

Q: Why is process part of BC/DR planning? It seems that if you take care of the technology aspects, you're set.

A: There are two reasons why process is part of BC/DR planning. First, you need to understand how normal business processes may be interrupted (or destroyed) by a disaster. Without understanding the potential impact on activities such as order processing, fulfillment, development, manufacturing, and so on, you won't be able to assess the IT needs during and after a disaster. For example, in some cases, the company may choose to use paper-based systems in the immediate aftermath of a disaster and transition back to computer systems after a particular milestone has been reached, such as regaining power to a key facility or building. Second, IT will need to develop processes for resuming or continuing business operations if some or all of the technology components are interrupted or destroyed. So, if your application server is up and running but your firewall and Web servers have been destroyed, you'll need to be familiar with what processes will be needed by IT and by users to resume business operations. Understanding business processes tied to technology helps you develop the most viable plan for various potential disaster scenarios.

Q: Our company doesn't really have any legal liability if sensitive data is lost during a major disaster, does it?

A: Yes, it does. If your company deals with confidential or sensitive data, it must be protected to the greatest extent possible, even during a disaster. If you have taken every "reasonable precaution" and sensitive data is lost, you may be cleared of any wrongdoing, but your firm may still face civil charges or stiff financial penalties. Your company should consult appropriate legal and financial counsel to determine the extent of your potential liability should a catastrophe strike.

Q: You've talked a lot about the financial impact on a company. Are there other factors that impact the company that I can use to convince senior management we need a plan?

A: Yes, financial impact is often the most easily determined and the one universally understood by most senior managers. However, the financial effect is just one aspect of the impact of a disaster. Other elements include human, environmental, public image, political, legal, and regulatory impact. The case study following this chapter highlights the legal, regulatory, financial, and public image problems faced by a credit card processing company that failed to protect customer data. In today's atmosphere where more and more data is stored electronically, you must be confident you have taken every reasonable (and all required) step to secure data. If this doesn't stir your management team to support a business continuity and disaster recovery planning project, you might ask yourself if you'd be better off working for a company that does want to plan for potential problems and proactively address them.

Q: A 40% chance of failure equates to a 60% chance of survival. It seems BC/DR planning is, potentially, just a waste of time and money, isn't it?

A: Not really. If you were to go to a bank and ask to borrow money and tell the loan officer there's a 60% chance you'll pay the money back, I doubt you'd find a bank willing to loan you the funds. If you were 60% confident you wouldn't get in a car accident, regardless of who's at fault, would you choose not to buy insurance? A 60% chance of things going right is just better than even—just better than flipping a coin on any given day. And remember, only 6% of companies that experience a significant event survive long-term. That's a 94% chance the company will fold. The lower percentages (40%, 50%) are short-term failures. So really, it's accurate to say that most companies face a less than 10% chance of long-term survival in the aftermath of a significant event. Most companies would consider those to be pretty dismal odds, but many companies or executives are unaware that they are even playing the odds. Those odds can be severely tilted in your company's favor through the fairly simple act of assessing, planning, developing, practicing, and maintaining a plan. Your job, as part of the IT team involved with BC/DR, is to bring these facts to management's attention in order to gain support for the BC/DR project. We'll discuss this in more detail in the next chapter.

Legal Obligations Regarding Data Security

Contributor Profile

Deanna Conn, Partner, Quarles & Brady, LLP

Deanna Conn is a partner at the law firm of Quarles & Brady, LLP. The firm provides broad-based, national legal services through a network of regional practices and local offices with over 400 attorneys nationwide. Ms. Conn practices in the areas of commercial and intellectual property litigation, intellectual property transactions, internet law, e-commerce, licensing and technology transactions, copyright, trademark, and trade secrets. Ms. Conn also specializes in advising direct selling companies on online distribution policies and regulatory compliance issues. She was Senior Editor at the Columbia Law Review where she earned her law degree.

Ms. Conn practices law in the Tucson, Arizona office of Quarles & Brady, LLP, and has worked closely with numerous clients on the issues surrounding electronic data. She was awarded the designation of Best Lawyers in America for Information Technology Law (2005–present).

> **WARNING**
>
> The information provided in this article is intended to inform readers of potential issues, responsibilities, and requirements of the law with regard to data security. It is *not* legal advice and should not be construed in any manner as such. The publisher, author, and contributor make no legal warranties of any kind and nothing in this case study should be taken as legal advice. For more information, contact your firm's legal counsel or an attorney who specializes in internet, e-commerce, and electronic data security law.

Background

Since 2005, data security breaches have exposed more than 100 million people to possible identity theft.[1] Incidents involving ChoicePoint, described next, as well as many others, have confirmed concerns among federal and state legislators that something needs to be done to address the problems of data security. In fact, security breaches had been occurring for years, however publicity regarding these breaches was more limited. In 2003, California was one of the first states to pass a law requiring companies to notify affected consumers regarding security breaches. Since then, many other states have passed similar notice laws. These public notice requirements have heightened consumer and public awareness regarding data security breaches.

The ChoicePoint Incident

ChoicePoint is one of the country's largest data brokers, reportedly having information on every adult living in the United States. ChoicePoint has amassed over 19 billion public records, including records from the Department of Motor Vehicles, credit card data, court records, and property records. ChoicePoint sells this data to direct marketers, such as credit card companies, for use in targeted mailings. In February 2005, ChoicePoint mistakenly sold the data of 145,000 people to a bogus group of companies that paid ChoicePoint to access its database. These bogus companies faxed applications to ChoicePoint using the same fax number, even though the requests purportedly came from different businesses. Approximately 1000 people whose data was sold by ChoicePoint have become victims of identity theft.

Following reports of the incident, the Federal Trade Commission (FTC) filed a lawsuit against ChoicePoint alleging that it had violated the Fair Credit Reporting Act (FCRA). The FCRA requires consumer reporting agencies to maintain reasonable procedures to limit furnishing of information for proper purposes. Pursuant to the FCRA, consumer reporting agencies must take reasonable steps to verify the identity of each new prospective user of the information. The FTC alleged that ChoicePoint failed to comply with these requirements.

As a threshold legal matter, it is unclear whether data brokers such as ChoicePoint even qualify as "consumer reporting agencies." As a separate claim, the FTC also alleged that ChoicePoint broke its own promises to its customers to safeguard data as reflected on statements made on ChoicePoint's Web site, such as: "ChoicePoint allows access to your consumer reports only to those authorized under the FCRA." The importance of taking care with representations made by companies on their Web sites is demonstrated by the ChoicePoint case. ChoicePoint ultimately settled the FTC lawsuit on February 1, 2006 by agreeing to pay an unprecedented $15 million, the largest civil penalty in FTC history. ChoicePoint also agreed to implement a security program that will be audited by a third party every two years until 2026. The FTC also is requiring actual on-site visits of potential subscribers who receive consumer information.

Although the FTC action has settled, more than 20 class action lawsuits were filed against ChoicePoint alleging violation of the FCRA, as well as invasion of privacy and negligence. Those lawsuits are still continuing. As a result, the total exposure to the company from this data security incident is well over $15 million dollars.

In addition to the ChoicePoint incident, there were at least 100 separate incidents of data loss reported after ChoicePoint. According to some reports, hacking accounted for approximately 50% of these incidents, and stolen or lost devices accounted for 25%. The remaining 25% were attributed to insider employee actions.[2]

State Laws Regarding Data Security

As a result of these many data breach incidents, state legislators concluded that the federal government was not acting swiftly enough to combat data security problems. A patchwork of state laws have been passed since, attempting to address the problems of data security. Whether these laws are successful, even if they simply shift the burden of security to companies, and ultimately, to consumers, remains unclear. Because each state's law is different, companies doing business with consumers throughout the United States must navigate many different laws. Most of the new state laws are vague, however, offering no clear guidelines on security requirements.

There are two new types of data security laws:

■ Laws that require companies to notify customers regarding personal data security breaches

■ Laws that require safeguarding of personal information

Notice of Security Breach Laws

As of January 2007, at least 35 states have enacted laws requiring companies to notify consumers if they suffer a data security breach. These laws require companies, and in some cases, state government agencies, to disclose security breaches involving personal information.

Definition of Personal Information

There is no established definition regarding what constitutes personal information. At a minimum, however, any individual's name in combination with another element of identifiable data, including social security number, driver's license information, account number, or credit card number, constitutes personal information.

What Triggers Notice Requirements?

Some states trigger notice obligations only if there is a "reasonable likelihood of harm." No guidance is offered regarding how to assess this issue. Companies may be likely to err on the side of avoiding disclosure. If, however, the failure to provide notice is later deemed unreasonable, the company will be subject to penalties for failing to comply with notice requirements.

A number of states make the further distinction of requiring notice only if there is a breach involving *unencrypted* data.[3] To the extent that companies have not already taken the step of encrypting sensitive personal data, state laws such as California's, which impose notification requirements only when there is a breach of unencrypted data, should encourage companies to encrypt data. Presumably, however, a breach involving encrypted data will also trigger notification requirements when the breach is linked to the encryption itself, such as when the encryption key is compromised.

Notification Procedure

Although the procedures vary by state, most laws require that companies provide consumers whose data may be affected with written notice, electronic notice, or "substitute" notice. Written notice is the standard method; however this can be costly if the breach involves thousands of consumers. Electronic notice is inexpensive, presuming the company has information regarding the consumer's e-mail address. The company must also comply with federal law regarding electronic contracts, which imposes certain additional disclosure requirements pursuant to the federal E-SIGN law (the Electronic Signatures in Global and National Commerce Act), 15 U.S.C. section 7001. Substitute notice may be used if the cost of providing written notice exceeds certain thresholds.

In California, substitute notice may be used if the cost of notice exceeds $250,000, or if more than 500,000 people are involved. In that case, substitute notice requires the company to conspicuously post notice on the company Web site. In addition, notice is required to

"major statewide media." As ChoicePoint learned, this requirement can cause a public relations firestorm. ChoicePoint initially limited its notice to 35,000 California consumers
because California was one of the few states at that time that had a notice of security breach
requirement. When media outlets found out that ChoicePoint was opting to provide notice
only to California consumers, there was a public outcry. Eventually, ChoicePoint sent
110,000 letters to people in other states. The mailing cost alone was $58,000.

The timeline for providing notice typically is not specified. Most states, such as
California, require companies to notify consumers in "the most expedient time possible and
without any unreasonable delay."[4] Some states, however, such as Florida and Ohio, have specific time periods that require notification within 45 days after the security breach.[5]

Penalties

Most states do not impose statutory penalties (penalties mandated by statute or law) for
breach of timely notification. In contrast, some states impose administrative fines for failing
to meet the timelines for notification provided by law. Florida law allows fines of up to
$1,000 per day for the first 30 days after the 45-day deadline. Thereafter, the fine can reach
up to $50,000 for each day that the breach goes undisclosed for up to 180 days. If the
breach is not disclosed within 180 days after the 45-day deadline, any person required to
make the notification under Florida law can be fined up to $500,000.[6]

Safeguarding Personal Data State Laws

At least 12 states have enacted statutes that require safeguarding of personal data. Most of
these laws provide no guidance on the steps that should be taken to protect personal information, instead simply providing that "reasonable measures" should be taken. As an example,
California law provides that: "A business that owns or licenses personal information about a
California resident shall implement and maintain reasonable security procedures and practices appropriate to the nature of the information, to protect the personal information from
unauthorized access, destruction, use, modification or disclosure."[7]

Some states are considering legislation that would provide at least some greater detail
regarding what qualifies as "reasonable measures." As an example, Massachusetts House Bill
4775 provides that "reasonable measures" shall include, but may not be limited to, implementing and monitoring compliance with:

1. Policies and procedures that are designed to protect personal data from unautho-
    rized access, acquisition or use.

2. Disposal policies and procedures that require the burning, pulverizing or shredding
    of papers containing personal data so that the personal data cannot practicably be
    read or reconstructed and destruction or erasure of electronic media and other
    nonpaper media containing personal data so that personal data cannot practicably
    be read or reconstructed.

3. Performing due diligence when contracting with a third party to dispose of papers or electronic or other nonpaper media containing personal data.

The Massachusetts' proposed legislation defines "due diligence" to include, but not be limited to:

1. Reviewing an independent audit of the third party's operation and its compliance with this statute and industry standards.

2. Obtaining information about the third party from several references or other reliable sources.

3. Requiring that the third party be certified by a recognized trade association or similar organization with a reputation for high standards of quality review.

4. Reviewing and evaluating a third party's security policies or procedures.

5. Taking other appropriate measures to determine the competency and integrity of the third party.

The foregoing suggests that companies need to develop and implement data security and destruction policies, audit those policies for compliance, review third party contracts, request audits of third party security policies, and conduct independent audits of the company's own security policies and systems.

One guide that may be helpful in achieving these goals is the best practices standard set forth by ISO 17799, accessible at www.iso27001security.com/html/iso17799.html. This standard has been promulgated by the International Standards Organization as a code of practice for information security management.

In addition, companies should keep track of what records they keep, where those records are stored, as well as contact information for their customers in the event the company has to provide notice of a security breach. Obtaining e-mail address information would be particularly useful to reduce the cost of complying with these notice requirements. Information regarding the state of residency is also important to ensure that the notice complies with that customer's state laws.

Federal Laws Regarding Data Security

To date, there is no comprehensive federal law governing data security. There is, however, a federal law governing data destruction policies. The Fair and Accurate Credit Transactions Act of 2003 (FACTA) is primarily directed to credit reporting agencies; however provisions of the act relating to data destruction have broader application. Section 682.3(a) of FACTA provides that:

"Any person who maintains or otherwise possesses consumer information for a business purpose must properly dispose of such information
by taking reasonable measures to protect against unauthorized access to
or use of the information with its disposal."

This law clearly has broader application and suggests that all companies must take care to develop data destruction policies, and undertake routine audits to ensure compliance with those policies.

Aside from FACTA, both the U.S. House of Representatives and the U.S. Senate have proposed numerous bills regarding data security. If passed, these bills would likely preempt state law, which would at least have the benefit of creating uniform requirements and protections. Some criticize that the proposed federal laws would serve only to weaken existing state laws[8]. In October 2005, 48 attorney generals sent letters urging Congress not to preempt the power of states to enact and enforce their own state security breach law.[9]

H.R. 4127 is representative of many proposed House bills. As currently drafted, it would fully preempt state law. In contrast, Senate Bill 1789 attempts to address this concern at least partially.

U.S. House of Representatives Proposed Bill

H.R. 4127 imposes data security safekeeping requirements, in addition to imposing notice requirements in the event of a data security breach. The proposed bill would require "the current state of the art in administrative, technical, and physical safeguards for protecting" personal information. No other detail is provided regarding what that standard requires.

The proposed bill requires notification to customers only if there is a "reasonable basis to conclude that there is a significant risk of identity theft." Finally, upon request, information brokers must provide a means for consumers to review personal information and dispute and correct the accuracy of the information. This bill does not impose any administrative penalties for violations. If enacted, this bill would fully preempt state laws.

U.S. Senate Proposed Bill

Unlike the House of Representative's proposed bill, Senate Bill 1789 would only partially preempt state laws by allowing states to impose requirements with respect to an individual's ability to access and correct personal electronic information. Otherwise, the bill would also require safeguards, including implementing a mandatory "Personal Data Privacy and Security Program," which would require risk assessments, training programs, and vulnerability testing. However, this safeguard would be required only for data brokers holding personal information of more than 10,000 U.S. residents.

The proposed Senate bill would also require mandatory notification to the U.S. Secret Service if there is a security breach involving over 10,000 individuals, with no exceptions. Finally, this proposed bill outlines penalties for failure to notify, ranging from $1,000 to

$55,000 per day depending on the degree of culpability in failing to comply with the law. Violators also could face imprisonment for up to five years.

Conclusion

To date, existing state laws regarding data security offer little practical guidance on what companies must do to comply with state laws. Proposed federal bills are similarly general in nature. As a result, companies must look primarily to standards and guidelines that are coming from within the data security industry.

At a minimum, companies should keep an inventory of the information they license or own, implement data security and data disposal policies, audit compliance with those policies, regularly update security systems, and finally, act quickly to notify customers in the event of a breach, recognizing that notification obligations vary based on the state of residency of the individual who is affected.

Footnotes

[1] See data collected by Privacy Rights Clearinghouse regarding the number of data security breaches at http://www.privacyrights.org.

[2] See State Security Breach Legislation, VigilantMinds, Inc., Feb. 2006, p.1.

[3] See Cal. Civ. Code §1798.82(a) ("Any person or business that conducts business in California, and that owns or licenses computerized data that includes personal information, shall disclose any breach of the security of the system following discovery or notification of the breach in the security of the data to any resident of California whose unencrypted personal information was, or is reasonably believed to have been, acquired by an unauthorized person.") (Underlining added).

[4] See Cal. Civ. Code §1798.82(a).

[5] See Fla. Stat. § 817.5681(b); Ohio Rev. Code § 1349.19.

[6] Fla. Stat. § 817.5681(b)(1)-(2).

[7] Cal. Civ. Code § 1798.81.1(b).

[8] See State Security Breach Legislation, VigilantMinds, Inc., Feb. 2006, p. 3.

[9] See State Security Breach Legislation, VigilantMinds, Inc., Feb. 2006, pp. 3–4.

Frequently Asked Questions

The following Frequently Asked Questions, answered by the authors of this book, are designed to both measure your understanding of the concepts presented in this chapter and to assist you with real-life implementation of these concepts. To have your questions about this chapter answered by the author, browse to **www. syngress.com/solutions** and click on the **"Ask the Author"** form.

Q: If our sensitive data, such as consumer's personal data, is encrypted and we experience a security breach, what's the best course of action to take?

A: If the security breach is linked to the encryption itself, such that the encryption has been compromised, state notification laws will likely be triggered. You must determine the state of residence of each of your consumers, and then follow each state's security notification laws. If the security breach is not linked to the encryption, many state notification laws will not be triggered; however you must again examine each state's law to confirm that only breach of unencrypted information triggers notification requirements.

Q: If our company experiences a significant disaster event such as a hurricane or tornado, and much of our data was on paper, which is now flying around a 10-square-mile area, what is our legal exposure with regard to personal information?

A: Many state security breach notification laws apply only to breaches involving electronic data. However, some state notification laws apply to both paper and electronic data. In addition, if a person's privacy is compromised, your company may still have exposure to common law claims for breach of privacy.

Q: If our company experiences a significant disaster event such as a hurricane or tornado and our computer equipment is stolen or compromised, how is that likely to be handled legally?

A: Any breach in security involving electronic data, regardless of the cause of the breach, will trigger state data security breach notification laws. Some state laws trigger notification only when the breach is "likely to result in harm" to the individuals whose personal information has been acquired. No guidelines are provided to assess this risk, however.

Q: If we contract with a reputable third-party company to store our data backups at a site we do not own or control and our backups are compromised, what is our legal responsibility and exposure? Does this change if the breach is the result of a natural disaster such as fire, flood, or earthquake?

A: The third party is obligated to provide your company with notice of the data security breach. Depending on the agreement between the parties, either the third party or your company must provide notice to the consumer of the data breach. If the third party fails to provide notice, your company must do so. The cause of the breach does not matter if there is a breach in security.

Q: If an employee compromises data security in the aftermath of a significant business disruption such as earthquake or flood, what are the best steps for us to take? What are the legal ramifications if the breach was intentional? Unintentional? Undiscovered for a period of months?

A: Breach typically is defined as "an unlawful and unauthorized acquisition of computerized data that materially compromises the security, confidentiality or integrity of personal information." If data is simply destroyed, security breach notification laws may not be triggered. In terms of steps to take, the company must determine in what manner the data has been compromised, and whether the breach is likely to result in harm to the individuals whose personal information has been acquired. An intentional breach may be more likely to result in harm. Notice requirements are triggered once a determination has been made that a breach has occurred. However, laws regarding safeguarding of data privacy may also be implicated if "reasonable" steps were not taken to safeguard data in the first place.

Q: "Reasonable precautions" can be expensive. Are the requirements the same for large, multimillion dollar companies as for small companies?

A: Most state laws impose the same data security breach notification requirements, regardless of the size of the company. However, with regard to data safeguarding requirements, "reasonable measures" are likely to vary based on what is reasonable given the resources of the company, and depending on the amount of data that is stored and its sensitivity.

Q: What advice do you have for IT professionals in small and medium companies with regard to data security, especially as it relates to business continuity and disaster recovery?

A: Where possible, data should be encrypted. IT professionals must try to keep up with industry data security trends because what is "reasonable" today is unlikely to be considered "reasonable" tomorrow. Finally, companies should keep track of what data they store, the state of residence of each consumer, and the contact information for each individual, including e-mail addresses, in the event security breach notification requirements are triggered.

Chapter 2

Project Initiation

Solutions in this chapter:

- **Elements of Project Success**
- **Project Plan Components**
- **Key Contributors and Responsibilities**
- **Project Definition**
- **Business Continuity and Disaster Recovery Plan**

- ☑ **Summary**
- ☑ **Solutions Fast Track**
- ☑ **Frequently Asked Questions**

Introduction

In Chapter 1, we discussed the relationship of disaster recovery and business continuity activities. You learned that disaster recovery plans are implemented in the immediate aftermath of a business disruption. Business continuity activities usually begin after the immediate impact of the disruption, event, or disaster has been addressed. It should be clear, then, that business continuity and disaster recovery (BC/DR) are two distinct plans that intersect. Each can be planned as a separate project using standard project management (PM) methodologies, or they can be planned as one larger, integrated plan. The steps needed to deal with the immediate aftermath of a disaster and the steps needed to ensure the business stays up and running may be one and the same for some companies. Only you and your project team will be able to make that distinction. However, regardless of the approach you take, you'll need to design your BC/DR projects as formal IT projects in order to avoid costly mistakes such as erroneous assumptions or gaping holes in your plan.

A project is defined as a set of tasks having a defined start and end point and specific objectives, requirements, and goals. Clearly, both business continuity and disaster recovery planning qualify as projects under this definition. The BC/DR planning process can, and should, be constructed as a project plan and each component (BC, DR) can then be implemented as a project. There is good cause for mentioning this. One of the reasons many companies fail to develop effective plans is that they do not approach BC/DR planning as a *project*. They see it as an all-encompassing, all-consuming, never-ending task; it becomes overwhelming or vaguely defined. No one can get in gear or stay motivated to complete a never-ending task. This also adds to the erroneous belief that BC/DR planning is just for large companies with deep pockets.

In this chapter, we're going to look at the process of creating a project plan for your BC/DR activities. If you're already familiar with IT project management, these steps will be familiar and should serve as a good reminder. If you're not familiar with IT project management, this chapter will help you become better acquainted with IT PM best practices and guide you through the process. You won't find an overly technical or detailed plan here; what you will find is a framework you can customize to the unique needs of your organization. If you're a certified Project Manager, you will find that these steps follow general guidelines, but may not adhere to the PM methodology of your choosing.

As with any IT project, there are numerous elements that tend to contribute to the likelihood of success. In this chapter, we'll begin by discussing those factors and how they relate, specifically, to your BC/DR planning efforts. We'll continue by looking at the elements you should include, how you might want to organize the project and your team(s), and how to develop success criteria so that you can mark your progress and recognize success.

Whatever you do, don't skip this chapter. It creates the framework for the rest of the book, and it will be relatively painless to get through this chapter without having nightmares or falling asleep. Throughout the remaining chapters, we'll refer to our progress diagram to help you keep a visual image of where we are in the overall process. Figure 2.1 shows we're in the first step, project initiation.

Figure 2.1 Business Continuity and Disaster Recover Project Plan Progress

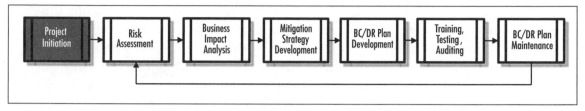

Elements of Project Success

Numerous studies through the years show there are a set of factors that, when present, tend to make projects more successful. The Standish Group International began researching project success and failure back in the 1990s. They've published a report called the CHAOS Report every two years. Each time the list is updated, the order of success factors changes slightly, but the same factors consistently show up in the top eight positions. These factors are:

- Executive Support
- User Involvement
- Experienced Project Manager
- Clearly Defined Project Objectives
- Clearly Defined Project Requirements
- Clearly Defined Scope
- Shorter Schedule, Multiple Milestones
- Clearly Defined Project Management Process

In this section, we'll discuss each of these factors as it relates to your BC/DR project.

Executive Support

Executive support for any IT project is typically the number one success factor. It makes sense that support from the top of the organization for an IT project tips the odds of success in your favor since executives have the ability to provide funding, resources, staffing, and political cover. If they are convinced there is a clear business need, they will go to bat for you and help ensure you get what you need to succeed.

How does that translate to BC/DR planning? As we discovered in Chapter 1, BC/DR planning has to be a comprehensive plan that covers every critical aspect of your business. In order for your plan to be successful, you *must* work with people from all key areas of your company. In order to do so, you clearly need the authority and reach that executive support can provide. You'll need to pull people away from other projects and tasks to participate on this project, and their managers are certainly going to be asking questions like—Who authorized this? How will this impact my other projects? Who is responsible for making this decision? If the senior managers or executives of your organization are behind you 100%, your authority to move forward on this project will be supported and others within the organization will typically fall in line. Granted, they may not be happy to shift their priorities to make room for this project, but they will if their boss and their boss's boss say so. Related to this, if you and your project run into resistance that you cannot overcome, you can escalate the issue through the proper channels and expect to find support as it reaches the executive ranks (assuming you're in the right).

Common Challenges

Gaining Executive Support

Some people in IT find it difficult to garner executive support for projects, and there are three common reasons and some fairly straightforward solutions to these challenges.

1. In many cases, IT staff get too wrapped up in the technology and fail to acknowledge and highlight the business need for the solution. In the case of BC/DR planning, you can cite the statistics provided in Chapter 1 as primary reasons for requiring a BC/DR plan. Throughout this book, you'll read about cases and examples of the business case for BC/DR planning that you can use in your discussions with executives.

2. Another common reason for failing to get support for BC/DR planning is a simple failure to communicate clearly, concisely, and convincingly. You don't need a huge presentation with graphics and Flash animation to make your point. You need to figure out who your audience is, what you want them to understand when you're finished, and how you should present the information. Some audiences want a PowerPoint presentation with bullet points; some audiences want a one-page written summary; still other audiences want a five-minute verbal presentation with a one-paragraph written summary. There is no one single correct way to present the information other than this: *Determine the preferred format for your intended audience and create a short, concise presentation using nontechnical language.* Most executives have heard of VoIP, NAS, SANs, and TCP/IP, but don't understand the underlying implications of these terms.

Continued

Therefore, keep it nontechnical and clear. Keep your objectives in mind and lay out your information in a logical progression.

3. Finally, a third common reason for failure to get executive support for IT projects is that IT projects sometimes appear (to executives) as bottomless pits into which time, money, and resources fall. You'll need to provide executives with a ballpark estimate for how long you believe the project will take and roughly how much it will cost (time and money) to complete. Here's the danger—until you've completed the initial project work, you probably won't know the answers to these questions. It becomes a circular problem because you can't give an estimate until you do some planning and you can't do the planning until you've provided executives with an estimate.

One way to short-circuit that problem is to tell executives you don't know how long it will take or what it will cost, but that it is critical that you be given the OK to find out. If you view your project as a two-part process where the first part is to create a rough estimate for time and cost so you can get approval for the larger planning project, you might gain support early. Alternately, you may be able to work with a few key people in the organization such as your financial person, one or more of your key Ops people, and your facilities person to come up with a ballpark estimate you're all comfortable with. Building this coalition early on may also help solidify your project team and could provide critical mass within the organization for the planning project.

Executives understand business and finance—they don't necessarily understand technology. Many are comfortable using technology and a vast majority understand the need to utilize technology effectively within an organization; few understand the terminology and the underpinnings of technology. Therefore, you have to speak business with them. Rather than saying "we need to investigate availability solutions for WAN and LAN support," you need to break it down and say "we need to investigate our options for providing data access across the entire company, both here in the building and across the globe." Both say roughly the same thing, but the second statement is business-centric, the first statement is techno-speak. If writing (or nontechnical writing) is not your strong suit, write up what you want to say and ask someone in a nontechnical department to work on it with you. For example, go to your Human Resources or Marketing department and ask someone to assist you. It would be better to get feedback from someone outside the decision-making loop than to submit your technical document to a decision-maker who doesn't understand it and won't take the time to ask you to explain it. Executives are typically very busy and they like when things are boiled down to their essence in plain language. It helps them quickly understand the situation and make a rational decision. You want to help them do just that.

From the Trenches...

When the Answer Is "No!"

The ideal scenario rarely matches reality. In today's IT world, budgets and capabilities are stretched to the breaking point and highly charged issues such as security and regulatory compliance garner the most visibility. You may very well find that your company is reluctant (at best) or unwilling (at worst) to devote time or money to developing a business continuity or disaster recovery planning project. Even though this is a suboptimal long-term business decision, the demands on a company's resources can sometimes be such that BC/DR planning is just not valued or appreciated. If you're an IT professional pushing for BC/DR planning in your company and you find you're hitting a brick wall, there are things you can do to help facilitate this process. Although you may not be able to set aside time (or money) to create a fully separate project for BC/DR, you can incorporate BC/DR concepts in all your other IT project plans. For example, if you're evaluating the implementation of a new server, new application, or new technology, include an assessment of the BC/DR concepts and include elements that will help you mitigate risk and plan for outages as you would within the scope of a formal BC/DR plan. The most basic BC/DR plan is to have redundancy and backups and most IT staff work to develop these attributes in the normal course of IT operations. There is a fair amount of coverage you can get if you begin adding BC/DR elements to your basic IT planning, projects, and operations. It won't ever substitute for a full BC/DR plan, but at least it will begin to move your IT operations in that direction without putting undue pressure on your time or your budget. So, if your company won't move beyond "no" when you discuss BC/DR plans, use the information throughout the remainder of this book in your IT operations.

User Involvement

You've probably been involved with projects that were going to dramatically impact end users, and yet no one talked with end users for months, if ever. These projects are almost always doomed to fail. User involvement consistently shows up in one of the top three spots on the list of success factors for IT projects. Many technology projects have failed because users were not involved and key decisions were made that were directly counter to user needs and wishes. Clearly, you can create any solution you want but you can't force users to use it. You can't force users to understand and accept convoluted processes for doing their once-simple tasks, to flex *around* awkward requirements of the technology. Although there can be compelling business drivers that force users to change their processes and methods, these should be created with user input and collaboration, not in the dark recesses of the IT department.

So, who are the users in a BC/DR planning project? There are essentially two sets of users. The first set includes those who will be involved in planning the BC/DR project itself. These folks may or may not be the same ones who will implement these plans should disaster strike. Therefore, you would do well to have both sets of users involved in this project. For example, you might want to have one team that focuses on defining the critical business processes that need to be addressed in the plan. A second (or subsequent) phase of the project plan could include a second project team that includes those people from around the company that would be responsible for implementing a BC/DR plan and would therefore define the implementation phase. If you work in a small company, this might be the same group of people. In larger companies, it might be two overlapping or two separate teams.

Whatever your approach, be sure to include the key people in the project from start to finish. Later in this chapter, we'll discuss who should be involved. For now, keep in mind that if you and your team create a great plan without input from those who will be the "boots on the ground" in an emergency, your plan is highly likely to fail under the stress of a disaster.

Experienced Project Manager

Experienced project managers bring a wealth of knowledge and skill to the table. They often have had some formal project management training or education and they may have achieved a standardized certification in one or more methodologies. Most importantly, though, they have been in the trenches managing projects, and have realistic understanding of what it takes to get the job done.

When we're looking at BC/DR specifically, an experienced project manager is likely to be more effective at working across organizational boundaries and in bringing together a diverse group of people and interests. Working effectively with people at all levels of the organization and in all areas of the company is critical to the success of a BC/DR plan. An experienced project manager is more likely to understand how to navigate the political waters as well as the organizational red tape that inevitably crops up during the development and implementation of cross-departmental projects.

In addition, an experienced project manager will utilize a defined set of steps, a methodology, to deliver consistent results. Most experienced project managers have developed a system of defining and managing projects that delivers positive results. Many have spent years honing their methods to generate an optimal outcome. Most adhere, in general terms, to standardized methodologies but each experienced and successful project manager undoubtedly will have customized those methodologies to suit their specific needs. This is key to delivering a successful BC/DR project plan. With the actual survival of your company at risk, it's imperative to have the most successful outcome possible. An experienced PM will increase your odds of such an outcome, though no single (or multiple) success factors guarantee success.

Clearly Defined Project Objectives

Clearly defined project objectives might sound incredibly obvious, but you might be surprised at how often projects are launched without clearly defined objectives. Clearly defined objectives are quite important because your BC/DR plan must be scaled to your organization's unique needs. Without defining the objectives, you and your team might spend a disproportionate amount of time planning and implementing a part of the plan that is less important, or you might short-change a very important area.

One way the task of defining objectives can contribute to BC/DR success is to develop a high-level list of functional areas of your company and invite key people from those areas to help define the objectives. This accomplishes two critical project objectives: it ensures that all functional areas are included and it brings together the people most able to develop appropriate objectives. As an IT professional, you are not in a position to develop objectives for BC/DR planning for, say, the Human Resources or Marketing department. You understand the technology but not the business or operational objectives, in most cases. In addition, you need to get these stakeholders together to agree on objectives because you will have to prioritize. During an emergency or disaster, many business operations, tasks, and objectives become secondary to the survival of people and the survival of the company. Determining these needs helps determine the project's objectives and this will help focus you and your team on the critical aspects of the business.

Clearly Defined Project Requirements

Related to objectives are requirements. Developing clear and complete requirements can also make the difference between success and failure, especially for an IT-related project. The requirements are those capabilities, attributes, and qualities that must be part of the final project deliverable. Defining these early in the project development cycle is important because going back to add them in later (called *rework*) is both inefficient, costly, and fraught with both errors and additional project risk.

Requirements are not the same as project objectives. The objectives should drive the requirements. Objectives are what you want to accomplish, requirements are how you will accomplish those objectives. For example, if an *objective* for your BC/DR plan is continuous availability of three key business applications and related data, your *requirements* would have to delineate this objective. Requirements may have to be refined or developed later in the project definition process as details about the project become clear. However, clear requirements before project work begins are absolutely critical to project success. Unclear requirements cause confusion, duplication of effort, rework, and wasted work. If your objective is continuity availability but you never specify which applications, which data, which users, which business functions, which locations, which customers (etc.) fall under that objective, you will undoubtedly find your project wandering off on its own.

Requirements typically fall into three categories:—business, functional, and technical requirements. *Business requirements* help you determine what the business needs to survive a disruption. This helps you understand the major building blocks of your company, how they work together, and what key areas should be prioritized. *Functional requirements* detail things such as which processes, methods, and resources need to be available during and after a business disruption. *Technical requirements* delineate things such as servers, network infrastructure, and business application requirements. The more specific your requirements, the more likely you are to have a successful outcome—a BC/DR plan that works when implemented. You can think of it this way. If you get in your car with a destination in mind but no particular route, you are more likely to take longer to get there than if you mapped out your route before you left the house. For those of you who immediately think "Yes, but I'd use my GPS navigation system," remember, there is no GPS equivalent in a project. To define your requirements during project work (the rough equivalent of using a GPS system) is to guarantee that you'll end up miles off course at the other end of the project. Though you may need to make several passes through your requirements definition phase to add detail and clarity as you *define* and *organize* your project, you should not begin the actual work deliverables (*Work Breakdown Structure* tasks) until you have clearly defined requirements.

Clearly Defined Scope

Related to clearly defined project objectives is a clearly defined project scope. *Scope* is defined as the total amount of work to be accomplished. Scope typically is defined through the project's objectives. Making sure payroll can be run during a disaster may be one objective; making sure your company can still take, fulfill, and invoice customer orders is another objective. If these are the only two objectives for your BC/DR plan, you can fairly easily determine the project's scope. Therefore, clearly defined objectives lead to a clearly defined project scope.

Scope creep happens to many projects, and BC/DR is a project type that is perhaps more susceptible to scope creep than many other types of projects. For example, it's not hard to imagine that you decide that being able to process payroll during a crisis will be critical to the well-being of your firm's employees, so you include a high-level objective to that effect. However, as the project planning stages progress, someone mentions that the Human Resources director also wants the ability to easily set up direct deposit for people during a crisis in the event they cannot come to their work location to pick up a check. Your scope has just experienced creep—as you may know or suspect, it's one thing to ensure that payroll can be processed per usual, but another thing entirely to suddenly add the ability to go to direct deposit in an emergency. Clearly, additional steps must be included in the project plan to enable this capability, especially if it does not currently exist or if it will need to exist in a different way. For example, it is likely that people who want to use direct deposit currently have to submit a form with a voided check from their bank account and it must go through one or two payroll cycles before it is effective. During a disaster, employees will not have access to the form, they may not have access to their checkbooks, and they won't know how to accomplish that without talking with someone from HR. Additionally, it might be unacceptable for it to take two payroll cycles for this to occur during an emergency. Therefore, your team will need to develop possible solutions or alternatives for this new requirement and address it in your plan. So, although the HR director may have wanted this to be accomplished, he or she may not realize what it will take to accomplish this. Be sure to have clearly defined objectives, make sure that each objective is necessary during an emergency, and have the people (in this case, HR) responsible for that business function develop the potential solutions to the objective. This helps reduce scope creep and helps manage clearly defined objectives and clearly defined scope.

Shorter Schedule, Multiple Milestones

Studies have repeatedly shown that shorter schedules with more milestones generate more successful results. How does this apply to BC/DR planning? In most cases, BC/DR planning is a comprehensive look at the business and its processes to determine critical functions and emergency procedures for those critical functions. You may choose to break your BC/DR planning project down into smaller projects—for example, one project plan for each functional area and one master plan that ties these all together. You may choose to perform your planning in an iterative mode so that during the first pass, you develop just the most basic, mission-critical solutions and each iteration after that builds on the one prior. Only you and your project planning team can determine the best approach to this, but keep in mind that long schedules typically just get longer. People lose focus, enthusiasm, and interest. Resources start being pulled away from the project the further out in time you go.

Milestones are, by definition, project markers that help you gauge progress. Milestones are checkpoints that can help you stay on budget, on schedule, and on scope as your project progresses. The more milestones (within reason) your project has, the more likely it is to be successful because you are consistently comparing where you stated you wanted to be with where you actually are. This regular course correction can certainly keep your project on target far better than an occasional checkpoint that may leave you wondering how you got so far off course.

Your company may be reluctant to undertake a full BC/DR project because executives might fear that it will go on forever and never produce a result or might cost far more than the company can afford in terms of staff time and money. If you run into massive resistance, you might want to parse your plan out into well-defined stages and get executives to sign off on the phased approach. Then, when you can show success with the first phase, you may more easily gain support for subsequent phases.

TIP

If you take a phased approach to your planning, be careful to clearly delineate what *is* and *is not* part of each phase. You want to avoid executives believing that Phase One will cover something in another phase. You also want to be sure that Phase One delivers a meaningful barebones BC/DR plan at the very least. The phased approach is most often useful when executives are uncertain about the potential costs and benefits, and success in the first phase can engender support for subsequent phases, but it can also lead to a false sense of security. You'll have to find the appropriate balance for your organization.

Clearly Defined Project Management Process

A clearly defined project management process typically goes hand-in-hand with an experienced project manager. As mentioned, an experienced PM is likely to have a set of methods, procedures, and associated documents that he or she has used successfully in the past. Most experienced PMs will hone those processes and procedures over time so that they become almost second nature. If you're an inexperienced project manager, you can increase your odds of success by using a well-defined project management process. Throughout the remainder of this book, we'll use standard PM processes that will help you develop a more successful BC/DR project plan. If you have a methodology that you've used successfully in the past, feel free to utilize that in conjunction with the material presented in this book. You'll find that the presentation of PM processes in this book follows a fairly standard format and should be compatible with any standardized process you choose to use. The key is to select a process and use it from start to finish so there are no gaps in the process, which inevitably lead to gaps in the plan.

As you can see from the success factors listed, achieving project success is not rocket science, but it does require a consistent approach and attention to detail. As business continuity and disaster recovery continue to show up in the news headlines, executives are becoming more aware of the need to invest in BC/DR plans. Clearly, regulatory pressure, shareholder requirements, and vendor initiatives also have pushed this to the forefront of many executive's awareness. Some companies' executives, however, are still a bit behind the curve, usually because they do not have regulators or shareholders pounding on their door asking for their BC/DR plans. If you work in one of these companies, you may still face a bit of an uphill battle, but by reviewing the project success factors and developing a strategy for approaching this planning, you are likely to find a solution that fits your company's needs.

From the Trenches...

Even Small Companies Need a Plan

If you're an IT service firm, you may be able to provide a great service to your customers by talking with them about their business continuity plans. There are millions of small companies out there, from sole proprietors to partnerships to small companies of five or ten employees. At a very basic level, they should have a solid data backup plan. If an author is working on a manuscript while traveling, what happens if the laptop is lost, stolen, or broken? Though this may sound very basic to you, be assured that millions of people don't give this much thought until they have an unfortunate incident that costs them thousands of hours, dollars, or headaches. Whether you work in a small company or in a service firm for small companies, you should keep BC/DR plans in mind when looking at how they work and how they should be protected. Most companies (including individuals) are thankful when you bring a solution to the table that helps them prevent a catastrophe. For the author, it might be as simple as backing up to a CD, USB drive, or online backup location to prevent a serious, costly data loss. Remember, keep it simple, especially for small companies and individuals. Find the minimum they'll need to stay in business and make sure that is what's protected. Everything else after that is just icing on the cake.

Project Plan Components

Now that we've reviewed the success factors, let's look at standard project management plan components. If you have a methodology you use, this should track (generally) with yours. If you don't have a methodology you use and you're not familiar with project management, this will give you a basic overview. Our goal is not to delve into the details of project management—for that, you can pick up a copy of *IT Project Management* from Syngress.

The basic steps in a project are:

- Project Definition
- Forming the Project Team
- Project Organization
- Project Planning
- Project Implementation

- Project Tracking
- Project Close Out

Let's look at each of these briefly, and as they relate to BC/DR planning. Keep in mind that project planning and project management are both linear and iterative processes. This means that there is a logical flow that defines the order in which steps are taken; at the same time, many steps are revisited over time to add additional detail that helps more clearly define the project. For example, some elements cannot be known at the beginning of a project, so an estimate is used until enough detail is developed to go back and refine the original estimate. Although this sounds like it could lead to interminable planning (and it could), the goal is to continually move forward and to refine and hone details as they become known. This approach actually prevents "analysis paralysis," in which planners feel they cannot move forward because they do not have enough information or detail yet. Instead, this approach allows you to use estimates or placeholders so you can move forward and develop the additional detail as quickly as possible. A good example of this conundrum is the budget for a project. In many cases it's difficult, if not impossible, to give a useful estimate for the cost of a project until you've defined the project's scope, objectives, and requirements. At the same time, it can be all but impossible to get the go-ahead for a project until executives have some idea of what it will cost. In some companies, this becomes a circular problem that causes projects to just spin in a loop, going nowhere fast. To overcome this, you may have to do some initial project definition work to develop a ballpark estimate to get the OK for the project to develop additional detail to give a more refined estimate of the cost. It may sound like a lot of rework, but it's actually refining instead of redoing, and this typically leads to forward progress.

TIP

If you're having trouble getting a budget approved for your BC/DR project plan because you don't yet know what it will cost or how long it will take to accomplish it, you may have to do some savvy negotiating with decision-makers. For example, in some companies it might be effective to ask for a specific budget to investigate the cost of a BC/DR planning project. If you can get some staff time and a small budget allocated to investigate this, you may be able to come up with a fairly realistic ballpark estimate for the actual BC/DR planning project. In other companies, it might be more effective to look at annual revenues and talk to one of the financial people in the company to estimate the cost, in terms of lost revenue, lost market share, and lost customers, if a disaster were to hit. Using some estimate from a financial analyst within the company will lend credibility to your estimate and should at least help you get a budget for the first phase or for an investigation into the potential cost of developing a full BC/DR plan. In still other companies, the only real constraint might be money, so you might get a go-ahead to use staff time and corporate resources as long as you don't spend any cash outside the company. This might be acceptable in the near-term and help you get your planning project underway with executive support.

You'll have to be creative and persistent in some cases, but the importance of creating a BC/DR plan cannot be overemphasized. Finding the right approach within your organization is probably the most important first step. The case study following this chapter highlights some of the financial considerations.

Project Definition

Project definition is the first phase of any project. In some companies, it's easier to get approval for a subproject plan whose only deliverable is a clear idea of what the project will be. The definition phase can include a variety of elements; we're going to look at this from both a BC/DR planning perspective and an IT perspective.

First, let's talk about project origins. In some cases, you may be approached by the CIO or another executive in the company about creating a BC/DR plan. In other cases, you may be pushing your organization to create a plan. If the former is the case, it's quite common that someone has given you marching orders to which you're expected to conform. The problem is that without doing the requisite research and planning activities, a directive to create "a BC/DR plan with x, y, and z" in it will likely be off-target. It will have gaps or will include things that are not needed. So, how should you proceed if this project is dropped in your lap? The first step is to talk with the person who handed you the project. He or she is most likely the project sponsor. Get a clear understanding of expectations and what the project should entail. Typically, you can extrapolate key elements that should be included in the plan and from there, you can check back with the project sponsor. This step should never be skipped, and unless your company is run as a dictatorship, you should always make sure you come back with questions, suggestions, and revisions. Rarely will a project be handed to you that is so well-defined that you can just put together a plan to accomplish the objectives and off you go. Remember that executive objectives are usually quite different than the objectives needed for a solid BC/DR plan. For example, an executive may need to show compliance with a particular regulatory requirement. Your job would be to develop project requirements that would, when implemented, result in regulatory compliance. Although closely related, you can see that the executive's objectives and the project's objectives are not always one and the same.

Problem and Mission Statement

Often the most effective starting point is a problem statement. In the case of BC/DR, it might be as simple as "Our company operates in two geographic locations and generates $25M in annual sales. We do not currently have a disaster plan for either location and the company is at risk as a result." Remember, you don't have to overengineer this, but a clear problem statement helps keep you focused as you move forward so you work on solving the *right* problem. A brilliant project that solves the *wrong* problem is useless.

Next, you should create a mission statement. This, too, can be a fairly simple statement (and should not require a three day off-site to accomplish it) such as "To create a business continuity and disaster recovery plan for both of our company's locations that will address the major risks to our company and that will provide a path to recovery of the basic, mission-critical systems including a, b, and c." When you define a problem and then define what the desired outcome (mission statement) looks like, you have created your start and end points. This helps you define the scope of the project and gives you a clear end in sight so you can begin planning a finite project.

Potential Solutions

Once you've defined your start and end point, you can develop a list of potential solutions. This is an important and often-skipped step in project planning. One potential solution to be evaluated in all projects is "do nothing." Although it's often not a viable option, it keeps things real and helps you evaluate all potential solutions against the do-nothing option. Even though it's highly unlikely that a do-nothing solution would be appropriate for BC/DR planning, it should nonetheless be included. You can then measure other options against the cost of doing nothing and use it as a reality check both with your project team and with your executive team.

Looking at all potential solutions will help ensure you don't pigeon-hole your project early on. The creative process of getting the project team together to brainstorm potential solutions may yield surprising results. For example, you might find that there is a vendor that sells solutions that will address all your BC/DR needs. You may find that you already have implemented many of the needed systems, and emergency procedures are all that are needed to create a solid BC/DR plan. Until you brainstorm all potential solutions based on the start and end point you've defined, you won't know what your options are.

That said, we're assuming that the solution will be to develop some sort of BC/DR plan. Though standard project management steps make sense throughout, the specific subject does need to be considered. After you've completed the risk assessment phase, which is part of the actual project work, you'll be in a much better position to determine the potential solutions. Then, you can select the one that meets project and organizational constraints and requirements.

Requirements and Constraints

Another important element is to understand project requirements and constraints early on. Every project will have a stated or implied budget, timeline, and expected deliverables. You *will* have to revisit these once you select a solution, but you often can't select a solution until you have some basic requirements in mind. If the project was handed to you, you should go back to the person for clarification about these expectations. If you're initiating the project, you should try to develop some realistic expectations that you can bring to your boss or the decision-maker for feedback. If you're way out of the ballpark, this is a great time to find out. If expectations are significantly misaligned, this is the best place to find that out and correct it. It only will get worse from here. For example, is your budget likely to be $10,000 or $10,000,000? The budget will have a huge impact on the solutions you may choose to consider. Is your timeline four weeks or four months or four quarters? Again, your project will need to meet time requirements. What about technical requirements? Does your company have multiple locations? Are there employees who travel or work in the field? Do your customers purchase your products online? Do they place orders online? Understanding some of the high level technical requirements will also be needed before you can select the optimal project solution.

Once you've developed your initial list of requirements and constraints, you should be able to identify which solution(s) best meet your project's problem statement and mission statement (problem and desired outcome). If more than one solution appears to be feasible and desirable, you may have to look at other factors to find the most optimal solution. What factors? Sometimes they're political—perhaps your CIO is partial to a particular vendor because they deliver the best customized solutions; perhaps your timeline would be better met by one solution; perhaps your budget constraint is really the most important aspect and one solution fits your budget better. The first major objective in a BC/DR plan is a risk assessment, which will drive many of your requirements. This is an excellent example of the iterative process needed in IT project management. You may have some requirements and constraints that are known at the outset of project planning; others may have to be developed later when more data has been collected.

Success Criteria

Another element often skipped in project planning is success criteria. How will you know your project is a success? If you know this, you have a better chance of creating a successful project plan. If you develop the criteria by which you'll judge or evaluate success, you're less likely to find yourself chasing after ever-changing definitions of success. It doesn't mean they won't change or that you won't have to work hard to maintain these success criteria once the project is underway, but it gives you a known starting point. If you're not familiar with success criteria, here's a rather mundane example that should help. Let's say you need to clean up your office because it's a mess. Your success criteria might be these: 1) all loose papers are filed in marked file folders, thrown out, recycled or shredded, as appropriate; 2) all books are stored in a bookshelf or stored in the company library in Conference Room A; 3) all writing utensils are stored in a pencil holder or in a desk drawer; 4) desk top is free of extraneous equipment not currently in use; 5) all computer hardware and software (other than desktop or laptop computer and associated devices) are restocked in the lab or an appropriate location (not your office or your desk). Even though this might seem obvious, it essentially tells you what your office will look like once you finish cleaning it up. Most of us don't need a list of success criteria to know if we've successfully cleaned our office, but you can probably see how this could be extremely useful in project planning.

Project Proposal

Once you select your optimal project solution, you'll need to put together a brief project proposal. Essentially, the project proposal should include all the elements we've just delineated. It should be submitted to the project sponsor (we'll discuss the project sponsor in a moment) or to your boss. If your organization has a process for submitting project proposals, use that format. If not, include the elements listed here (you can modify to meet your specific needs):

- Business case (can include the problem and mission statements)
- Financial analysis (if appropriate)
- High level scope, timeline, budget, and quality metrics
- Requirements, constraints, assumptions, exclusions
- High-level resource needs
- Phase schedule (if a phased approach will be used)
- Success criteria
- Risks, mitigation, and alternatives (risks to the project, not BC/DR risks)
- Recommendations

The project proposal can serve several purposes simultaneously. First, it can be used to convince executives that a business continuity and disaster recovery plan is needed and that you've given serious thought to the subject. It can be used to rally others in the organization around the need for a BC/DR plan and begin the process of gathering support and critical mass for such as project. Finally, it documents the beginning of the project so that you don't have to needlessly and repeatedly revisit these decisions and this data in the future. This can help the project move forward instead of sideways and help momentum build instead of just spinning in circles.

Estimates

Although it's outside the scope of this book to discuss estimating in detail, it is worth talking briefly about estimates. First, estimates are dangerous. More often than not, they become targets. The vice president catches you in the hallway and asks how much this BC/DR plan will cost. Without thinking too hard about it, you reply that you think it shouldn't really take more than a month and about $5,000. Congratulations, you've just committed yourself to two targets! Though you may preface your response with "well, it's really early but I'm guessing..." you may still have a problem. The VP may simply remember one month and $5K. She's a busy person, after all, and she remembers just the facts. So, use extreme caution when tossing around estimates, especially ones that are just wild guesses.

Estimates can be generated based on past experience of similar projects, and these estimates, called *parametric estimates*, tend to be fairly accurate. When you can say "the ABC Project was very similar in scope and requirements and it took six months and cost about $50,000" you're in much better shape because the odds are good, if the projects really *are* similar, that estimates should be close. If you don't have a similar project you can use for comparison, you have to develop an estimate from scratch. These can be created top-down or bottom-up. The *top-down* approach is the fastest but least accurate method. The bottom-up is the slowest but most accurate method. A top-down estimate would start with an estimated budget, let's say, $100,000. From past project experience (this is where the "experienced project managers" as a success factor comes into play), you might know that designing the requirements is usually about 10% of the budget, so you could estimate that designing requirements will cost $10,000. You might also know that planning the project is typically 18% of the project budget, so planning should cost about $18,000, and so forth. If you don't have a lot of project experience or a lot of experience at this company, you generally cannot create realistic top-down estimates. *Bottom-up estimates* typically are developed after you've created your detailed Work Breakdown Structure, after you know what your tasks and deliverables are. Once that is known, it's a simple matter of adding up the cost of each task and developing a total. The biggest problem with bottom-up estimating is that you won't end up with an estimate early in the planning cycle. If you're trying to get project approval based on an estimate, a bottom-up approach won't work because it will require you do a fair amount of planning before you ever develop an estimate. In fact, the result of bottom-up estimating is usually close to a real number you can use as a target or commitment (as opposed to an estimate that will need additional refinement).

We've covered the basics of project definition very briefly. If there are elements you're accustomed to using or elements required by your company, certainly use them. If you're not familiar with these or other project definition elements, there are plenty of great resources on IT project management you can use, including one by this author mentioned earlier.

Project Sponsor

A project sponsor often is the person that hands the project to you and assigns you as project manager. In other cases, especially situations in which you are initiating the project on your own, you'll need to identify a project sponsor. A project sponsor is someone in the organization who has the authority—both organizational and political—to help you accomplish your project goals. He or she should have enough knowledge about the company and the project to help you make sound decisions, create a budget and timeline (or approve them), help remove obstacles, rally resources, and support and evaluate choices and decisions all along the way. If you haven't yet identified your project sponsor, you should do so early. Finding the right project sponsor can make all the difference to you and your project's success. If your project was handed to you, don't assume that person is also the project sponsor. Ask. The question that usually gets to the bottom of the issue is "are you the person who will approve the budget and schedule?" If no, you haven't found your project sponsor. He or she should be authorized to approve project documents, budget expenditures, and schedules. Some companies may operate differently, but the project sponsor normally has the authority and responsibility to approve expenditures, assign resources to your project, and remove obstacles to your success.

Keep in mind that your project sponsor for a BC/DR plan may be different than a typical IT project sponsor. The sponsor might be an executive vice president responsible for regulatory compliance; the project sponsor might be someone who oversees facilities. It's hard to say who your project sponsor will be for a BC/DR project, but it should be someone who 1) understands the importance of a BC/DR project and 2) who has the organizational and political power to help a BC/DR project get off the ground.

Common Challenges

Project Sponsor MIA

Some project sponsors go MIA (missing in action) and become impossible to contact. If you're in search of a project sponsor, be sure to select someone you're confident will be available to you on a regular basis. You shouldn't have to run to the project sponsor for approval of every single item, but you should meet with your project sponsor regularly (via phone, net meeting, or in person) to discuss project progress. Should your project sponsor be someone who is not good at returning e-mail or voice mail or regularly travels and is unavailable, your project will likely run into significant delays. Find a sponsor who will be available and then make sure you use your time with the sponsor productively. Be prepared, have an agenda, and get on with it. Some project sponsors go MIA if they find their time is not being used productively. Don't expect to

Continued

chit chat over a *double tall nonfat half caf vanilla latte* with your project sponsor—stay focused and your project sponsor may actually look forward to your meetings. Waste their time, and they'll dread (and then avoid) your meetings.

Forming the Project Team

You should form a project team pretty early in the project cycle for numerous reasons. You're unlikely to have all the knowledge you need to develop a sound plan by yourself and you will eventually need the input from various subject matter experts during the course of the project. Studies repeatedly show that those who help plan a project are more likely to contribute positively to the project because they feel a sense of control and ownership. Therefore, you would be ill-advised to create the plan on your own and unveil it to your project team. Instead, decide who should be on the project team, form the team, and create the plan.

There are times when it's not initially clear who should be on the project team. This may well be the case with a BC/DR project because it crosses so many organizational boundaries. One approach is to create a preliminary project team whose sole job it is to create the basic project definition and then determine who the right team members should be. In some cases, the original team might be suitable; in other cases, the team may require a few new members or the removal of a few existing members whose long-term participation just doesn't make sense. For example, it might be that the initial planning meeting(s) include area directors or vice presidents—especially if this project has high visibility, is related to ongoing regulatory compliance, or has been assigned by an executive. Once the high level definitions are created, directors or VPs may then select members of their teams to join the project team and they, themselves, may bow out. This is entirely appropriate. The key is to develop your project team with an eye toward covering all your organizational bases. You want neither a team that's too narrow in focus nor a team that is so large as to be difficult to manage and coordinate. You can choose to look at your company in many different ways. One is to include the basic functions, which typically are operations, human resources, finance, facilities and security, logistics/purchasing, public relations and, of course, IT. Another way to look at it is by reviewing various categories. When forming your team, you could also consider these elements, which we'll touch on briefly next.

- Organizational
- Technical
- Logistical
- Political

Organizational

The first place to look is at your company's organizational chart. This will help you identify geographic locations, functional departments, or divisions of the company. These should all be included in your BC/DR planning process. Clearly, there will be some overlap that can help you pare down team members. For example, you may have an HR manager or director at each of 10 worldwide locations. You may need all of them to work as a subteam on HR needs during a crisis or you may select two of the most experienced to represent HR on our BC/DR team. If there are significant differences among your various worldwide locations, you'll need to look at the best way to approach HR needs, especially IT functions. For example, how is payroll processed in the Netherlands compared to how it's processed in Brazil or Australia? If it's all done through a single payroll processing company, you have different options than if each country processes its own payroll. There is no single right approach, so your best bet may be to form a preliminary team with the understanding that it probably has too many representatives and should be pared down once roles and responsibilities are known and understood.

WARNING

Things can get very political very fast when discussions of mission-critical areas of the business begin. Everyone wants to think of their part of the business as being critical to the company (i.e., if it's not mission-critical, why does it exist at all?). Many people become quite concerned when their area of business is not tagged as "essential to operations." Therefore, be on the lookout for power plays and bruised egos here. It's a delicate balance. You may need to resort to scenario-based questions to help people understand what BC/DR planning *is* and *is not*. For example, you might ask, "If this building caught on fire, what's the bare minimum we would need to get back up and running again?" This can help focus people on the project's objectives rather than on worries that their department or job function isn't included. Also, it often helps to indicate that *work* will be required of team members—that usually rids the team of the people who want to feel important but do not want to be held accountable for deliverables.

Technical

Which different technical specialties should be included? You may be unable to answer this question until you understand which areas of your business are mission-critical. The technologies used in areas of the business not considered mission-critical may not be represented in your planning. Alternately, they may be included in business continuity planning but not in disaster recovery planning since these technologies may not be needed immediately but will be needed in the long-term recovery of the business. Remember that technical issues involve not just the IT department but the facilities functions as well. How the building is heated, cooled, powered, maintained, and more impacts the IT function. If you have power but no heating or cooling, you may not be able to get systems up and running again. With the integration of the various technical components of your business, whether that's desktop computers, large server clusters, manufacturing systems, healthcare devices, and others, there are numerous technological factors to be considered in addition to IT-specific systems. Understanding how these work together and what elements truly are mission-critical in the aftermath of a disaster is an important area to explore in your project planning work. Understanding how you'll transition from disaster recovery to business continuity also requires a close integration of IT and non-IT technology management.

Logistical

There are two logistical components—one related to disaster recovery and one related to business continuity. In the immediate aftermath of an emergency or disaster, the most important tasks are related to stopping the impact of the event. If there's flooding, one disaster recovery task might be to move servers or computers to higher ground or to contact your company's other locations and let them know they will need to pick up the slack. These steps "stop" the effect of the flood. After those tasks are underway, business continuity activities typically begin. How can the business get up and running again? To continue with the flood example, it might be locating a temporary office building or arranging with a contractor to come in and begin pumping water out and start repairs. These all involve logistics of various kinds. Those in your company responsible for logistics and/or purchasing should certainly be included in the BC/DR planning activities.

As mentioned, it's often quite helpful to contract for emergency services before an emergency. You can lock in better rates before demand spikes the cost. You can calmly and rationally order what you need rather than ordering whatever comes to mind during the emergency. You can develop and maintain a relationship with a variety of firms to provide those services; as an existing customer, you're likely to get preferential or priority service during an emergency. Your logistics staff needs to be involved with negotiating contracts for these mission-critical needs before disaster strikes.

Political

One element often overlooked is the political aspect of managing a crisis. Your company may need to communicate publicly after a crisis to assure stockholders and key customers that the company is intact and prepared to maintain or resume operations. This was the case with many of the financial institutions in the days following the September 11th attacks on the World Trade Center in downtown New York City. In addition, there may be internal political ramifications related to managing a crisis or emergency in your firm that should be addressed as part of your BC/DR plan. As an IT professional, you may not be aware of these political needs, so you should include people on your team who will be up to date on internal and external political requirements for your plan.

Project Organization

The project organization includes how you will organize and run your project. It begins with identifying the right project sponsor. Assuming you have a project sponsor, one of your first organizational tasks will be to define the elements that must go to the project sponsor for approval. Typically, the project sponsor approves the project scope, budget, and schedule, and any significant changes therein. Avoid having to get sponsor approval for every single expenditure and change or you'll forever be chasing your project sponsor trying to get requisite approval rather than getting project work done. Other organizational elements are discussed briefly next. Again, it's not an exhaustive or comprehensive list but just a reminder of the top level items and how they relate to BC/DR planning. In most cases, these are developed with your project team.

Project Objectives

You've developed the problem statement, the mission statement, high level requirements, constraint, and an optimal solution through the project definition stage. Now, you need to develop project objectives. The objectives for a BC/DR plan can be very narrow or very wide, depending on your company's specific situation. In this section, we'll list some of the types of BC/DR plans companies create. From there, you can develop specific objectives suitable to your company.

Business Continuity Plan

Focuses on sustaining the company's business activities, particularly those related to revenue generation and management of corporate obligations (employee payroll and health care insurance are two notable examples). A business continuity plan can be written for a specific business process or for all key business processes. As mentioned earlier, projects with smaller scope and more milestones tend to be more successful, so it might make sense for you to break your BC plan into smaller subplans, each addressing a specific set of mission-critical business processes. In most cases, a BC plan addresses the long-term recovery processes needed for the company to resume normal operations. Clearly, almost every company's BC plan will have an IT component since almost every company operating today (in the industrialized nations) utilizes IT to some extent.

Continuity of Operations Plan

This plan focuses on restoring mission-critical operations at an alternate location and performing these functions for an extended period of time. This type of plan addresses a wide variety of companywide concerns and typically is developed independent from a BC/DR plan, but clearly there are overlapping components that may need to be coordinated across the projects. A continuity of operations plan might be developed as the overarching project and the BC and DR plans may be considered subplans. IT operations and alternate IT arrangements clearly are an important part of any continuity of operations plan.

Disaster Recovery Plan

This plan focuses on restoration of key business processes immediately following an emergency or disaster. Unlike a BC plan, the DR plan typically does not include processes and procedures for ensuring the continuing and ongoing operations of the company long-term. Within the DR plan, there should be an IT section devoted to the technological needs of the company during an emergency and a section on the initial steps needed to restore all affected IT systems including business applications, servers, network infrastructure, and computer facilities to normal operations. It will also include the identification and specifications for an alternate operations site following an emergency. In some companies, noncomputer technologies such as communications equipment (cell phones, walkie-talkies, Blackberries) fall under the purview of the facilities manager rather than the IT department. As an IT professional, you are in an excellent position to understand how all corporate technology can be effectively deployed and utilized during an emergency, so your input on this topic, even if it's ultimately managed by some other department, can be extremely helpful.

Crisis Communication Plan

A crisis communication plan should be developed, either as part of your BC/DR plan overall, or as a separate but related project. Communicating effectively during a crisis can make a difference in determining whether the company ultimately succeeds or fails. It also can help employees maintain a sense of calm and order by giving them important information they might not have access to otherwise. During normal operations, employees gather information about the company from a variety of informal sources—instant messaging, chat, e-mail, and hallway conversations with coworkers. During a crisis, employees often are cut off from one another or unable to use normal communications channels. Planning for this and providing consistent and clear communications for employees and external parties helps organize recovery efforts and helps reduce some of the anxiety employees face when a disaster hits their company. Questions such as "was anyone injured?", "will I still get paid this Friday?", and "where should I report to work on Monday?" can be answered through the processes developed in advance.

In addition, you may have a large enough community or market presence that the media is swarming in the aftermath of a serious event. Even if you're a small company and you experience a large fire, the local media may show up asking questions about how and when the company will resume operations. Without a clearly articulated crisis communication plan that kicks into gear when a disaster hits, you could end up with an uninformed employee talking inappropriately with the media, sharing unflattering information rather than having an official company spokesperson answering questions in an intelligent and thoughtful manner.

Cyber Incident Response Plan (CIRP)

This is certainly a plan that should be developed within the IT department and is most likely separate from the BC/DR plan. Although we typically think of disasters as large, physical events such as earthquake, fire, or flood, a security breach into a corporate network's critical areas can be a huge disaster on a number of levels. Therefore, your firm will need a CIRP as part of the DR portion of the BC/DR. The CIRP establishes procedures to immediately address cyber attacks against the organization. Procedures to identify the nature and extent of the attack, the ability to mitigate and stop the damage, and to recover from and resume IT operations should all be included. Clearly, cyber threats are ever-evolving and your CIRP will need to continue to evolve and be updated on a regular basis. Incidents such as unauthorized access, denial of service, data theft, data alteration, and unauthorized system reconfiguration are all examples of problems to be addressed by a CIRP. Like the BC and DR plans, the CIRP plan requires a specialized team that is trained and at the ready. A Cyber Incident Response Team (CIRT) should be formed and trained to quickly and effectively take action immediately upon discovery of a cyber incident. Typically, the faster and more decisive the action taken, the less damage done to the company's computer assets.

Occupant Emergency Plan

This type of plan is related specifically to the building's occupants. It includes how to safely exit the building in the event of a fire; where to gather outside the building or where to congregate inside a building if the best option is to *shelter-in-place*. It also details procedures for contacting emergency personnel including fire, police, and medical assistance. This plan is typically part of the Facilities department, but in the absence of such a department, you should include these types of plans in your BC/DR plan, either as part of the overarching BC/DR plan or as a subplan that ties in. Clearly, if the building experiences structural damage as part of the effects of an earthquake or bomb, it will impact other aspects of the company's operations and a BC/DR plan will be implemented. Developing fire drills and evacuation procedures (for example) may or may not be considered part of your company's BC/DR plan, though these details should be managed by someone at your company. In the absence of a Facilities department or manager, these functions often are handled by the Human Resources department.

We've run through six different types of plans that are all related, in one way or another, to business continuity and disaster recovery planning. Clearly, they all overlap to some extent. It's important that you evaluate your company's emergency and disaster readiness and determine which of these types of plans will most suit your company's needs. Some of these plans may be developed independently from the BC/DR plan. In other cases, you may form a steering committee or designate a Program Manager to oversee the development of a number of related subplans.

Project Stakeholders

Broadly defined, the stakeholders for a BC/DR plan can include the government, various regulatory agencies, financial markets, public shareholders, private shareholders, employees, vendors, suppliers, contractors, and the community at large. That's a large list and clearly, it's not appropriate to invite them all to your planning sessions. However, it is important that your project team consider the broadest scope possible so that you can ensure that various stakeholder interests are considered and included when appropriate. For example, if your firm is subject to financial, legal, or environmental regulations, these must be considered as part of your overall BC/DR plan. If your firm is subject to these regulations, you will most likely require input and assistance from specific subject matter experts such as your financial or legal counsel. If you have in-house experts responsible for maintaining compliance to various regulations, these folks can be invaluable resources as part of your planning team.

However, let's narrow the scope for a moment. *Stakeholders* are, by definition, those who have a stake or interest in the outcome, results, or activities of the project. This is important because we're not only talking about the *results* of the project but the *activities* of the project. That means that if you need to pull people from other departments or off of other projects, those department or project managers are stakeholders in your project—they have a stake in the activities of your project. Their interest may be limited to when they can get their personnel back in the department or project. As you develop your BC/DR planning project, keep in mind all potential stakeholders. If your project is going to pull resources from other departments, for example, be sure to include those department managers in high-level project progress reports (if appropriate) so they are aware of the project's progress and have a contact person they can go to for status update or timelines. In small companies, this probably won't make much sense, but in larger companies or companies that are geographically dispersed, it can make the difference between gaining support for your project and irritating a bunch of people who have the distinct ability of making your life as project manager difficult.

The usual suspects should also be included in the list. Executives are stakeholders since they have a vested interest in the outcome (and effectiveness) of the project. They may have legal obligations as well, so they may be very closely tied into the project's objectives and outcomes. Even if there are no legal obligations, the executives or senior managers of the company should certainly care about the project's outcomes and objectives since the very survival of the company may depend on how well you define, create, implement, and maintain your BC/DR plan. Other stakeholders include facilities management, Human Resources (representing both HR functions and employees as a whole), operations management, marketing/sales/PR management, financial/legal management, and of course, IT management. If there are other departments in your company not represented by the preceding categories, include them if appropriate.

At this point, you may have a long list of project stakeholders. That's OK. At this point, you need to be *inclusive* rather than *exclusive*. As you move through your project planning process, you will develop communications plans to address the various *categories* of stakeholders so that you don't have a list of a thousand people to whom you have to report every day. Stakeholders' interests and concerns must be addressed, but it doesn't mean they need to participate in the project itself.

Project Requirements

Poorly defined project requirements can cause project failure, so what can you do to develop better project requirements? It begins with project definition, which we discussed earlier. What problem are you trying to solve? What is your mission statement or big-picture outcome? When you know what problem you're solving and what you're trying to achieve, you have defined the boundaries of your requirements.

Another activity that adds to the success of project requirements development is involving the right people early. If you wait until your plan is 50% complete to bring the Facilities Manager into the loop, you're likely to have missed something important or you have to rework much of your plan. Failing to bring the Facilities Manager into the loop at all would be even worse. It's not inconceivable that a plan would be created by the IT folks and completely omit facilities issues such as heating, cooling, drainage, and other issues. IT folks are smart but, like any other specialists, can become myopic and miss things that are obvious to others. So, word to the wise. Bring in the right experts early, and have them help develop the project requirements. It's easier to pare down requirements than to try to add to them when you discover an omission or gap later on.

Let's look at some examples of project requirements for a BC/DR plan. Clearly, you'll need to create your own list and modify it as you move through your project in order to address the specific needs of your company, but this should give you a running start. You should start by delineating requirements known at the outset of the project, but accept that project management is an iterative process and you will have to revisit your requirements as more information becomes known and details become clearer. So, let's look at some samples.

- E-commerce functionality (*define which functionality this includes*) must remain up 99% of the time, enabling customers to place and manage orders and to receive order status. Functionality includes product presentation, price and product information presentation, search, shopping cart, payment, order processing, credit card processing, customer order notification, warehouse order notification, warehouse pick tickets, shipper notification, customer shipment tracking notification, and inventory management.

- E-commerce customer service must remain available 85% of the time, enabling customers to interact with company representatives to answer questions and resolve problems.

- Customer order fulfillment must remain in place, regardless of company warehouse status.

- Employees must be paid on a regular basis or on the normal schedule during an emergency.

Notice that the first requirement describes functional and technical requirements. Sometimes they're closely intertwined, other times you can list technical requirements and functional requirements separately. In this case, the e-commerce functions are required and the technical requirements of that functionality are described. Don't get caught up in whether something is a functional or technical requirement at first, just be sure to capture the requirements. You can always move them around later once you've capture the requirements.

This example shows a company that does 100% of its business via a Web site (or Web sites). Clearly, Web site uptime is critical and will be the primary focus of BC/DR planning efforts. However, there are a lot of backend functions required to ensure that the Web site does more than electronically keep track of orders. Product must be in stock in inventory somewhere, someone must pick it, pack it, and ship it, and notify the customer it's been shipped. The on-hand inventory must be updated, credit cards have to be charged, income accounts must be updated, and so forth.

Well-defined project requirements will help you ensure that your project works once it's implemented. Although people rarely feel there is adequate time to plan (including creating project requirements), they will be forced to find the time to deal with the aftermath of a disaster. Thus, the choice is to take time to plan now to reduce the time to recover later, or extend the time to recover later with the high likelihood that recovery will fail and the business will close its doors.

Project Parameters

Project parameters are scope, budget, schedule, and quality. The scope of a project typically is defined by the objectives and resulting technical and functional requirements. However, you can also create scope statements. One method that is particularly helpful is to state what *is* and *is not* included in the project. Often by creating paired statements, you generate a clear picture of what the project work will entail. Although it may seem redundant to state what is not included, it helps avoid making incorrect assumptions. We all know that 100 people can read a statement and come away with 100 different interpretations of that statement.

When you include both IS and IS NOT statements, you help narrow down the interpretations so that you are more confident that everyone is literally and figuratively on the same page. Scope is defined as the total amount of work to be accomplished; budget is the total cost; timeline is the schedule or total duration of the project; and quality is the number of defects you're willing to accept. In this case, defects might be total hours of downtime per incident.

The budget and schedule for the project will certainly require multiple refinements. However, you may be handed a deadline or a specific budget amount to which you must manage the project. In most companies, there is neither unlimited time nor unlimited money, so you will be required to limit one or more parameters. Just as a quick review, Figure 2.2 shows the relationship of scope, budget, schedule, and quality.

Figure 2.2 The Relationship among Project Parameters

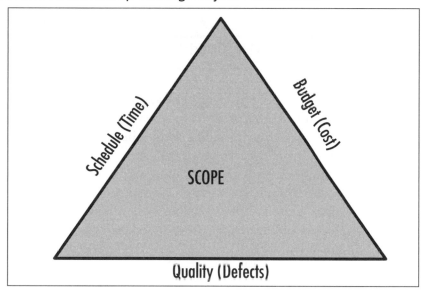

The project's scope should be defined at the outset by the amount of time and money you can devote to the project and also by the level of quality you require. Any change to the project's schedule, budget, or quality requires a change to the scope or to another parameter. For example, in Figure 2.3, you can see that if you reduce the budget and keep the schedule the same, the scope is reduced by a corresponding amount.

Figure 2.3 Reducing Budget Reduces Scope

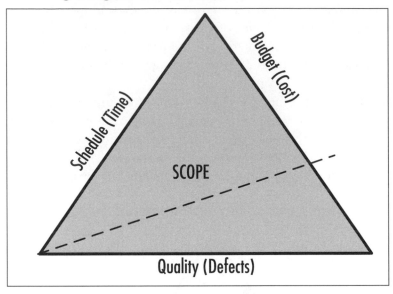

Another alternative to reducing scope is to increase another parameter. If you have to reduce your budget by 30%, you may be able to increase your schedule by some amount to offset the reduction in budget. If you have to reduce your schedule (meet a tight deadline), you really have only four choices: increase your budget, reduce your quality, reduce your scope, or cancel the project. In the case of a BC/DR project, cancelling the project is not a viable option (though often done), but in many other types of IT projects, it may actually be a viable alternative to consider. Figure 2.4 shows the impact on other project parameters of reducing the schedule to meet a deadline. In this case, the option is to cut the scope significantly. In other cases, you might choose to keep the scope the same but reduce quality or keep the scope the same and increase the budget. The bottom line is that you cannot reduce the schedule without impacting one or more of the other parameters.

If you decrease the schedule and don't want to reduce the scope, you typically have to increase the budget. Sometimes you can reduce the quality, but that gets tricky. Quality usually is defined as the number of defects. In a BC/DR plan, it might mean the difference between requiring 90% critical systems availability and 75% critical systems availability. Although 75% might be considered a lower quality metric, it doesn't mean you're actually accepting a "poor" quality project deliverable. There are various areas in your BC/DR plan where you may say that a certain number or percentage is acceptable. This typically is done to address the interplay between planning for business continuity and working within known constraints.

Figure 2.4 Reducing Schedule to Meet a Deadline

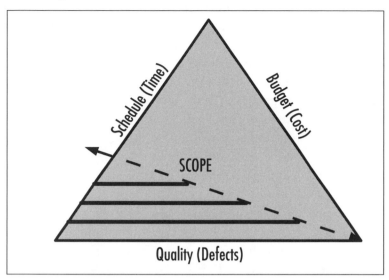

If you have a specific budget or timeline, you'll have to compromise throughout your project planning process. Although it's best to start with the most optimal solution and scale it back from there, you may also start out knowing that you have significant constraints within which you have to work, and you begin your planning process there. Remember, it's unlikely that any realistic plan you devise will be perfect. The goal is to create a plan that is workable and that ultimately will keep you in business after you experience a business disruption such as a natural disaster or a security breach. In many cases, the cost of the project will be predetermined—an executive might say, "We understand this is important but our funds are stretched. We can devote $5,000 to this project over the next three months and not a nickel more." Therefore, your budget will be your *least flexible* parameter. This means that everything else will have to flex around your budget. Often this forces you to take longer to complete a project or to accomplish less work than desired. Regardless of whether or not you feel the parameters are optimal, you will have to manage your project to them.

Project parameters need to be ranked from least flexible to most flexible. The least flexible item is usually the one to which you must manage the entire project. In this case, we're assuming that the least flexible parameter will be your budget. Another parameter must be designated as most flexible in order to provide your team with the flexibility it needs to develop and implement the project successfully. The most flexible parameter is the one that will change when things go wrong (and they will go wrong). The other two project parameters may flex or not, depending on your company's constraints and the project's needs. As you develop these parameters, you will have to come back and make modifications once more detail is known. This typically occurs during or after the development of the Work Breakdown Structure (discussed later in this chapter). Once you know what tasks are involved in your project, you'll be able to create a better timeline, a tighter budget, and a more realistic view of the total project scope. Parameters can then be marked as less flexible, most flexible, or not labeled. By default, any project parameter not labeled as least flexible will flex only if needed.

Keep in mind that the project parameters are all interrelated. You can't change one without impacting the others. It's a certainty that something will go wrong in your project—the project management adage says "Things are more likely to go wrong than likely to go right" sums up how projects tend to run. Therefore, one of the project parameters must be able to flex. In a BC/DR project plan, this could be the scope, the timeline, or the quality. If your budget is least flexible and your scope is most flexible, you would do less project work if you find yourself straining your budget. If your timeline is most flexible, you would push your schedule out into the future to accommodate a tight budget. This might mean that you avoid overtime or that you schedule this project work around other higher priority projects. Finally, your quality might be the most flexible item and you might decide that your organization can put up with longer outages than originally identified. Sometimes a company will look at the cost of a BC/DR plan and determine that if the cost of creating the plan is X, then it is willing to put up with an outage of 0.25X at any given time. A case study following this chapter discusses how to create financial estimates for the cost of planning versus the cost of outages versus the cost of potential failure.

Some people think that designating a parameter as the one that will flex is a crafty way of saying you don't have to make a solid commitment and hit real targets. No so. The reality is that you will find that as the project moves forward, unanticipated things happen to which you must respond. When you have your project parameters labeled as most to least flexible, you know how to make appropriate decisions. If you go into your project saying that you will hit all four project parameters on the nose, you have very little chance of success. It doesn't mean that you won't give your absolute best effort to meeting the metrics to which you've committed, but it does give you (and your project sponsor) a solid framework in which you can make decisions when things change or go wrong. Also keep in mind that as you move through your project definition and project planning stages, your project parameters will become better defined. You may have to settle for estimates in the early stages and redefine these parameters as detail becomes known. Once you have a project plan in place, you will have to commit to those parameters as targets, but you'll still need to understand the interplay of these parameters to make appropriate adjustments as project work progresses.

Project Infrastructure

Project infrastructure refers to the tools and resources you'll need to have at your disposal as you develop your BC/DR project. This might include computers, software applications, testing labs, communications equipment, and more. For example, if you're working with a team located in several countries around the world, your project infrastructure might include some sort of collaboration software, net meeting capabilities, along with e-mail and instant messaging. Since you'll likely be working at different hours, depending on the time zones involved, you'll need tools that allow you to manage a cohesive team while working within the local constraints. Your infrastructure needs for your team should also be viewed with an eye toward your BC/DR planning. Which of these tools might be useful if one location were to experience a natural disaster or business disruption? Which of these tools would be completely useless or inappropriate? Whenever possible, you should look at the technology and infrastructure tools you use in your planning process to determine whether it would also be a good tool during a disaster or event.

The infrastructure for a BC/DR planning project may be as simple as e-mail, instant messaging, and a shared folder on a network drive. It might be quite a bit more complex, especially if you work in a large, geographically dispersed organization with multiple sites or many different business units. Clearly, the infrastructure for a 10-person software development company or a 50-person nonprofit social agency is very different than the infrastructure needed for a 10,000-person, 40-worldwide-location, 10-business-unit type of company.

Defining the infrastructure you'll use to plan and implement your project is important at the outset so that you know what you will have access to, what you can use, and what's off limits. Defining your infrastructure needs may impact your budget if you don't have the infrastructure in place, but in most cases, it's a matter defining what infrastructure you'll need, how you'll acquire it, and how you'll utilize it to accomplish project objectives.

Remember that geographically dispersed teams may have varying levels of access to infrastructure such as conference calling, video conferencing, net meetings, e-mail, instant messaging, shared team intranet web sites or portals, and so on. For example, some members of the team may have access to wireless Internet access, others may have to be at a corporate location to connect. Some members of the team may be able to make international phone calls, others may not. Some members of the team may have reliable communications lines, others may not. Keep these considerations in mind as you plan your project's infrastructure.

Project Processes

Processes needed for developing and running a BC/DR plan are similar, if not identical, to running any other sort of IT project. If you're an experienced project manager, you probably have notes on processes and procedures you've used in the past, which you can pull out, dust off, and review. In most cases, these processes and procedures can be reused, though usually with slight modification. If you're not an experienced project manager, you might be wondering what sorts of processes you'll need. There are numerous resources you can reference; we've included a few of the basics here to give you a jump start.

Clearly, you'll need to adjust existing processes to address the unique needs of your BC/DR team. In many cases, an experienced PM has a set of documents he or she uses for every project, modifying the processes slightly each time to fit the circumstance. If you're not an experienced project manager, you will need to define all the project processes you plan on using. Some may argue against documenting these processes, stating they are a waste of time to document. It may not be useful to document every process, but it is well worth your time to define and document (and later archive) these processes for two primary reasons. First, it will reduce problems later if you have a set process for handling common project tasks or processes. Consistency in project work typically yields higher quality results, so defining processes that can be reused during the project (such as the process for generating progress reports or the process for running team meetings) will help save time and lead to better results. For example, if you're stumbling around during each team meeting because you don't have or aren't following a set process for running team meetings, you're wasting everyone's time. You'll find that people begin to skip team meetings because they view them as unproductive. This leads to other problems that ripple through the project. A second compelling reason for creating these processes and documenting them is that at the end of your project, you can review these processes, make adjustments, and file them away. When your next project comes along, you simply have to pull out the files, review them for appropriateness, make minor modifications, and move along in your project work. Using processes that have been used and tested over time usually leads to higher productivity and lower stress for you and your team. You should take advantage of any opportunity to save yourself time and aggravation.

Team Meetings

Project processes in a BC/DR planning project are basically the same as those used in any other type of IT project plan, with one exception. BC/DR plans typically span across organizational and functional boundaries. Many IT projects do not. You'll need to determine how the team works together and how information will be shared, stored, and archived. You'll also need to address the logistical aspects such as determining how, when, and where the team will meet. Letting team members know how meetings will be set, at what interval, how they'll be notified or reminded, how agendas are determined, and how meetings will be run is all part of defining the team meeting process.

Reporting

You'll need to develop a two-tiered reporting process. One level of reporting will be you, as IT Project Manager, reporting to your project sponsor regarding the overall status of the project and any challenges or roadblocks you encounter that impact the project or with which you need sponsor assistance. The other level of reporting will be reporting from your project team to you, as IT Project Manager. You'll need to develop reporting methods using processes and procedures familiar to those in your company. Whenever possible, you want to avoid reinventing processes and procedures for two reasons. First, developing new procedures often leads to resistance because team members are unfamiliar with them. Second, new procedures can cause unintended effects that can ripple through the team or the department's productivity. Whenever possible, keep reporting requirements focused. Avoid dragging a wide net gathering every last detail. There may be an appropriate place to create a repository, such as creating a project-related *wiki* into which team members can dump all their collective knowledge. Keep formal reporting to the elements that actually impact the project's progress and ultimate success. Your teammates will appreciate your lean operating style and you won't have to pour through reams of documents to find pertinent details.

Escalation

How will problems within the project or external problems that impact the project be escalated? It's important to define this process before it's needed. You'll need to know which problems require sponsor notification or sponsor approval versus problems that require escalation through a different channel than the project sponsor. If you have a problem with a team member that cannot be resolved, how will that be handled? By you? Your boss? The team member's boss? If you have a major problem getting payment for an approved expenditure to a vendor, how should that be escalated? If you run into a serious project roadblock and all work has to come to a stop until it's resolved, how is that escalated? Your basic escalation procedures may include a division of internal and external. Problems that are internal to the team will be treated in one manner, problems external to the team (i.e., within the company at large) will likely be treated in a different manner. Similarly, problems internal to the company will be addressed in one way, problems external to the company (i.e., with vendors or regulators) will most likely be handled in a different manner.

The escalation procedures for each category should include *boundaries*—which issues qualify for escalation? If possible, create specific and quantifiable metrics for deciding which issues are escalated. In most cases, these escalation procedures should be provided to the team so they understand how issues will be resolved. There may be a subset of the escalation procedures you choose not to share with the team, but in general the more the team knows about how things will work, the more effective they can be in making project-related decisions.

Escalation procedures should also indicate the chain of command for various types of problems so that you know which is the appropriate resource to utilize should a problem arise. Which types of issues should be brought to your manager, to your sponsor, to someone else's manager, to the executive team?

Finally, delineate how escalated issues will be tracked and closed. Some project managers like to dispatch issues as Closed–Resolved, Closed–Unresolved, Open, and Deferred. Whatever categories you choose, define how they'll be used and use them consistently.

Project Progress

How will the progress of project work be tracked? You might choose to have task and project progress tracked through team reporting, through an interactive team web site, or through other tools you may have used in the past or have available to you for this project. Again, as with reporting, keep it simple. It's human nature to fabricate reports and progress notes if the requirement for producing such documentation is overly burdensome. In plain English, encourage your team to give you the real data you need with as little effort as possible. Think about how you'll need to report project progress to the executive team and to your project sponsor, then determine the bare minimum you need to know from your team. Start from there and add to the tracking requirements only as needed. Also, just because certain data may be available to you or your team, it doesn't mean that data helps you monitor project progress or help move the project forward any better. Keep it streamlined and simple whenever possible and be sure that you're asking for data that actually is useful in monitoring or managing the project.

Change Control

Most projects encounter the need for some sort of change along the way. In a BC/DR project, one of the most likely sources of change is that a new technology is being implemented. For example, if your firm decides to implement a customer relationship management (CRM) solution, you'll need to add that to the scope of your BC/DR plan so that you can address the specific needs related to this application. The business will not stand still while you're planning your BC/DR project, so there's a good likelihood some sort of change will be required as you move forward. What process will you use to control change?

Though it's outside the scope of this chapter (and book) to discuss change control in detail, a few reminders will probably help. First, though it may sound clichéd, it's still true: Control change or change will control your project. Define a change request process so that needed changes must be formally requested. This provides you and your team the opportunity to evaluate the requested change. In too many companies, someone (usually an executive) with clout will demand a change to the project and everyone on the project team will scurry around trying to incorporate that change. Without a formal process for managing it, change will be introduced into your project in a random fashion. Once you have a defined process for requesting change, you should develop a defined process for evaluating change. The evaluation should include the risks involved with making the change, the risks involved with not making the change, and the impact to the scope, schedule, budget, and quality of the project if the change is implemented. Finally, you should have a change tracking process that indicates that a change was requested, reviewed, accepted (or rejected), and incorporated.

Quality Control

Quality control is a topic that can fill volumes on its own, so we'll limit our discussion to quality control as it relates to BC/DR projects. Clearly, one error in the project plan could mean the difference between getting a critical system back up and running in one hour versus one month. However, it's unrealistic to expect that you will have zero defects in your project plan and implementation. How much quality is enough? It's always interplay between how much time and money you have and how critical the systems are. For example, you might decide to spend a disproportionately greater amount of time testing your CRM or ERP BC/DR functions and a disproportionately lesser amount of time testing some other part of your plan. If your CRM or ERP application is the most critical application from a corporate perspective, you may choose to devote the most planning, implementation and testing to this area to ensure as close to 100% quality as possible. Although specific quality metrics for BC/DR plans may be difficult to develop or measure, you can use qualitative methods to determine what level of quality is required for each element of your plan.

There are many other project processes that you'll no doubt develop or use, but these are some of the basics to get you started thinking along these lines. If it's something you do repeatedly, you should consider developing a process for it so that you can perform more consistently and not have to think about (or reinvent) the process over and over again.

Project Communication Plan

Technically speaking, the communications plan may be considered part of the project processes. However, communications are so vital to the success of any project that it's listed as a separate entity because it has a unique level of importance and awareness. With a BC/DR plan, your need to communicate across departmental, positional, and geographic boundaries may be greater than in any other IT project plan you've worked on. That can be a tall order for even the most seasoned project manager. How can you be sure you're communicating to the right people with the right language in the right frequency and the right medium? One of the best ways is to check with your subject matter experts, who are typically the founding members of the team. If you're going to be working with technology plans that touch the entire organization, you will probably need to communication with all corners of the organization. Those people on the team who represent the various areas of the company are probably in the best position to provide input as to how, when, and in what format to communicate with their constituents. It will ultimately depend on where your BC/DR plan falls in the overall list of priorities. If your plan is low on the list of priorities, timely and positive communication might help boost its rating. On the other hand, there are times when you might want to communicate only as needed in order to keep your project flying below the radar (to avoid it being canceled or reduced in scope, for example). Typically, communicating project progress is good for the project and a bit of positive PR can go a long way. However, in BC/DR projects that may not be too popular to begin with, you may decide your best move is to keep your team and your project sponsor fully informed but avoid more expansive communications. A discussion with your manager or project sponsor on this topic will help you develop an effective communication plan that fits in with the overall political climate of your organization.

Common Challenges

Communicating Results of a Low Priority Project

We can emphasize the importance of business continuity and disaster recovery planning until the cows come home, but the truth of the matter is that some companies just don't care. You might find IT professionals creating BC/DR plans almost on the sly or as part of another IT initiative so they don't have to formally announce a project that will get stonewalled from the start. Depending on how your company runs, it may be possible that you're working on a BC/DR plan without a formal charter—without formal recognition or approval from existing corporate authorities. As we've mentioned several times, some basic BC/DR planning can be incorporated in other IT activities. Though this is not an optimal situation, it is better than doing no BC/DR planning at all. So, how do you communicate to the organization if you have no formal BC/DR plan or if you know that your BC/DR activities might be thwarted if they came to light? Your communication plan, in that case, may be limited to those individuals who will be impacted by your plan. You may choose not to use the terms *business continuity* or *disaster recovery*. Instead, you may ask, "What would happen if this system went down?" This is clearly something that must be addressed in any organization, so if using the phrase BC/DR planning will get you off-track, don't use that language.

For those of you who might think this is a bit underhanded, remember that IT staff have to deal with the reality of system outages every day, whether a server fan goes out and the unit overheats or whether a vital cable is disconnected from a router. The point is that business disruptions will occur, regardless of whether you plan for them or not and regardless of what you call them. So, you might as well build in some BC/DR planning to your IT activities and avoid an all out communications blitz if that's what it takes to protect your company, your employees, and your job. Granted, if you don't have executive level authorization or support for your BC/DR project, you will not be able to count on wildly successful results, but that doesn't mean you can't still make a positive impact... quietly.

Project Planning

Planning the BC/DR project involves all the typical steps you'd undertake in any IT project planning process. However, there are two key elements in the project planning process worth discussing here. Developing the Work Breakdown Structure for your BC/DR plan essentially defines the scope of the project. By definition, the Work Breakdown Structure (WBS) is the list of outcomes that must be accomplished in order to successfully complete the project; it describes 100% of the required work. Therefore, a well-developed WBS helps you deliver a successful project. The *critical path* is, by definition, all tasks in the project that if delayed will delay the completion of the project. Why is this important? If you are trying to meet a deadline for completing your BC/DR plan, you'll need to understand which tasks in the WBS are on the critical path and which are not. Those that are on the critical path can delay project completion; those not on the critical path cannot delay project completion. As a result, tasks not on the critical path are more flexible as to when they can be scheduled. Let's look at WBS and critical path in more detail as they relate to creating your BC/DR plan.

Work Breakdown Structure

The top level of your WBS in your BC/DR plan will most likely follow the structure of this book: Risk Assessment, Business Impact Analysis, Risk Mitigation Strategy Development, Plan Development, Emergency Preparation, Training, Testing, Auditing, and Maintenance. As we move through the remaining chapters of this book, you can compare your WBS structure to the material in this book. You may have additional elements for your WBS that are specific to your company's BC/DR needs; feel free to include those.

Remember, the completed WBS should describe the total amount of work to be accomplished. If there is a mismatch at this point, you need to reassess your WBS or your project's scope. These two items should be aligned so that the scope is fully described in the WBS (or that the WBS fully describes the desired scope). The elements of your WBS may vary from those outlined in this book, but the overall elements should match fairly closely. If there are any discrepancies, be sure they are by intent and not by accident.

Critical Path

The critical path in your BC/DR plan will describe exactly how long the project will take and which tasks will delay the project if you run into problems on those tasks. Remember, tasks not on the critical path are the tasks that provide you and the project team with a bit of flexibility. As you probably know, tasks not on the critical path, by definition, have some float. *Float* is (essentially) the time flexibility of a task—it can be completed this week or next week without impacting the overall timing of the project. Tasks can move on or off the critical path. A noncritical path task that is delayed long enough may end up on the critical path. A critical path task may, for some reason, move off the critical path if you discover there is some flexibility in the timing of the task. The key is to understand the tasks that are on the critical path and make sure you keep a close eye on these tasks if you are working on a tight timeline. If your schedule is your least flexible element (as discussed earlier), then you will have to manage to your critical path more than to your budget. You might face this situation if you are required by a customer to provide a BC/DR plan by a certain date in order to close a large deal. You may also be required by certain regulatory or governmental agencies to complete and submit a BC/DR plan by a certain deadline in order to meet compliance requirements or to avoid heavy financial or legal penalties. In these cases, your critical path will define your project's timeline and will be the key to successfully managing to a required schedule.

Looking Ahead...

Strategically Planning Your Project Schedule

In planning your business continuity and disaster recovery activities, you're likely to find that the risk assessment and business impact analysis are the most tangible aspects to the process and are therefore the phases that move along smoothly. Conversely, when you get into the actual risk mitigation and emergency response strategies, you may find the project getting bogged down. It's easier for most people to sit around and discuss theoretical issues (what happens if the building catches fire, how would that impact the business) than to devise practical solutions within given constraints. In scheduling your project, you may want to account for this and move the early phases along more quickly and allow more time for strategy development rather than giving equal time to each phase or section of the planning process.

Project Implementation

Since we're focusing on the IT aspects of a business continuity and disaster recovery plan, we need to state the obvious: IT is a moving target. Unlike an area such as facilities, which is usually a fairly stable and static area, IT is always changing. From reconfigurations to new security threats to moving a data center, there's no shortage of change in the IT department. How does this impact your BC/DR project implementation? It means that you will need to build in a process for monitoring change in the IT department and assessing what should be incorporated into your BC/DR plan. Here's an example. A nonprofit agency decides it needs an IT BC/DR plan. The IT director, with her staff of four, creates a plan. In the meantime, they have funds that must be spent before the end of the fiscal year that are earmarked for a case management application. The application must be implemented and it requires the use of Windows Server 2003 and SQL Server 2005. These should certainly be included in the plan so BC/DR issues can be addressed. If the agency's location has a major fire, how will the server hardware be affected? Is it on-site? What about backups? What arrangements should be made to have alternative server hardware available? What arrangements should be made for off-site backup and data recovery? How will this be shaped by the implementation of the new application? These are the kinds of questions that may come up throughout the lifecycle of your BC/DR planning process because other IT projects are in various stages of planning, design, implementation, and testing. Each of these must be assessed for their impact on your BC/DR planning and each must be assessed for the BC/DR impact on these projects. The interaction between your BC/DR planning activities and all your other IT projects is a two-way street—each ultimately impacts the other.

Ideally, you should strive to add BC/DR types of assessments and considerations into your standard IT processes. Over time, you'll discover this makes keeping your BC/DR plan up-to-date a bit less onerous. You might also discover ways to save time and money or streamline processes along the way.

Managing Progress

Managing project progress is often a matter of organizational and project manager traits. As unique as each company and each project is, there are a few things that you should keep in mind for your BC/DR planning project. As mentioned previously, you should have some method of ensuring that current IT changes and initiatives are evaluated in your BC/DR planning. It would make no sense to develop a risk mitigation strategy for a technology that is being phased out or to purchase some solution that is easily rolled into a current or planned initiative. As simple as this sounds, it's often complex in practice. It's easy to suggest that various technologies be evaluated as part of your BC/DR plan; it's another matter entirely to make sure that happens. Your project processes (described earlier in this chapter) should include some method of keeping an eye on what's going on in the IT department across the enterprise. You should develop a method of evaluating and incorporating techno-logical components. We'll look at risk assessment in detail in the next chapter and that will provide you with some of the tools to evaluate the risk to various technologies. Once you understand the risk and the potential strategies to reduce that risk, you will have a better sense of which IT initiatives underway in your organization will likely come into play. It might be at that point that you talk with the entire IT team to see how their work impacts your project and how your project might impact their work.

In addition to keeping an eye on changing technology, you'll need to make sure you use a consistent method to monitor and measure project progress including standard tools such as reporting, dashboards, and whatever tools you are accustomed to. Since this type of project spans beyond the scope (and authority) of the IT department, you'll also need to use your best people skills to keep the project moving forward since it's likely you won't have the authority to require that project work be accomplished. This is true of many types of IT projects, and it's especially true of BC/DR projects that definitely cross all organizational boundaries and that often have a low priority in the day-to-day scheme of things.

Managing Change

Any experienced IT project manager will tell you that managing a project is managing change. Despite our best efforts, our plans are always subject to change in the dynamic world of business. Although there must be certain areas that are nonnegotiable (these are typically the least flexible project parameters), there must be some room for change in a project in order for it to have any chance for success. Earlier we discussed the importance of managing expectations regarding changes to your project. Without requiring the consistent use of your defined change management process, you'll end up with a project scope that is all over the map and four times larger than originally defined. As an IT project manager, you're probably familiar with managing change, but perhaps not across corporate boundaries. Let's review a few pointers for managing change effectively:

1. Define a simple, easy-to-use change management process.

2. Require that all requested changes go through the change management process.

3. Evaluate each requested change as to how it will impact the current project plan.

4. Evaluate the risk of each requested change and determine if it increases or decreases risk in the plan.

5. Incorporate the requested change into the plan and update all parts of the plan impacted by the change.

6. When possible, incorporate one change at a time so each can be properly evaluated.

7. Do not allow random or informal change requests to become incorporated into the plan.

8. Communicate with those requesting change as to the status of their request.

9. Communicate the rationale for rejecting a change request and be willing to listen. It may be possible you overlooked a critical factor or misunderstood key data.

10. Keep track of all requested changes and how they are ultimately handled (accepted, rejected, postponed) and why.

Project Tracking

The metrics used to track projects vary with each company and with each IT PM's systems of project tracking, so there is no single, universal method for tracking a project. One of the keys to a successful project is to create multiple milestones so that you can easily see where you are versus where you said you would be. At minimum, you should create milestones for each phase of work. Using our framework as an example, you should have a milestone at risk assessment, business impact analysis, and at each additional phase. However, if these phases or sections are going to span months, you should create interim milestones. If you have a small company and your risk assessment activities will span a week or two, one major milestone is probably sufficient. Milestones will keep you on track but you shouldn't create so many that you drive yourself and everyone around you nuts.

As an experienced IT PM, you also know that tracking involves monitoring the budget, keeping an eye on the scope and quality, and making sure change is being properly handled. Depending on how your company operates and what tools are available, you can use any one of a number of programs to track project progress, schedules, expenditures, and such. With BC/DR planning, you'll need to make sure that whatever you ask of team members to enable you to track the project is available to them. Remember that the Facilities Manager may not have access to the same tools you do or that they may not be as comfortable with project tracking tools as you are.

There are more detailed ways to evaluate and track project progress including Earned Value Analysis, Schedule Performance Index, Cost Performance Index, and Estimate At Completion, to name several of the more commonly used tools. Discussion of these tools is outside the scope of this book but you can certainly learn more about these methods through a variety of sources.

Project Close Out

In most IT projects, project close out involves a hand-off to some other organization. It might be that the new wireless infrastructure is now managed by a different subset of IT staff or that the new application is now maintained through normal application maintenance procedures. The same holds true for your BC/DR plan. Once you have completed the plan, it must be kept up to date through some sort of maintenance procedures. In many organizations, this is as simple as an annual review of the plan and a paper walk-through of the business continuity and disaster recovery steps delineated in the plan. In other companies, it might involve actually testing out some of the BC/DR procedures and even testing some of the defined recovery processes. Maintenance of the plan is important if for no other reason than the plan took a lot of time and effort to create and it takes far less effort to keep it current. As we've mentioned, an outdated plan can be worse than no plan at all because incorrect assumptions might be made. For instance, if a legacy system was slated as a fail-safe backup technology and that legacy system has since been decommissioned, your fail-safe backup is gone. If a disaster strikes now, the assumption that there is a bottom line fail-safe option is incorrect and could spell the difference between the company surviving the incident or not. Be sure your project close out activities include handing off the BC/DR plan in such as way as it will be maintained. It can be helpful to schedule the annual check up at the close of the project so that it will be on the calendar, though it's also likely that in 12 months, that calendar will change and your review plans may be impacted.

Also as part of best practices in project management, it's a great idea to have a post-project review session to review what worked, what went wrong, and how project processes could be improved in the future. Taking away lessons learned and best practices from each project you participate in (or lead) can help improve organizational results and your personal effectiveness as an IT project manager. It will also save you time and money in the future to avoid the obvious mistakes. The goal is not to completely avoid making mistakes—mistakes will always happen—but to avoid making the same mistake twice. Capturing best practices and lessons learned from across the organization might help you fine-tune other IT projects in progress and might also help streamline your BC/DR project review process.

Key Contributors and Responsibilities

Thus far, we've talked in fairly general terms about the business continuity and disaster recovery planning process. In this section, we'll discuss key contributors to your BC/DR plan and what the roles and responsibilities should be in an ideal scenario. While we outline the ideal scenario, you'll need to take a look at your company specifically and make modifications as needed. For example, your company may be so small there is no Facilities manager and that task falls to the Human Resources director or the IT director. Your company may be so large that there are multiple levels of Facilities management up to and including a vice president. You'll need to scale the information in this section to your company's size and needs, but this will give you a good overview to use as your starting point.

First, let's list the roles and contributors and then delve into the details of each:

- Information Technology
- Human Resources
- Facilities/Security
- Finance/Legal
- Warehouse/Inventory/Manufacturing/Research
- Purchasing/Logistics
- Marketing and Sales
- Public Relations

Information Technology

Since we're focusing on information technology in this BC/DR planning process, you clearly need representatives from the IT group on this project. Which members of the team should participate? Well, that largely depends on the size and scope of your IT department. If it's a three-person IT department, all three of you probably have to participate. If you have a team of 40, you will need to select those people best suited to this project. Some of the factors you can consider in making your decision can be:

- Experience working on a cross-departmental team
- Ability to communicate effectively

- Ability to work well with a wide variety of people

- Experience with critical business and technology systems

- IT project management leadership

Experience Working on a Cross-Departmental Team

Having IT people on your BC/DR team that have successfully worked on a cross-departmental project can help facilitate the success of your BC/DR project. They may have established positive working relationships with key people in other departments that you can
draw upon for your BC/DR project. They may have learned how to navigate some of the
tricky political waters in your organization or they may simply have developed a broader
organizational perspective that can help as you work across departmental boundaries to
develop a successful plan.

Conversely, you want to try to exclude those who have developed a reputation for being
difficult and not working well with others (especially in other departments). Though you
don't always have a choice as to who's on your team and who's not, avoid troublemakers
from the start. If you need their expertise on the project, try to contain them and restrict
their interaction with others to a bare minimum. For example, if you have a difficult person
on the team who is the subject matter expert for a critical business application, try to have
that person work with as small a subset of the project team as possible. It's better to annoy
three people than 30.

Ability to Communicate Effectively

The business continuity and disaster recovery project planning process is all about being able
to discuss risks, alternatives, and strategies that work for a variety of stakeholders. Without
the ability to communicate effectively with all kinds of people—technical, nontechnical,
executive, management, and front-line—your team and your project will suffer. It's not
uncommon to find some of the best technical people are the most challenged in terms of
interpersonal communication skills. If that's the case in your IT department, you may want
to include a few generalists who understand the technology and can also communicate
effectively with a variety of people. These folks can act as translators, taking very technical or
detailed information and paraphrasing it for other non-IT members of the team. If you don't
have at least one person on the team who is an excellent communicator with regard to IT,
you may well find that you have some serious miscommunications that occur through the
course of the project.

Ability to Work Well with a Wide Variety of People

The ability to work well with a wide variety of people often accompanies the ability to
communicate effectively. IT members on the BC/DR team will have to interact with end-

users, facilities people, financial people, and many others during the course of the project. The last thing you need is someone from the IT department representing the BC/DR project in a way that alienates the rest of the company. An example that probably jumps to mind for many of you might be the stereotypical "know it all" IT person who talks down to those who don't understand the intricate details of the technology. These people tend to simply make enemies where none existed and you don't need the added problems this type of person brings to a cross-functional team project.

Here's one more tip: Because your team may need to communicate with corporate vice presidents, department heads, or front-line staff, make sure that you, as project manager, or someone on your team is comfortable talking candidly and appropriately with people in all positions. Some people get nervous talking with those who are much higher up in the corporate hierarchy and are unable or unwilling to speak candidly. Others don't understand what constitutes "appropriate" communication and they tend to drone on or disclose inconsequential or embarrassing information. Make sure you or someone on your team is comfortable with and capable of interacting at all levels of the organization into which this BC/DR project may take you.

Experience with Critical Business and Technology Systems

Having people from the IT department with experience in critical business and technology systems almost goes without saying, but we'll say it anyway so it's not overlooked. Clearly, the critical business systems from IT's perspective may not be the same critical business systems from the CEO's perspective or from the financial analyst's perspective. You'll have time to test out your assumptions as you move through your project. If you're a small IT shop, this is probably a moot point. In larger IT departments, you'll need to look at your staffing needs and determine who should participate. This almost always runs into problems because your IT staff have other roles, responsibilities, day-to-day tasks, and project tasks they need to address. Often you are forced to choose between the person who is your first choice and the person who has the time or bandwidth to deal with your project. Unfortunately, the person who gets things done and delivers the best results is probably the person with the least time and bandwidth available.

In addition, you'll need to balance the needs of your project with the needs of the IT department's ongoing activities. If you're in the middle of a roll out of a new technology, it's going to be tough to find the time and resources to also participate in BC/DR planning. This may impact your overall BC/DR schedule or it might force you to work creatively with those subject matter experts. One potential work around is to flag the technologies you think are critical but wait until you've confirmed this information with the rest of the organization. Once you've identified critical systems from an organizational perspective, you can then tap necessary resources. This small work around might provide a bit of flexibility so that you can still utilize the subject matter experts you need without reworking your schedule or theirs.

IT Project Management Leadership

We're assuming you're the IT project manager and you're heading up this project. However, whenever you can tap others with leadership ability, you'll get that much more mileage from your project. Clearly, you want people with leadership abilities who can also work effectively as part of a team. You don't need four or five people all trying to manage the project. Good leaders usually know how to follow and to head up smaller groups or initiatives within a project team. Tap into these resources to take some of the burden off your shoulders as long as these folks won't be the cause of messy political maneuvers and power plays. Remember, a good project manager willingly delegates to others on the team and spends more of his or her time monitoring project progress. When you have competent people on the team who will take leadership roles, your job of delegating and monitoring will be that much easier.

Human Resources

Depending on how your company is organized, you may find that Human Resources (HR) encompasses many of the other functions listed in this section. In small companies, HR ends up being the catch-all for all the miscellaneous functions that have no home. In large companies, the miscellaneous functions typically are large enough to require a distinct department such as facilities, security, or public relations. We'll stay focused on traditional HR functions in this section, but as you read through subsequent sections, keep in mind they may be housed in HR. Regardless of how or where the function is managed, they all need to be represented in the BC/DR planning process.

Human Resources typically is responsible for helping to recruit new employees; managing the legal issues around hiring, firing, and performance management; and managing the payroll process including periodic paychecks, contributions to retirement accounts or saving accounts, increases due to promotions or raises, and managing other required paperwork such as verifying citizenship or the right to work in the country.

How will these functions be impacted by a fire in the building or a flood or earthquake in the area? The best person to answer that question is someone (or several people) from the HR department. What computer systems do they use to process payroll and other HR functions? What applications do they use? What forms are required by law? What forms are required by the company? These are questions your HR specialists will have to answer when you get into the BC/DR planning process. For now, think of whom in your company is responsible for these functions and add them to the list of team members.

Facilities/Security

As mentioned, the function of facilities and security may be handled by your HR staff if you're in a small company, or these functions may be handled by someone in operations. Regardless of where these two functions are housed, they are a critical part of the BC/DR planning process and should be represented by subject matter experts. Facilities typically handles the management of the office, warehouse, manufacturing, or storage spaces including cleaning, maintenance, set up, build outs, and remodels. They deal with occupancy and tenant issues as well as other legal or licensing requirements related to business operations in the facility. The department (or function) also usually handles the installation, management, and monitoring of key utilities such as electricity, gas, water and, in some companies, communications equipment (Internet connections, telephones, cell phones, mobile radios, mobile devices, walkie talkies, etc.).

Security includes controlling and monitoring access to the building, facilities, and grounds, dealing with imminent dangers such as structural, mechanical, or electrical failures, and sometimes dealing with employees who have been terminated or who are behaving in a manner contrary to acceptable and safe means. Security provided by most companies is to secure the assets of the company, not to provide policing services. Therefore, you can't assume that your security staff deals with all aspects of security. Nevertheless, both facilities and security functions should be well-represented in your BC/DR planning project. You'll need to answer questions such as these: What is our fire drill policy? How is the building evacuated in the event of a fire or other internal disaster? What plans are in place for protecting staff from external dangers such as nearby chemical spills, noxious fumes, or railroad derailment? If something happens to the building, how will access to and from the building be controlled? What tools, supplies, and equipment will be needed by the company's emergency staff to communicate with each other and manage the initial impact of a business disruption such as a fire, explosion or earthquake? These questions should not be answered at this juncture, but these are the types of questions that will be asked and answered later, within the scope of your BC/DR project plan.

Finance/Legal

In some companies, it is the finance department, not Human Resources, that handles the payroll processing function. Finance also handles the tracking and managing of accounts payable (money owed to others) and accounts receivable (money owed to the company). Clearly, the company's very survival depends on being able to keep track of what is owed and what is due. Without the ability to manage the income and outflow of funds, the company cannot survive for very long. Certainly, the bank with which your company does business has records of recent transactions, so the cash in the bank is relatively secure. The problem clearly is with the receivables and payables. No company can survive the loss of information related to these types of transactions, so understanding what systems track that data as well as where and how that data is stored, backed up, and archived will be critical to your BC/DR plan.

The legal aspects of data security in the event of a disaster is a topic too broad to discuss at length in this book. However, there are important legal considerations when looking at your company's responsibilities for data security and integrity in the normal course of business as well as in the face of a major or minor business interruption. As you learned from the case study presented prior to this chapter, "Legal Obligations Regarding Data Security" by Deanna Conn, partner at law firm Quarles & Brady, there are definite legal requirements to be dealt with. Although the legal aspects may not be your responsibility to address, as IT project manager your job is to ensure the project meets all requirements, and some of those requirements may be legal. You should contact your company's legal counsel and discuss your BC/DR plans with them. You can then follow their recommendations as to how to ensure your BC/DR plans meet minimum legal requirements.

As with other corporate functions, you'll need to be sure the financial and legal requirements are met within the scope of the BC/DR plan. This may include compliance with financial or legal requirements related to data security and integrity; it may include simply being able to process payroll after a significant business disruption or understanding how you'll recreate your accounts payable and accounts receivable files so you can resume the task of collecting money due and paying money owed.

Warehouse/Inventory/Manufacturing/Research

If your company maintains warehouses or facilities with inventory, manufacturing, research, or other similar activities, you'll need representatives from these areas to answer specific questions related to the BC/DR process. Certainly some of these areas are covered by facilities and security such as access to secure areas or repair of broken or ruptured pipes, for example. However, facilities staff will not be able to tell you which equipment is mission critical, which research is vital to the ongoing success of the company, or which inventory is most important to the company. These are questions that can be answered only by experts in those areas. Because this may well be the heart of corporate operations, you may need to use a phased approach to BC/DR planning with this group. For example, what would it take to get back up and running in a minimal manner? Next, determine what it would take to get back up and running in a normal manner. These are often two distinct phases that must be planned out. We'll address this in more detail later in the book.

Common Challenges

Using the "What If"...

Remember, it's likely that just about everyone in every corner of your company will view their work as vital to the ongoing success and operations of the company. In reality, there are critical functions and important support functions, but not every-

Continued

thing that goes on is actually mission critical. Unfortunately, you'll be caught in the middle of the stream. As IT project manager for this BC/DR project, you won't necessarily have the expertise to say which are and are not critical functions, so you'll be dependent upon subject matter experts to make that determination. At the same time, they are the very ones with the vested interest in their areas or activities being labeled "critical" or "vital" to ongoing operations. One method to sort this out is to ask specific questions rather than vague, values-based questions. Instead of asking, "What's the most important function here?", you may need to ask, "What if this system went down? What would you do?" or "What if this room caught on fire and charred everything to bits, what would you do?" Asking specific scenario-based questions can help you move past egos and agendas to the underlying issues. If you're extremely lucky, you might actually get a team of people who are willing to clearly identify mission-critical, important, and support activities with the proper perspective.

Purchasing/Logistics

Purchasing and logistics may be two distinct functions within your company or they may be handled by one group. They may go through finance or HR or they may be tied to other departments such as warehouse or research. These functions may even be much more informal than that, such as departmental managers having the authority to purchase supplies as needed. Regardless of how this is handled in your company, you'll need to ensure that these functions are addressed in your BC/DR plan.

Purchasing and logistics are involved in three distinct ways. First, if your company regularly purchases things like equipment, inventory, and supplies used in the normal course of business, you have to deal with the potential disruption of this function. Second, you may need to arrange for the purchase of services and supplies related to disaster readiness. For example, you may need to contract with a remote data center for backup computing services in the event of a disruption at your primary place of business. In this case, the purchasing folks may get involved in terms of preparing or reviewing contracts, developing requests for information (RFI), or requests for quotes (RFQ) or requests for proposals (RFP). Emergency services provided by third-party vendors are typically less expensive and more reliable when contracted for outside the scope of a business disruption and your purchasing or logistics folks may be involved. In addition, there may be other emergency supplies your company chooses to keep on-site or available such as emergency medical supplies, food, water, office supplies, tents, clothing, whatever is appropriate to your company and its unique needs. Third, you'll need to incorporate emergency methods for purchasing and logistics in the event of a disaster. If you urgently need to purchase three new servers, how will this purchase be authorized and completed if your company is temporarily running out of the CEO's garage?

Marketing and Sales

Marketing and sales activities rely heavily on customer information. Marketing staff typically mine corporate customer data or target market data to determine the best approach for marketing, advertising, and related activities to generate sales for the company. Sales data, whether generated online, by phone, or in a brick-and-mortar setting, is used to determine inventory levels, purchasing requirements, manufacturing lead times, and a whole host of other corporate decisions. What would happen if your CRM system went down or the building in which the server hosting the CRM application caught fire? What would happen if your marketing and sales staff did not have access to key customer data, sales history, or order history? Marketing and sales are the engines that drive the revenues that make everything else in the company possible, so these deserve a special place in your BC/DR planning. These are the activities that get revenue coming in the door to help your company resume or continue operations after a business disruption. So, while IT systems and physical facilities rank high on the list of priorities in a BC/DR plan, pay special attention to what the marketing and sales folks have to say and be sure you include them in your planning sessions.

Public Relations

You may ask "What does public relations have to do with your BC/DR plan?" In February 2007 JetBlue (NASD: JBLU), a low-cost airline, experienced systemwide problems due to bad weather across much of the United States. In some cases passengers were held on planes away from the gate for up to eight or ten hours. Although this was not an IT failure, it was a serious business disruption caused by bad weather. This is exactly the type of scenario a BC/DR plan for an airline should account for. "What would happen if bad weather forced us to cancel or delay flights across the United States?" is the question the airline should have asked and answered prior to a weather event (and they may well have done so). However, they failed to address one of the most basic concerns. Many trapped passengers could not understand why the plane did not just taxi back to a gate and allow passengers to get off, or why portable stairs were not rolled up to the plane on the tarmac to allow passengers to disembark onto buses or other vehicles to take them back to the terminal. This was not done. News channels on TV and radio were quick to broadcast the news of these delays and passengers or those waiting for them alerted the media to these issues and the media frenzy began. When it was all said and done, all JetBlue could do was mop up the damage through its marketing and PR departments. In this case, they announced they were instituting a passenger's bill of rights and would compensate passengers who were delayed or rerouted. Was it too little, too late? It's too soon to tell what the long-term impact will be but the stock tumbled from a high of $16.62 per share on January 16, 2007 to a low of $12.56 on February 20, 2007, a drop of almost 25%. Can the company recover from the loss in its market capitalization (price per share x number of outstanding shares)? Can the company recover from the loss in consumer confidence? Time will tell, but remember the not-too-cheery statistics on recovering from a major failure. Even though this may not qualify as a major or catastrophic failure, it was a significant event that had an immediate impact on market capitalization and revenues—that's "major" in most people's minds.

Public relations can go a long way in smoothing over the public image of the company and soothing bruised and battered customer. There are numerous examples of companies owning up to a problem immediately, taking steps to stop or resolve a serious problem, and recovering consumer confidence over time. PR can be extremely helpful to your company should you experience a major disaster or outage. Suppose you are an e-commerce company and there is a breach of your Web site and customer names and other personal data are stolen? How will you deal with this? Certainly from an IT perspective, you'll lock the virtual doors and investigate. From a corporate perspective, however, you may need to engage in some serious PR to help manage the company's image.

If you're in a small company, this function may be handled by the owner, the general manager, or the HR staff. However, don't discount the need for these kinds of activities in the face of a business disruption. Even if you don't have a dedicated department or even a dedicated staff member for marketing and PR, you can identify a firm that specializes in PR and perhaps develop a relationship with them prior to a business disruption. At the very least, you may want to identify two or three firms in your area (both in and outside your immediate geographic area) that could be resources for you in the event of a business disruption. Perception impacts how your shareholders, stakeholders, vendors, suppliers, customers, employees, and community view your company. There is an opportunity to shape perception in the immediate aftermath of an event, even if it is to calm frayed nerves or reassure the public that immediate and decisive action is being taken. Don't miss this opportunity because you dismissed the value of PR in a BC/DR plan. For more on this, be sure to read the case study that discusses crisis communications, following Chapter 6.

From the Trenches...

Marketing, PR, Spin—What's the Difference?

Many IT people are focused so intently on staying up to date with technology that they are not familiar with the difference between marketing, public relations, and spin. Let's take a brief detour to discuss these. Marketing includes the activities a company undertakes to make consumers aware of the products and services it offers or to create demand for those products and services. For example, Coca Cola airs commercials on television to make consumers aware of the attributes of the beverages it offers for sale in order to increase consumer awareness of the Coca Cola products as

Continued

well as (hopefully) to increase demand for the products. That's marketing. Public relations or PR is done to announce pertinent corporate news and information to the public, typically through news outlets. For example, a press release may be issued when a company hires a new CEO or when it executes a new union contract or decides to implement a new enterprise software application. Any time the company wants to disseminate information about the company (and not specifically about a product or service), it may engage in PR activities. Clearly, there is some cross-over. If a company discovers a new process that is used in a new product or service it is offering, there is a mix between marketing and PR activities. "Spin" is a term often used in conjunction with PR, and it has negative connotations. It typically means that the company is putting a positive or company-biased slant on the information. PR activities may attempt to put the most positive face possible on a situation, whereas *spin* often implies disingenuous or dishonest motives. Unfortunately, some companies that experience a problem attempt to put spin on the situation rather than address it in an honest and forthright manner. To use an analogy, PR would say, "The glass is half full" rather than half empty. The statement is true, but it points the listener to the positive aspects of the situation. *Spin*, on the other hand, would say "It's dangerous to fill the glass more than half way, it could overflow and hurt someone." It's not true and they're trying to make you think anything more than "half full" is bad simply because they can't provide more than "half full." Keep this in mind if you're in charge of developing your crisis communication plan. It's fine to put your best foot forward—just make sure that foot doesn't end up in your mouth.

We've delineated the more common functions found in small, medium, and large companies alike, but we know this list is not exhaustive. Take inventory of your company—look at your company's Web site, intranet, or phone directory to make sure you have all the subject matter expertise from all the departments represented. Also keep in mind that during your project, you may interview these folks but they may not be a formal part of the project team. In some cases you might find it more productive to create a list of interview questions to ask each subject matter expert or departmental representative. You can then compile that information and rank system and information criticality based on your own assessment or (more ideally) the assessment of a small subset of subject matter experts. For example, you may gather and collate all this information and bring it to three corporate vice presidents or to the senior management team for discussion. You may want to avoid trying to be the referee among competing interests and you may not be in the best position to make some of those calls. You *are* in the best position to ensure you gather the relevant data from the subject matter experts and present it to those who can and should make those decisions. If it is left to you and your project team to cull through the data and make those decisions, be sure to present your conclusions and decisions to your project sponsor for formal, written approval before moving forward. This will cover you in the event of a turf war and will hopefully help you avoid any organizational or political gaffes as well.

Project Definition

OK, let's take a moment to regroup. So far in this chapter, we've reviewed the basic process of creating an IT project plan with an eye toward the specifics of business continuity and disaster recovery. We've looked at the key resources and stakeholders you'll need to consider including in your project planning process. While we were looking at those key resources, we also discussed the various business functions and what kinds of questions you might ask to ascertain the criticality of those functions. So, we understand what the project plan should look like and who should be involved.

Now let's turn our attention to the project definition. As we discussed earlier in this chapter, first you'll need to define or understand what the basic project parameters are—the scope, budget, schedule, and quality for this project. In most companies, one or more of these parameters are assigned to you, but not in all cases. Next, you'll need to define the business, functional, and technical requirements. These need to be determined before you can begin your risk assessment, which is the first "work" phase within the BC/DR project plan and which is discussed in the following chapter. The business requirements define the scope of the project. Will you be addressing the top three critical systems, all critical systems, or all business systems? The functional requirements tell you what the plan must do or accomplish to meet the business requirements. Do these critical systems need to have full redundancy or do they need to be available within 72 hours of any serious business disruption? The technical requirements tell you how these business and functional requirements will be met. Will you use a backup data center or will you rely on another location of your own company to provide the backup services or redundancy you need? What are the requirements to set up your critical applications in another location? How exactly will your data be backed up, verified, and archived? As you can see, by starting with your business requirements, you can develop functional and then technical requirements.

As IT professionals, the temptation is strong to begin with the technical requirements, but you clearly could miss some critical data with that "bottom up" approach. That said, you no doubt already have some technical solutions in place, and these should be incorporated into your planning. For example, if you already have a solution in place for backing up and archiving data in a secure manner (i.e., data is secure while being backed up and backup itself is off-site in a secure location), then this should be rolled into your BC/DR plan. So, if you have the backup scenario figured out, you can then look at your business requirements and ensure your backup plan adequately addresses backing up the four critical business applications defined in the business requirements. You can also look at your backup process in light of your functional requirements to determine whether your backups meet those specifications. For example, perhaps you realize through developing functional requirements that the period between backups is too long. Perhaps trying to recover from an outage that occurs the day before a scheduled weekly backup would be too difficult or time consuming, or even impossible if it was due to a severe business disruption. You might conclude that it would be better to do a twice per week backup to not only meet your current business needs but to provide the kind of business continuity and disaster recovery capability called out in your business and functional requirements. As mentioned earlier, you may choose to have different subsets of your project team for different phases of the project. You may want to get all the major business units together to develop the business requirements. You may want to have subject matter experts from each of the business areas help develop the functional requirements, and you may then have just IT staff work on the technical requirements.

One last note here before we look at the details of business, functional, and technical requirements. You may find that it makes more sense for you to put together your BC/DR planning team and perform the first major task, the risk assessment. The risk assessment phase, discussed in the next chapter, will help you look at the potential risks to your systems and to your business. You might choose to create several deliverables from your risk assessment phase that help you develop or refine your project requirements. You may decide that it makes sense to create general business and functional requirements so you can ensure your project falls within required parameters. That is, you might choose to create high level business and functional requirements that fit into your budget, your timeline, or your overall objective for the scope of the project. You might then go into the risk assessment phase and come back to your requirements with more specific information. If you do this method, it is more likely that your business and functional requirements will define a project scope (budget and schedule) that meet the goals or objectives you've been handed. From there, you can continue to refine. If you go into the risk assessment phase without a general understanding of the business and functional requirements, there's a good chance your scope is going to balloon quickly. So, keep in mind that both methods (define requirements then assess risk or assess risk to define requirements) have their own set of benefits, risks, and limitations. Be aware of these as you make your decisions and remember that the key is to use a process that works best for your business.

Business Requirements

Business requirements are the first step in developing BC/DR project requirements because you must first understand the critical areas of your business. What questions should you ask to ascertain which are the critical business functions? As you know, if you ask users what the most important systems are, they'll give you a list a mile long. Rather than ask that type of question, many experts advise using scenario-based questions to help focus attention and elicit useful information. This may take a bit longer in the short-term but will save you time and headaches later. Keep in mind, too, that the first major deliverable for your BC/DR plan is likely to be the risk assessment, which may lead you back to modifying your business, functional, and/or technical requirements. Each iteration should move more quickly and inject less change into the prior iterations. So, if you create your business requirements, and then do your risk assessment, you may find the priorities for the business requirements change or that a critical system was omitted during the first go-round. This is a normal part of project planning. However, if you find that each iteration is injecting more change and more uncertainty or more confusion to the process, you need to step back and assess what's happening. It might be that the project is beginning to manage you (rather than you managing the project), or it could be some key assumptions were incorrect or that your organization is in the midst of a significant change. Be aware that a project plan that feels like shifting sand beneath your feet is in danger of getting out of control and failing. Let's look at some questions you can use to elicit the type of business information you need to create a useable BC/DR plan, understanding that some of the answers to these questions may change slightly over time. You can tailor these questions to the specifics of your organization but this should give you a good start.

- What would happen if the server room caught on fire and the sprinkler system went off?

- What would happen if there was a fire in the building and we had to evacuate the building immediately? What would happen if we were not able to reenter the building for three weeks or a month?

- What would happen if a security breach was discovered in our customer database?

- What would happen if we discovered that our Web server had been hacked?

- What would happen if an earthquake (hurricane, tornado, flood) destroyed this building and many of our employees' homes in this area?

- What would happen if a major snow storm made it impossible for employees to get to work for a week or two?

- What would happen if a chemical spill from a nearby plant or railroad forced us to evacuate this building for a week? A month? Six months?

- What would happen if electricity to this site were cut or unavailable for half a day, one day, one week, one month?

- What would happen if our high-speed connection to the Internet were to go out for half a day, one day, one week, one month?

- What would happen if a bomb went off in this building and we could not get back into it, ever?

- What would we do if major transportation routes (air, rail, road, sea) were shut down or disrupted?

- What would we do if key people were killed, injured, or missing?

As you can see, these questions elicit information because they create "what if" scenarios to which team planners have to respond. It gets people thinking in very concrete terms and you, as project manager, can help step them through this process. Again, since there may be aspects to the BC/DR planning process that do not fall under your direction or management, you may not be responsible for managing this process. However, whether you're heading this up or simply participating as part of the team, you can bring your skills and expertise to the team process and help step people through this process. By envisioning "what would happen if," you can help craft a realistic view of what the next steps would be. Immediately, people will begin thinking about what they would do without a server or without an application or without the resources at their desks—and this helps you begin to determine what the technological priorities and needs are for your BC/DR plan.

Functional Requirements

Once the business requirements are developed, you can begin to craft the functional requirements. Functional requirements state what is needed, not necessarily the technology that will fulfill that need. For example, the sales department might say "We'll need a way to contact our customers. If the Internet connection and e-mail are down, will we have phones? If we have phones, we can call customers but how will we get their phone numbers?" The functional requirement might be to always have access to the most current customer contact information. For example, some small businesses might keep a printed copy of their customer list at their attorney's office, which has branches in four cities other than the one in which you're located. Although this might seem extremely low-tech, the only requirement is that the solution meets the needs of the organization. This can include the need to have low-tech, low-cost solutions available to the most common business disruption scenarios. Only you and your company can determine what's most appropriate. The functional requirements describe what functions or features must be available. When you delve into the technical requirements, you can define the technological counterparts to the functional requirements.

When asking and answering the questions listed earlier in the business requirements section, listen carefully. Chances are good the functional requirements will start forming there. Your planning team will say, for example, "If our servers are down, we'd need a way to contact our customers." This is a business and a functional requirement. The business requirement is that customer contact is a vital aspect of the business; the functional requirement is that there needs to be some method for accessing current customer contact information in the event of a server outage. If the team ponders the question, "What if our entire server room caught on fire and was destroyed? What would you do without the tools you currently use to do your job?", you're likely to begin to understand that they might be able to get by without the billing system as long as they had a printout of the current status of accounts payable and accounts receivable, but they would not be able to continue business operations in the near-term at all without the Web server being up. That's good information. You now know that the Web server functions are critical, the billing system functions are important but not critical.

As you work through these scenarios, you may want to create a system of ranking these requirements so that you can gain agreement with the team of subject matter experts as to the relative importance of these requirements. You'll have to listen carefully if you're in charge of this process because people may describe one priority and then assert another. Using the billing system example, when you ask about what they would do in a particular situation, they may indicate that they could get by without the billing system because if they had current data, they could keep track manually; even though that might be a major task, it would be doable. However, if you then move to a ranking system and indicate the Web server function would be mission-critical but the billing system would not be, you might get some disagreement. You may need to manage the situation by helping people understand the point of the ranking system—which is to determine what you'd need within hours of a disruption versus what you'd need within days or weeks of a disruption. Table 2.1 provides a few suggestions for ranking systems that you might find helpful, but feel free to create whatever works best for your organization that will lead to clarity and agreement.

Table 2.1 Sample Ranking Systems

Sample One	Sample Two	Sample Three	Sample Four	Sample Five
Mission-Critical	Very High	Red	One	Revenue
Critical	High	Orange	Two	Support
Important	Normal	Yellow	Three	Maintenance
Support	Low	Green	Four	Other

The Sample One ranking system is fairly self-explanatory. The ranking is related to how critical a system is to the ongoing company operations. The Sample Two system uses commonly used word to describe priorities. Sample Three uses the color system, similar to the threat level system the U.S. Department of Homeland Security uses. Sample Four is obviously a simple priority ranking system using numbers. Sample Five is an example of a customized list you might create to indicate that anything related to revenue generation is the most important priority; anything related to supporting business operations and revenue is second. Tasks that maintain business operations come in third and the "Other" category is the catch-all. You can create any ranking system that is appropriate for your organization, but spend some time defining the categories you choose to use so that you will know where the boundaries are for each category. This will help everyone have a shared understanding of the categories and will help everyone to consistently rate and rank priorities as a team.

Technical Requirements

The technical requirement will define how the business and functional requirements will be met. As the IT professional on the team, you will have the unique vantage point of understanding how these current business and functional requirements are being met with technology. As stated earlier, it's entirely possible that some or many of your business and functional requirements for BC/DR planning are being met with current technology strategies. As you review the business and functional requirements, you can begin assessing how close or far your technology is from the desired state. In most cases what you'll find is that your technology solutions in place meet some but not all of your BC/DR requirements. This is essentially a gap analysis that tells you where your technology meets business and technology requirements for BC/DR and where it partially or completely misses the mark.

Technical requirements are important for reasons stated earlier, but they are also important because these will form the foundation of specific tasks related to defining the required technological solutions so you can go out to bid for these products or services; so you know what would need to be replaced; so you know what resources you'll need at a remote/off-site data center, and so on. When it's all said and done, you should have a complete list at the end that defines server types, server capabilities (number of processors, speed, RAM, disk space, network connections, etc.), applications, application requirements, application configuration needs, user configuration needs, and more. If this is currently part of your standard IT operations, then you may have all the configuration and technical needs documented already. If this is the case, you'll need to make sure everything is up to date and accounts for any new or changed technology implementations (current or upcoming). Most importantly, you'll have to determine where and in what format this data should be stored so that it is available to you in the event the server room, building, or location is destroyed or inaccessible.

As you can see, your BC/DR planning process ideally will help you make better use of existing technological solutions and help you implement new ones to meet your needs. You may discover through this process that one or more of your existing solutions is a perfect fit for your BC/DR needs. You may discover that some existing solutions are only being partially utilized or that they can be utilized in new and different ways. Finally, you may discover large gaps in your BC/DR technology readiness and this should give you the information (and ideally, the organizational support) you need to implement the right solutions for your company.

As you can see, developing the business, functional and technical requirements are all inter-related and none happens in a vacuum. As you develop your BC/DR plans, you should continually assess your current practices, processes and capabilities. In some cases, you'll find that what's already in place will work perfectly with your "best case" BC/DR plans. In other cases, you'll find that you want to modify existing processes and methods slightly to address your optimal BC/DR requirements. In still other cases, you'll find areas of the business (and technology) that is critical to ongoing operations that are completely exposed. This is the good news—the BC/DR planning process should help you assess these areas from the top down (business, functional, technical) and from the bottom up (technical) and determine the status of your current business operations as it related to BC/DR. From a known state, you can make intelligent choices about how much work you need to do to provide the level of readiness required for your business.

Business Continuity and Disaster Recovery Project Plan

Those of you familiar with project planning know that the project plan itself will be comprised of several major elements. The first element includes the various project definitions, which we've covered at length in this chapter. The project parameters (scope, budget, schedule, quality) should be defined; the project requirements must be delineated so they fall within the project's parameters. Once the project definition stage is complete, you create your Work Breakdown Structure (WBS). As you know, the WBS defines all the major and minor tasks of the project that, when taken as a whole, describe the total amount of work in the project, or the project scope. The WBS we're using as the framework for this book and for our BC/DR plan is as follows:

1. Project Definition
2. Risk Assessment
3. Business Impact Analysis
4. Risk Mitigation Strategies
5. Plan Development
6. Emergency Preparation
7. Training, Testing, Auditing
8. Plan Maintenance

Project Definition, Risk Assessment

We've discussed project definition at length in this chapter and we've also linked it to the risk assessment to be discussed in detail in Chapter 3. The risk assessment is the phase in which all potential risks to the business are listed and then evaluated both for likelihood of occurrence and impact in the event of an occurrence. As a company and as a project team, you'll need to create a cut-off point so that risks that fall below the line are not addressed. This is one way the scope (and as a result, the budget and schedule) of the project are managed. We'll look at how to perform this phase of project work in the next chapter.

Business Impact Analysis

Business impact analysis, covered in detail in Chapter 4, looks at how the business would be impacted if the major risks were to occur. In order to make this process productive, it occurs after the risk assessment so that only the risks that fall above the cutoff point, or above the risk line, are addressed. This, too, contributes to your ability to manage the scope, budget, and schedule of the project.

Risk Mitigation Strategies

Tying the risks with the business impact analysis together yields your BC/DR priorities. Clearly, you want to address only risks that have a high likelihood of occurrence and a medium to high impact should they occur. If a risk has a low risk of occurrence and it would have a low impact on your business, you may choose to not plan for that particular risk. Every company has to make that call individually—there is no single right answer, though there are real world limitations to the value of planning too far down the risk/impact ladder. It probably isn't of any benefit to spend two weeks and 300 staff hours planning for something that probably won't happen, and if it did happen, would impact only six of your company's 148 employees and none of your company's top 100 clients. In Chapter 5, we'll look at how to develop strategies to manage the risk including ideas on how to reduce, avoid, and transfer risk.

Plan Development

Once you've assessed your risk and the impact of those risks, and developed strategies for mitigating those risks, you'll need to start working on putting those strategies into action. That means developing a set of tasks that will deliver the required results. Plan development will include creating the project plan's WBS tasks related to actual BC/DR activities (as opposed to plan activities) as well as all owners, deliverables, and success criteria. We'll look at this in Chapter 6 in detail.

Emergency Preparation

Part of every BC/DR project plan should be the actual emergency preparations that a company should undertake, and we'll look at this in detail in Chapter 7. If your job is limited to IT-related functions, you might find that your role here is limited. Emergency preparations include the specific steps to take in the immediate aftermath of a disaster and the definition of when business continuity activities should begin. Though this might be outside the scope of your responsibilities or authority, we'll cover the basics so you can be a knowledgeable contributor to the overall BC/DR planning process. If your job as the project manager for this BC/DR project includes all aspects of BC/DR planning, then this section will help you rally the resources you need to create an effective emergency response plan.

Training, Testing, Auditing

Chapter 8 covers the tasks you'll need to include in your BC/DR project plan related to training staff for emergency response and for implementing the BC/DR plan should that be necessary. Testing is something all IT professionals are familiar with and this takes on significance when you look at testing from the BC/DR perspective. Finally, you'll need to audit and assess strategies after you've trained staff and tested the plan. This is part of the iterative process you'll use throughout the project management process. It is here where you discover key gaps or broken processes; it is here you have the opportunity to fix these gaps, errors, and omissions so that your BC/DR plan is as solid as possible within the given constraints of the organization.

Plan Maintenance

As we've discussed, an out-of-date plan is sometimes worse than no plan at all because it allows staff across the organization to make assumptions about BC/DR readiness that may simply be wrong. If your plan was crafted several years ago, there's a high likelihood it is no longer current. If you have a plan that you believe may be relatively current, you can short-track some of your planning processes by reviewing the plan against the steps delineated throughout this book. For example, you may choose to perform the risk assessment and business impact analysis with fresh eyes then compare the results to the plan you have. If there are significant gaps or disconnects, you may choose to scrap the old plan altogether or modify, update, and test the existing plan. The choice is yours. Whichever path you choose, whether you have an existing plan or are creating one for the first time, you should build in tasks that allow the plan to be periodically reviewed and updated. In Chapter 9, we'll discuss some of the methods companies use to do this so that you can create a maintenance plan for your BC/DR plan that makes sense for the way you do business today and in the future.

From the Trenches...

Perfect World versus Reality

Throughout this book, we'll discuss perfect-world scenarios as well as real-world realities. Project planning rarely follows the prescribed methods, timelines, and order we've discussed. It's useful to understand best practices and preferred methods so that you can strive to mirror those in your work. However, it's highly likely that there are one or more mitigating factors that come into play with your project planning process. To assume that things will follow the defined order and work out perfectly is to set yourself up for disappointment and failure. The goal should be to strive to follow the predefined processes and steps to the greatest degree possible and to diverge from those only with intent and conscious choice. Anytime you find yourself diverging from best practices, make sure you ask yourself if this is by accident, by choice, or by necessity. That should help keep your project on track while still giving you the flexibility to deal with the specifics of your organization. As long as you're aware that you're taking a side road or alternate path, you can still arrive at the same destination. It's when you close your eyes and hit the gas pedal that you're likely to get yourself (and your project) into trouble.

Summary

Planning your BC/DR project is similar to other IT projects you've defined, developed, and managed in the past with the possible exception of the reach of the project. A BC/DR plan reaches into every corner of your organization and tests your assumptions about what you do, how you do it, and how important those things are to the viability of the company in the short- and long-term.

We looked at the elements that drive success in any IT project and looked at the specific factors that you'll need in order for your BC/DR project to be a success. Clearly, having executive support means that those further down the organizational ladder will be compelled or required to participate and to give this project the appropriate priority in the corporate scheme. Other success factors discussed include involving the right people early in the process, having an experienced project manager, clearly defining project objectives, requirements, and scope, as well as setting a shorter schedule with multiple milestones and using a well-defined project management process.

The project plan components are standard IT project plan components that should be incorporated into your planning process. We looked at the elements including project definition, forming the project team, organizing the project, planning the project, implementing the project, tracking project progress, and closing out the project. Although there should have been no surprises here, we did cover some of the more important elements in each of these categories related specifically to BC/DR project plans.

A business continuity and disaster recovery project plan requires that key contributors throughout the organization participate with you to identify the bottom-line needs of the organization. As an IT professional, there's simply no way you can know all you need to know about how business operates on a day-to-day basis to create an effective BC/DR plan without key contributors from your various business units or functional groups. Including people from facilities, accounting, legal, human resources, engineering, and others will be crucial to success, and with the information discussed in this chapter, you should now have a better idea of who needs to participate and what you can expect them to contribute.

Another important aspect of the BC/DR project plan is the definition of the project including the business, functional, and technical requirements. As you learned, this is likely to be an area that requires additional detail or refinement. Most companies find that it makes the most sense to create initial high-level definitions and then come back after the risk assessment and business impact analysis are complete to refine those definitions. The business requirements define what is needed to operate the business in the aftermath of a disaster (disaster recovery) and in the longer-term (business continuity). The functional requirements delineate what is needed by way of functionality so that the business can get up and running as quickly as possible should a business disruption occur. Finally, the technical requirements, with which you may be best acquainted, determine exactly what the technical specifications are that will meet the requisite functional and business requirements.

Lastly, the BC/DR project plan elements are the elements specific to the BC/DR activities. We looked at the framework that will be used throughout this book as an example of one potential approach to the Work Breakdown Structure for the project. This includes risk assessment; business impact analysis; risk mitigation strategy development; plan development; emergency preparedness; training, testing, and auditing; and plan maintenance. Each area will be discussed at length in the upcoming chapters.

Solutions Fast Track

Elements of Project Success

☑ There are numerous well-known, time-tested success factors that contribute to the likelihood that your IT project will be a success. Executive support is always at or near the top of success factors for an IT project.

☑ User involvement is key to success because any IT project that does not ultimately meet user needs will likely be seen as a failure. User involvement must be well-managed to ensure it remains focused and productive.

☑ An experienced project manager is another success factor, which makes sense. Anyone who's had to manage a project with all the challenges that come with it is likely to be more successful in subsequent projects.

☑ Clearly defined project parameters (scope, budget, schedule, quality) as well as project definition and requirements are also key to success. The more detail you develop to describe what you're trying to accomplish and how you'll accomplish it, the more likely you are to hit the target.

☑ Shorter project schedules contribute to project success. In some cases, it might make sense to break down various parts of your project into subprojects.

☑ Projects with more milestones also tend to be more successful because milestones help keep you focused on deliverables. Milestones are checkpoints, and the more often you check your progress against plan, the more likely you are to be able to make small changes early that help you stay on track.

☑ Having well-defined project processes also contribute to project success for fairly obvious reasons. If the team knows what to expect and how to go about accomplishing and reporting project progress, the project has a better chance of moving forward. If you get bogged down in convoluted, illogical, or unnecessary project processes, your team will find ways to circumvent the processes or fail to complete project deliverables on time (or at all).

Project Plan Components

☑ The BC/DR planning process follows standard project management steps. The components used are ones you are likely familiar with, especially if you are an experienced project manager.

☑ Project definition serves to define what work needs to be accomplished. The definition includes the project parameters (scope, budget, schedule, quality), project objectives, and the business, functional, and technical requirements.

☑ Forming the project team is unique in a BC/DR project plan because the people that need to be involved with the project will likely come from every corner of your organization. Involving the right people at the right time will boost the chances of success for your BC/DR project.

☑ Project organization includes defining how the project will be run from team meetings to task status updates to overall project progress reporting. It also includes key processes such as change management and escalations. Having well-defined process improves the project's chance for success by providing standardized, easy-to-use processes for the team.

☑ Project planning includes creating a Work Breakdown Structure that, when complete, defines the scope of the project and meets the project's requirements and objectives. A well-crafted WBS includes tasks, dependencies, timelines, milestones, and success criteria, among other things.

☑ Project implementation is where the "rubber meets the road" and is where all the planning efforts take effect. Once the project is underway, change must be actively managed and project progress must be actively tracked and managed.

☑ Project tracking consists of all metrics you, your team, and your project sponsor have agreed to with regard to project progress. This may mean tracking time, costs, labor hours, percent complete, and other metrics that help bring the project to a successful conclusion.

☑ Project close out includes all tasks required to bring the project to an end and shut things down methodically. It typically includes reviewing change requests, project deliverables, and lessons learned.

Key Contributors and Responsibilities

☑ Key contributors in a BC/DR project plan may differ from key contributors you've worked with in the past.

☑ BC/DR activities require the coordinated efforts of key people throughout the company from facilities management to security to human resources to finance and accounting.

☑ Identifying the key contributors to your plan and what areas they will be responsible for in the BC/DR planning process will help ensure you have the people you need on the team to create a solid, workable BC/DR plan.

Project Definition

☑ Project parameters define the scope, budget, schedule, and quality of the project. There is an interrelationship among these parameters. If you choose to increase the scope, you must also increase the schedule or budget. If you reduce the budget, you must increase the schedule, reduce the quality, and/or reduce the scope of the project.

☑ Understanding the relative flexibility of the project parameters helps you make decisions in line with the business requirements. For example, if budget is fixed, it is least flexible. Therefore, you will have to modify your schedule, scope, or quality if things change later on.

☑ The business requirements for the project must be understood before the project can get underway. Business requirements may be modified or revised after the risk assessment and business impact analysis is completed.

☑ Functional requirements for the project indicate what functions or abilities must be present in order to continue business operations after a major (or minor) business disruption. The functional requirements indicate what must be present or available, not how those functions will be made available.

☑ Functional requirements may be revised or modified after the risk assessment and business impact analysis phases of the project work have been completed. Information may come to light that suggests (or requires) your functional requirements be modified to better suit the company's needs.

☑ Technical requirements can be established only after the business and functional requirements have been developed. However, it is also true that your organization already has many technical solutions in place (or in progress) related to BC/DR.

☑ Your current (or in progress) technical capabilities should be taken into consideration as you assess business and functional requirements. There will be areas of complete alignment, areas that overlap, and areas where you discover large gaps.

☑ The BC/DR planning process will help you utilize existing solutions more effectively as well as address any gaps or future needs.

Business Continuity and Disaster Recovery Plan

☑ The BC/DR plan will follow a standard project management framework and rely upon standard project management processes.

☑ The BC/DR plan, at this highest level, should have these sections (or work phases): definition, planning, organizing; risk assessment; business impact analysis; risk mitigation strategy development; plan development, emergency preparedness; testing, training, auditing; and plan maintenance.

☑ Risk assessment and business impact analysis may induce changes to the business, functional, or technical requirements for the project.

☑ The plan development should include the development of the work breakdown structure including tasks, dependencies, task owners, success criteria, and milestones.

Frequently Asked Questions

The following Frequently Asked Questions, answered by the authors of this book, are designed to both measure your understanding of the concepts presented in this chapter and to assist you with real-life implementation of these concepts. To have your questions about this chapter answered by the author, browse to **www.syngress.com/solutions** and click on the **"Ask the Author"** form.

Q: My boss doesn't really see the point in creating a disaster recovery plan. He says we've been in business for 30 years and never had a disaster, so it would be just a waste of time and money. Any suggestions for how I might sway his opinion?

A: You're facing a very common challenge in business today, so you're not alone in feeling like you're fighting an uphill battle. If your boss's argument is that things have been fine for 30 years, there are two compelling counterarguments that jump out. First, logic would argue that if you've managed to run for 30 years without any major business disruptions, the odds are getting stronger with each passing year that something will happen. Can business run for generations without major business disruptions? Sure, but the odds are not in your favor. Second, and perhaps more important to you and your boss, is the fact that the last 30 years of business did not include technology to the extent it now does. A company in the food distribution business 30 years ago was scheduling routes with maps and rulers; it was tracking customer orders, inventory, and staff hours by hand on paper. It verified orders or resolved problems using the telephone. You may want to discuss with your boss how your business has changed in the past 30 years and how the role of technology in your company has changed over time. When he realizes that there are new and different considerations, he may come around. And, with the increasing use of technology, there are new business threats that did not exist before technology was as pervasive as it is. Data integrity, data security, and data availability are three areas that were handled very differently in the precomputer age. Talking with your boss about how these things have changed and what some of the new threats were may also help. Thirty years ago, a virus meant several staff members had to stay home from work for a week; now it means your data may be corrupted, compromised, or stolen by friend or foe. Your boss should understand these fundamental changes and hopefully these arguments will help him understand the importance and necessity of BC/DR planning.

As a last resort, you may be able to come to some agreement with him about how much time and money he *would* be willing to spend to allow you to perform some level of BC/DR planning. Sometimes the objection comes from a fear (whether real or imagined) that the BC/DR planning process will take a long time and cost a lot of money. If you allow your boss to set one or more reasonable parameters, you may find you've removed the roadblock. For example, you might ask if you could take three people and work on this for two months and come back with a risk assessment, impact analysis, and recommendations. You will have removed some of his potential concerns about time, money, or scope, *and* you'll be able to come back with real data that may open your boss's eyes to the value you've added and the value of continuing on this path. Finally, look around to see if your company is required by law or regulation to comply with various BC/DR requirements. Financial institutions, healthcare providers, and others are required to have some plans in place. Check to see if any of these requirements open the door for you to create a full BC/DR plan for your company, regardless of industry, size, or segment.

Q: I've never managed a cross-functional project team before. Any suggestions?

A: There are several important aspects to managing a cross-functional team. First is having executive support so when you begin crossing organizational boundaries, you don't get stopped by the first person who says, "Who are you to ask me to do that?" Second, you need to make sure you ask a lot of questions and listen to the answers. You will learn a lot that way and you'll avoid making common mistakes such as including or excluding the wrong people from the team. Third, it's vital that you understand (or learn) the organizational culture of each cross-functional area. For example, don't assume everyone is comfortable using e-mail, instant messaging, net meeting, and the like. Don't assume that everyone runs on the same schedule or cares about the same things. Listen more than you talk and you'll likely find your way through the cross-functional maze.

Q: My boss has said, flat out, that we are not going to create a BC/DR plan. Should I just give up?

A: Interesting question. If you simply "obey orders" then you have no choice but to avoid creating a BC/DR plan. However, unless you've been given strict orders *not* to create a plan, then you may have some flexibility. First, you might ask your boss directly, "What would happen if this building caught on fire because the tenant downstairs violated the fire code?" This might help him or her realize there are things beyond your company's control that can be planned for in advance. If, for example, your boss said, "I don't want to waste any time or money on creating a BC/DR plan," you can begin building risk assessment, business impact analysis, and mitigation strategies into your normal IT processes. For example, it would be reasonable to look at how at risk your customer database was if the server it ran on went down. That's a normal part of IT operations management. You might ask who would be impacted and how would it impact them. Even if you can't go to your subject matter experts, you may be able to have informal conversations so that you understand the relative impact of that database. If you find that it's at significant risk, you may opt to use a different backup strategy or begin sending backups to a third-party, off-site storage facility. These are the kinds of things that can be built into your everyday IT operations. As you proceed through this book, you'll find plenty of opportunities to incorporate many of these BC/DR concepts into your normal operations. Although it may not be optimal and it may not result in a robust BC/DR plan, it will give you better protection than not thinking about it at all. Of course, we're not encouraging you to disobey directives—we're just encouraging you to do what you can within the scope of standard IT operations to help bridge the gap.

Q: We're comprised of about 20 different business units spanning 10 different countries, though the overall company size isn't huge. This project seems overwhelming. Any suggestions on how to approach a project with this type of reach?

A: You do have a big job in front of you. However, it would be surprising to find out that nothing in the way of BC/DR has been done yet, so your first step should be to do some research to find out if something has been done in the past or is currently underway. Many times, organizations have initiatives underway that other departments or business units don't know about, so make sure you ask before launching into a project. If you've determined that there is no plan or no initiative underway and you've been given the project charter (or authority to begin this project), start by asking division heads who should be on the initial project team. Be clear about the charter for this first team (defining the project, defining requirements, perhaps performing the risk assessment and business impact analysis) so the department or division heads can assign the right resources to you. Remember that your job is to *coordinate* project efforts, not to be the subject matter expert for all topics. So, start by casting a wide net and narrow your focus as you go. It may take some adjustments early on to ensure the right people are involved to the right extent, but continually ask yourself, "what else do I need to know, who knows that, and who else do I need to talk to?" These three questions should help you cover all your bases. Once you have identified the right people, you can assign large subsections of the project to them. For example, you might have a project lead for each of the countries or a project lead for each business unit. It doesn't mean they'll all be participating fully throughout the lifecycle of the project, but they can help ensure you create a thorough project plan and identify the go-to people for each phase of the project.

Case Study 2

The Financial Impact of Disasters and Disruptions

Introduction

You don't need to be an accountant or a financial analyst to understand the financial aspects of a business disruption. In this case study, we'll look at it from an outsider's view—someone who has little knowledge of accounting—and look at the various costs to a business both in the short- and long-term. If you're not a financial type of person, please be sure to read this case study. You'll find it's far easier and less boring that you might imagine and if you can take away one or two new concepts, it will help you in your BC/DR planning and in your career as well.

Financial Aspects of Business Disruptions

The financial impact of a business disruption can be significant and long-lasting. In the past several years, disruptions from data security breaches have cost numerous firms millions of dollars in recovery costs and legal fees. The disruption to businesses impacted by the hurricanes in 2005, including Hurricane Katrina, is still being felt. How many of the businesses in the path of Katrina's fury are open today? Exact numbers are not available, but anecdotal evidence clearly indicates that many firms went out of business as a result of the impact not only of the hurricane itself, but the continued lack of emergency services, recovery services, and infrastructure in the aftermath of the storms.

From the Trenches...

The Cost of Disruption

In the legal case study presented earlier, you learned about ChoicePoint, one of the country's largest data brokers, that mistakenly sold the data of 145,000 people to a bogus group of companies that paid ChoicePoint to access its database. To settle the matter, ChoicePoint paid a $15 million civil penalty and agreed to implement an audit program that will be reviewed by a third party every two years until 2026. What do you think the cost of complying with the audit requirement will be for ChoicePoint through 2026? Certainly more than it made selling data to companies that had no legitimate right to that data. The cost of planning and prevention is almost always less than the cost of remediating the problem after the fact, especially when external, auxiliary, or related costs such as the cost of audits, the cost of fines, and the cost of the damage to the company's reputation are factored in.

Cash Flow

Cash flow is the lifeblood of businesses, just as it is for most individuals. When you get your paycheck every week or twice a month, you pay bills from that income and hopefully have enough left over to buy groceries, go to the movies, buy new clothes, and put some away in a savings account. Some months, your bills exceed your income and you charge things on a credit card (which is just a high interest loan) or pull from your savings account. Other months, your income exceeds your bills and you're able to put a bit more into savings or pay down some of those credit cards. If you're like an increasing number of Americans, your bills start to creep up and exceed your monthly income on a regular basis and you start finding yourself taking out new credit cards or payday loans to make it through to the next paycheck. This short-term solution is not sustainable. Eventually, you will falter if you use this financial model. The same is true for companies.

Each day, week, or month, companies generate revenues. During those same timeframes, they have expenses including employee payroll, health insurance, business insurance, rent, utilities, travel costs, supplies, and more. In any given month, the company typically plans to have its revenues exceed its expenses and to have some extra cash that can be put into the company in terms of a cash surplus sitting in the bank or capital improvements to the company (buying new equipment, upgrading the offices, etc.).

If the company's expenses exceed its revenues, like you it must find an additional source of capital. For businesses, this might be the cash cushion sitting in the bank that they've been able to amass during times when revenues exceeded expenses. In other cases, the business might draw upon a line of credit established at a bank, which is a loan that must be repaid with interest. It might rely upon credit terms with key vendors to delay paying invoices for 30, 60, or 90 days. If a company knows in advance it will have unusual expenses, such as during an expansion to new countries or the building of new plants, it can get a loan from banks, financial institutions, or investors, or it can sell stocks or bonds (depending on the type of company). These are all long-term financial strategies that allow a company to raise capital for growth because most companies cannot fund significant growth through existing revenue streams. However, the assumption is that by taking on this debt, they will be able to generate new or larger revenue streams and thus remain profitable or even become more profitable.

So, now let's relate this to business continuity and disaster recovery planning. Companies have revenues that they usually can predict to some degree. They manage expenses based on expected revenue streams. When a disaster hits, whether it's a data security breach, a hurricane, or a chemical spill, the company experiences a downturn in revenues and an uptick in expenses.

Lower Revenues

Revenues are generated through sales activities. These activities include order taking, order fulfillment, and order shipment. Each of these is needed to generate sales revenues. Revenues can be impacted by a disruption in a variety of ways.

Sales Activities

The most obvious impact of a business disruption is that your company is unable to conduct sales activities. If sales are done through the Web and a Web site is hacked, breached, or goes down, sales stop. If sales are conducted through customers contacting the company, as might be the case with selling software or large equipment, for example, customers may not be able to contact sales representatives if the building is destroyed or uninhabitable. If customers place orders through fax or purchase orders sent to the company, mail or faxes may not be properly received. Customers visiting a physical sales location such as a retail or wholesale store may not be able to access the building for a variety of reasons. Have you ever been in a grocery store check-out line when the power went out? If so, you know that all sales activities halt because few, if any, large grocery stores have manual sales systems in place as workarounds.

Order Fulfillment

If your company can still take orders, you may end up with a backlog in the order fulfillment area. If orders are shipped from a warehouse that has been impacted by fire, flood, hurricane, earthquake, or other natural or man-made disasters, either products may not be available to ship (if inventory is damaged) or the processes needed to pick, pack, and process orders may be damaged—from automated warehouse systems to IT systems that track inventory. Orders awaiting processing create a backlog, which in turn can create frustrated customers. If they need a part or product in a certain time frame and their order is delayed, they may cancel the order and go to your competitor. These kinds of lost sales are difficult to recover because you don't necessarily know which customers have abandoned you or if they've temporarily or permanently gone to your competitors.

If your company sells software or other intangibles such as services, you may have difficulty fulfilling customer orders because your staff is busy dealing with the aftermath of the disaster. So, although you might not see your company as being impacted by this type of issue, the inability to deliver services in the aftermath of a disaster is related to "order fulfillment" of more tangible goods. In both cases, revenues are impacted by the inability to provide the product or service to the customer in a timely manner.

Order Shipment

In some cases, you may be able to take the sales order and pack the order for shipping but not be able to complete the transaction due to infrastructure issues external to your firm. If an earthquake damaged roadways in and around your area but left your facility unharmed, you may not be able to get trucks in and out to ship products to customers.

Accounts Receivable

Some companies extend terms to their customers by allowing them to pay later for goods and services received today (this is typically referred to as *offering terms*). In these cases, companies send out invoices on a periodic basis. Customers typically have some defined period of time in which to submit payment to the company. If your company's ability to sell goods is not impacted but its processes for tracking and invoicing sales and tracking customer payments is impacted, you have a different kind of challenge to your revenue stream. You clearly don't want to send products out to customers and not bill them at all, bill them for the wrong amount, or bill them far past the sales transaction date. You also don't want to have income coming in from customers that is not properly credited to their accounts and entered on the books. So, even if your company can still generate sales, fulfill and ship orders, you may have other problems that impact your revenue stream.

Higher Costs

At the same time you have a disruption to the sales and order processing part of your business, you have increased expenses. First, you have expenses that have been scheduled as part of normal operations such as payroll, loan payments, or quarterly tax payments. If things were humming along as usual, these expenditures would be covered by revenues. However, in a disruption, revenue streams can slow to a trickle or be cut off completely in an instant. These recurring costs must still be paid in most cases.

In addition, there is the potential for additional costs such as setting up alternate work locations, providing emergency cash for employees (payroll advances), hiring contractors, buying replacement or emergency equipment, paying for prearranged services (such as firing up an alternate IT site), paying overtime, paying for nonproductive time (keeping employees on the payroll even if there is no work for them to do), and paying to temporarily relocate employees to an alternate work location. These are expenses that very few companies anticipate or plan for—unless they develop a business continuity and disaster recovery plan. When these unexpected expenses are coupled with reduced or nonexistent revenues, the company can quickly get into a deep financial hole. If the company is on shaky financial ground to begin with, this could cause the company to fold.

Impact on Cash Flow

If the company has lower revenues and higher expenses, it must either have a solid cash reserve on hand or it must have a line of credit available to it to meet higher expenses for the near- and mid-term. The amount of cash needed to make it through a business disruption depends on several factors: nature of the disaster, extent of impact on business, length of time to recovery, and other business-specific factors. Let's look at each of these briefly.

Nature of disaster. Clearly a large scale natural disaster will have a greater impact, both short- and long-term than will a small, localized event such as localized flooding or a fire in the building. Large natural disasters typically disrupt the infrastructure, which takes more time to restore. Although electrical service *may* be restored quickly, roadways, bridges, or other infrastructure elements may be damaged, requiring a longer recovery period. Therefore, the nature and extent of the disaster has a significant impact on the cost to the company.

Extent of impact on business. A disaster can impact your company fully, partially, directly, or indirectly. An earthquake may not damage your building but may damage nearby buildings or the homes of your employees, who then must deal with their own personal issues before they can refocus on work. A flood may make nearby roads impassable but leave your building intact, or your building's pipes may burst, causing a flood in your building that doesn't have any impact on the surrounding area. If the company is directly affected, the impact on cash flow is likely to be far higher than if the company is indirectly affected. Though this may seem obvious, it's important to think through these scenarios when looking at the financial impact to the company.

Length of time to recovery. In Chapter 4, we discuss the *maximum tolerable downtime* (MTD) as the maximum time critical systems can be down. The shorter this time frame, the greater the cost to recover (or to avoid the downtime in the first place). This is a critical assessment each company must make and it's vital that you understand the relationship between the MTD, the cost, and the company's cash flows. It's not always a simple straight line, either. For example, the longer you're down, the more it costs your company in terms of lost revenues. However, it typically costs more to facilitate a faster recovery, so quick recovery costs more in terms of higher expenses. How will this impact your cash flow? Although your company's financial analysts or accountants may be responsible for doing the actual calculations, it helps if you understand that there are trade-offs to be considered. In a perfect world, you would spend whatever it costs to get up and running quickly to begin restoring your company's ability to generate revenue. In the real world, this is rarely such a clean decision. Chapter 4 discusses some of these trade-offs.

Other. There may be other financial considerations, such as the type of employees the company employs to the types of financial obligations your company has. If employees are salaried, they must be paid regardless of whether they work two hours or 20 hours in a day. Hourly, temporary, or contract employees can be sent home without pay if there is no work for them. However, your company may choose to continue to pay these hourly workers in the aftermath of a disaster as part of the corporate philosophy to "do the right thing" with regard to employees and community relations. How long you can afford to pay nonproductive employees, whether salaried or hourly, is a matter of cash flow and corporate policy. Your company may also have complex financial obligations that must be serviced in a timely manner and these may not be normal, ongoing expenses. Therefore, any unusual or nonrecurring expenses or obligations must be considered as well. For example, if your company is embarking on an aggressive expansion or acquisition plan and is raising capital for doing so, your BC/DR plan should be modified to reflect the additional financial risk due to these unusual expenses. This is part of the BC/DR plan maintenance (reviewing current business operations), but these can be factored in if they are occurring as you create your BC/DR plan. Other business-specific impacts on revenues and costs should be included in your review.

Impact on Valuation and Ability to Raise Capital

Reduced revenues and increased expenses reduce the company's value in the market. Standard valuation techniques typically look at cash flows over time. As a simple example, if a company is publicly traded, it usually trades at a multiple of earnings. This is called the Price-Earnings (P/E) ratio. The price of the stock divided by its earnings (per share of stock) is some number, often called the *multiple*. That number can be compared to other companies to determine whether the price of the stock is high, middle, or low in comparison to other stocks in that category or other stocks of any kind. The P/E ratio impacts a number of investor decisions. For example, value investors look for low P/E ratios, believing the cost of the stock is low compared to its earnings and therefore provides price growth opportunities.

How does this impact your company? Regardless of whether your company is publicly traded or privately held, the *value* of the company is related to the earnings and revenue streams. The value of the company dictates the amount of money your firm can borrow and often the interest rate at which those borrowed funds must be repaid. Let's take a simple example. If your company has monthly average revenues of $2 million, it means your company earns $24 million annually. The value of the company would be $24 million if it had no expenses (there are other factors, we're keeping it very simple). However, if you have $23 million in expenses, what's the company worth? About $1 million. Now, suppose you need a loan after a disaster. You go to the bank and say, look, we typically earn $24 million a year and we need a $4 million loan to patch things up. The bank may look at your financials and say, "Yes, we see that you *used to* do $24 million a year, but what was your revenue last month or last week and what were your expenses last week?" The bank knows as well as you do that your ability to generate revenue may be impaired and your expenses are higher. If you have a solid business model and a strong financial track record, the bank will happily lend you the funds. On the other hand, if your company's finances are tight, a bank might decline to lend you the funds or it could charge you a significantly higher interest rate than it might otherwise.

In addition, getting a loan after a disaster might mean waiting in line behind hundreds of other companies. Bank or lender resources may be overwhelmed trying to respond to a sudden increase in demand. Internally, your company's staff may have its hands full trying to implement and manage the disaster recovery and business continuity activities. It may not have the time or capacity to develop the appropriate documents often required to obtain outside funding. These documents may include the last three years' worth of tax filings, financial statements, and such, which may be difficult (if not impossible) to obtain in the immediate aftermath of a disaster.

The company's ability to raise capital, whether in public or private markets, is tied directly to cash flow and valuation. Therefore, a disaster or business disruption can significantly impact your company's ability to raise capital or it can impact the cost of that capital. This has a long-term impact on the company because more expensive capital reduces cash flow and profits.

Summary

As you can see, there are numerous factors to consider when looking at the financial impact of a disaster or business disruption on the business. Revenues usually go down, expenses usually go up, and the methods for filling in the gaps can be as simple as using cash reserves and as complex as going to outside funding sources for additional funds.

The bottom line is this: Whether you're responsible for the entire BC/DR plan or just the IT section, you need to understand the basics of how companies finance operations and the details of how your company, in particular, manages finances. The disruption of revenues, the introduction of higher costs, and the resulting impact on cash flow and valuation can have a devastating and long-term impact on the company's financial health. Though you may not be involved with any of these elements or decisions, they impact your approach to business continuity and disaster recovery planning, and they certainly impact your decisions regarding risk management, business impact, and risk mitigation strategies. Understanding these principles can help you whether you're developing the plan or working with your financial resources within the firm on this or any other project.

Chapter 3

Risk Assessment

Solutions in this chapter:

- **Risk Management Basics**
- **Risk Assessment Components**
- **Threat Assessment Methodology**
- **Vulnerability Assessment**

☑ **Summary**

☑ **Solutions Fast Track**

☑ **Frequently Asked Questions**

Introduction

In this chapter, we're going to discuss the concept and practical application of risk management. We'll look at the broad business perspective, the practical business continuity and disaster recovery planning perspective, and the IT-centric perspective. We'll look at risk management to understand the overall process, then delve into the risk assessment process. This is where the first phase of project work begins.

To help you keep track of where we are in our overall planning process, you'll see an image similar to that shown in Figure 3.1 at the outset of each chapter. As you can see in Figure 3.1, we've completed the basic project initiation steps (see Chapter 2) and we're moving into risk management. Clearly, we can't create a viable BC/DR plan until we know which specific threats the company faces. Every company faces numerous common threats such as the potential for a server failure or power outage; but each company also faces numerous threats that are either unique to the organization or unique in their potential impact. Throughout this chapter, we'll discuss risk management from a BC/DR perspective, but there may be risks your business faces that are not mentioned. In Chapter 1, we provided a fairly extensive list of potential threats to be addressed, but the list is not exhaustive and you'll need to look at your own business with other knowledgeable members of your company to determine what risks you'll need to assess. We'll cover many of those threats in more detail in this chapter and the information provided may be new or surprising to you. At the very least, it should serve to prompt you to think about these events in light of business continuity and disaster recovery planning processes.

Figure 3.1 Business Continuity and Disaster Recovery Project Plan Progress

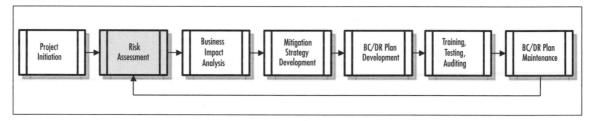

One of the common objections to BC/DR planning is that there are just too many things that could go wrong to plan for them all. That's partially correct—there are thousands of things that *can* go wrong, but fewer things that actually are *likely* to go wrong. Just because something may not happen doesn't mean that it shouldn't be planned for. Let's use driving to work as an example. There are hundreds, if not thousands, of threats to you as you drive to work each day, but you still took some basic precautions to reduce, remove, or avoid certain risks. First, you probably took a driver's education class when you were learning to drive so you could learn the basic rules of the road. That's risk reduction. Second, you probably obey traffic controls and signals such as stop signs and signal lights. That's another risk reduction strategy. Third, you probably have car insurance so that if something does go wrong, you won't face exorbitant costs related to repairing the car. That's a risk transference strategy. Still, it doesn't prevent someone from plowing through a red light or losing control of their car going 75 miles per hour on the highway. So, you've done what you can to reduce your risks and you also realize there are risks that you can do nothing about. The only way to take your risk profile to zero would be never to go in a motorized vehicle anywhere and never to go near roadways containing motorized vehicles. Not very practical, but effective. It may be effective but it is neither feasible (for most) nor desirable as a risk mitigation strategy. Just because you can't manage all the risks involved with driving a car, it doesn't stop you from getting in your car and driving to and from work five days a week. You can't control all the risks but most people will still do what they can, within reason, to limit their risks—like take a driver's education course, follow traffic laws, and buy insurance. So, it makes sense to try to address the risks we can in business, recognizing there will be risks we do not or cannot address. We'll also discuss risk concepts including avoidance, reduction, acceptance, and transference of risk. These are four general methods that can be used to manage risk and we'll discuss how these apply to your BC/DR planning process. We'll look at the risks to your company and to your IT operations to help you determine which are acceptable, which must be mitigated (reduced or avoided), and which can be transferred, all within the constraints unique to your company. Risk management must fit within the financial and time constraints of the company to be viable; in other words, they must be reasonable. When you finish reading and absorbing this chapter, you'll have information you can use to go back to your executive team to gain support for your BC/DR project if you've been unable to gain that thus far.

Risk Management Basics

Risk management is a general topic that looks at how all risks are managed across the enterprise. The number and type of risks companies face in today's world are many and varied. For example, companies face risks to the value of their company through gyrations in the stock market, they face shareholder lawsuits for mismanagement of the company, and they also face risks based on currency fluctuations in an international trading environment. These are just three risks companies face with regard to the value of the company. The value of the company impacts its ability to raise additional capital, the interest rates it receives on loans, and the rating of any bonds the company may try to issue (or has issued). This is just one type of risk management focused on financial risks to publicly traded companies. Let's look at some other types of risks. There are risks associated with union contracts, labor agreements, or outsourcing agreements. There are risks associated with products such as product tampering, product malfunction, product contamination, or product failure. These are risks companies face related specifically to the products they make or sell. We're not going to delineate every possible risk a company could face, nor are we going to have an in-depth discussion of risk management here. However, it should be clear to you that risk management is a large undertaking at any company and there are risks beyond business continuity and disaster recovery that your company has probably already addressed or is aware of.

Let's begin with looking at the risk management process visually. Figure 3.2 contains a flowchart that indicates the four basic steps in risk management:

- Threat assessment
- Vulnerability assessment
- Impact assessment
- Risk mitigation strategy development

Figure 3.2 Risk Management Process Overview

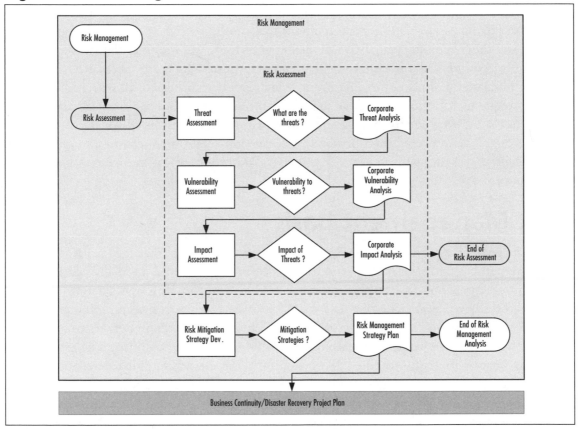

We're going to focus on threat and vulnerability assessment later in this chapter. In the next chapter, we'll discuss the impact assessment process in more detail, though we will mention it throughout this chapter as appropriate. In Chapter 5, we'll discuss risk mitigation strategy development in detail, but again, we'll also touch upon it as we discuss threat and vulnerability assessments in this chapter. As you can tell, these four areas are intertwined and it's difficult to discuss one aspect without also touching upon the others. However, we'll save our in-depth discussions of impact assessment and mitigation strategies for later chapters.

We've used shading in Figure 3.2 to indicate the general boundaries of risk assessment versus risk management. However, we're far less concerned with the boundaries than the actual work products. In each of the phases, there is an assessment and an analysis that should result in a report or written document. This helps you move from one phase to the next in an orderly and coherent manner. You can also use these phases as part of your Work Breakdown Structure to delineate tasks, deliverables, timelines, and deadlines. Let's begin with a brief look at each of these four areas just to be clear about definitions and boundaries.

From the Trenches...

Risk Management Certification

There are numerous organizations that offer risk management courses and certification programs. If this is an area of interest to you, a quick Web search will yield a variety of resources. There are also numerous links found in the risk management section on Wikipedia at http://en.wikipedia.org/wiki/Risk_management. At the bottom of the listing, you'll find links on certifications, degree programs, organizations, institutes, and training. Some of the certifications and programs listed are industry specific, notably insurance and finance. However, there are numerous general-purpose certifications and courses you may find of interest as well.

Risk Management Process

The process of managing risk includes assessing potential and also analyzing the trade-offs, or opportunity cost. Imagine a company that says we need to make sure our systems never go down. The potential for systems to go down occasionally is very high; most systems go down for one reason or another from time to time. The cost of those system outages varies, usually in direct correlation to the time the system is down. If the system is down for 10 minutes while it's rebooted due to an emergency patch installation, the cost may be negligible. If the system goes down for days because the database is corrupted by a hacker and restoring back to the previously validated database data experiences a few problems, the cost is much higher. Now, let's offset that with the opportunity costs. If we spend $5,000 on various systems to keep that server up and running, that's $5,000 we *couldn't* spend on something else, such as marketing materials, advertising, or employee wages. In addition, there's the cost of the downtime versus the cost of the solution. What does one hour of downtime for that server cost your company in lost sales, lost productivity, lost reputation, or lost consumer confidence? That's the opportunity cost of downtime.

The point is not to get into a detailed financial discussion regarding business costs but to understand that for every activity that occurs, some other activity cannot occur. For every dollar spent doing something, that dollar cannot be spent doing something else. Clearly, if you have $50 left at the end of the week, you can only spend that $50 once. If you choose to spend that $50 on dinner and the movies, you cannot also spend that $50 on the latest electronic toy. Every choice made excludes other choices not selected. Understanding opportunity costs within the risk management process is important because you can't manage every risk to zero. In some cases, it simply can't be brought that low; in other cases, the opportunity cost of doing so is disproportionate to the benefit. These assessments require some level of qualitative assessment (an assessment made without hard data, a value judgment). Understanding all aspects of the decision-making process will help you and your team make better decisions based on the unique requirements and constraints of your company.

Two other useful concepts in this process are *magnitude* and *frequency*. For example, an earthquake's impact to business operations would have a high magnitude, meaning the impact would be extreme. However, in many places, even those prone to earthquakes, the frequency is relatively low. California does experience fairly regular earthquakes, but the frequency of large earthquakes that impact business operations is relatively low.

Finally, each threat and potential mitigation strategy has a cost and a benefit. As we discuss various threats later in this chapter, we'll look at costs—in dollar figures and in the cost to human life and business operations. There's also the benefit of the mitigation, which ideally should more than offset the cost of the event. Let's look at a concrete example. The cost of installing fire suppression systems in a building may be $15,000. Typically, fire suppression systems cost about $5.00 per square foot; specialty fire suppression systems may run as high as $10.00 per square foot. Compare the total cost of installation with the cost of a major fire in terms of 1) building damage, 2) equipment damage (desks, computers, carpet, decorations, files, records, inventory), 3) IT equipment damage, and 4) human injury and death. $15,000 looks like an excellent investment because the cost of installing the system is far lower than the benefit it provides by way of risk reduction. These are the kinds of assessments you'll need to complete for your BC/DR plan. Let's look at each of the phases briefly so you understand the framework for the entire risk assessment process.

Threat Assessment

We've used the words "risk" and "threat" several times, almost interchangeably. Although this is correct in a general context, it's not quite accurate in a specific risk management context. *Business risk* is defined as:

> The total process of identifying, controlling, and eliminating or minimizing uncertain events that may affect businesses. It includes risk analysis, cost benefit analysis, selection, implementation, and testing of selected strategies, and maintenance of those strategies over time.

The key words here are "identifying, controlling, eliminating, or minimizing uncertain events." Risk management is about trying to manage uncertainty. We can't ever completely remove all risk all the time, but we can find ways to reduce or eliminate many risks to some degree. The process of risk management is the process of determining which risks should be addressed and how they should be addressed.

A more IT-centric view of risk management was defined by Joan S. Hash in the Computer Security Division of the Information Technology Laboratory at the National Institute of Standards and Technology (http://csrc.nist.gov/publications/nistbul/itl02-2002.txt):

> "Risk is the net negative impact of the exercise of a vulnerability, considering both the probability and the impact of occurrence. Risk management is the process of identifying risk, assessing risk, and taking steps to reduce risk to an acceptable level. The objective of performing risk management is to enable the organization to accomplish its mission(s) (1) by better securing the IT systems that store, process, or transmit organizational information; (2) by enabling management to make well-informed risk management decisions to justify the expenditures that are part of an IT budget; and (3) by assisting management in authorizing (or accrediting) their IT systems on the basis of the supporting documentation resulting from the performance of risk management. Risk management encompasses three processes: risk assessment, risk mitigation, and evaluation and assessment."

Both business risk and IT-specific risk must be addressed using the same methodology; only the details will differ. We can use the following equation to define risk as well.

Risk = Threat + (Likelihood + Vulnerability) + Impact

Thus, risk could be viewed as the combination of the threat itself, the likelihood of that threat occurring, the vulnerability of the organization or system to that threat, and the relative or absolute impact of that threat on the organization or system. Likelihood and vulnerability are shown in parentheses simply to indicate that some people prefer to assess these in one pass or as one value. For example, vulnerability could be construed to include likelihood; others may want to specifically break out the likelihood from the vulnerability. Either method is acceptable as long as you account for both factors in your equation. Although this might seem like splitting hairs, it's important to define these various elements so that we can discuss them at the level of detail needed to perform a thorough and meaningful risk assessment. We'll discuss *threats* and *threat sources* in depth later in this chapter.

Vulnerability Assessment

The *vulnerability assessment* analyzes how vulnerable, susceptible, and exposed a business or system is to a particular threat. It should include an assessment of how *vulnerable* a particular system is to a threat as well as the *likelihood* of that threat occurring. The likelihood portion of the assessment can be part of the vulnerability assessment, though you could also break it out as a separate process if desired. As long as your risk assessment includes vulnerability and likelihood assessments, you should be in good shape. Clearly, it is useful to know that a system is *vulnerable* to a threat that has a 90% *chance* of occurring, a 50% *chance* of occurring, or a 1% *chance* of occurring. The vulnerability and the likelihood of the event are closely related and the results are used as inputs to the impact assessment. Certainly, a server that is outside the firewall is far more vulnerable to external attacks than a server inside the firewall. This is an example of relative vulnerability since both servers are vulnerable but one more so than the other. How likely is it that either server will be attacked? Probably 100% for the server outside the firewall and perhaps 90% for the server inside the firewall in today's attack-laden environment. As you can see, creating relative assessments for vulnerability and likelihood result in different risk profiles for the two servers. We'll look at this in greater detail a bit later in this chapter.

Impact Assessment

The impact assessment analyzes how great or small the impact of a threat occurrence will be on the business or system. An earthquake has an enormous impact on a business that is in or near the epicenter of the quake; it has a lesser impact on businesses further from the epicenter; it may have a slight impact on other companies around the country if infrastructure fails or if key suppliers or vendors are located in the region impacted by the earthquake. Therefore, impact varies based on numerous factors. Clearly, a fire contained to the lunchroom has a much lower impact than a fire that engulfs the entire building. We'll look at the impact of various threats in detail in Chapter 4 but also in conjunction with our discussion of the risk assessment process throughout the remainder of this chapter.

Risk Mitigation Strategy Development

We mentioned four distinct strategy types of risk mitigation earlier in this chapter. You can reduce, avoid, accept, or transfer risks. Each strategy comes with an associated cost. It's far more expensive in many cases to completely avoid a risk than it is to reduce the impact of the risk. Most businesses are more likely to build in state-of-the art fire suppression systems rather than construct a building with absolutely no flammable materials. The cost of building a completely fireproof building is far higher than installing a high-quality fire system. However, each company has to make that assessment. There are certainly situations in which building a fireproof facility is not only cost effective, it may be the only viable option for a particular type of company.

Some risks are worth accepting. As we discussed in the introduction to this chapter, we all accept risks in our everyday lives. We drive cars, we cross a busy intersection on foot, we eat unhealthy food, we buy high-risk stocks. These are all risk-laden activities but we accept these risks. We may find ways to reduce our risk such as obeying traffic signals when driving and crossing streets; we may limit our intake of junk food to some extent; or we may put 25% of our investment funds in a savings account. These are all attempts to reduce risk but there is also an element of acceptance. It's like saying, "I'll accept 35% of this risk," meaning I'll obey traffic signals but I'll still drive my car. You've accepted that even if you obey traffic laws, there's still a chance, albeit a smaller one, that you could end up in an accident.

Risk mitigation strategy development is the process of deciding which risks you should address and in what manner. The inputs to this are the risk assessment analysis or reports, which delineate which threats exist, how vulnerable your systems are, and how likely the threat is to occur as well as the impact of these occurrences on your business. The compilation of this data will help drive sound business decisions because you'll be able to look at your entire risk profile and decide how to proceed. Since there are rarely perfect solutions in business, your job during this phase is to make intelligent decisions and trade-offs in light of the data collected. We'll discus this in detail in Chapter 5. For now, let's turn our attention to the risk assessment components.

People, Process, Technology, and Infrastructure in Risk Management

Earlier in the book we introduced the framework of "people, process, and technology, as a framework that works well for IT projects. However, for business continuity and disaster recovery planning, a fourth category needs to be included: *infrastructure*. As IT professionals, it's relatively easy to look at technology and assess the various risks. It's a bit more difficult to assess the risks to people and the processes they use to run the businesses, but it's part of normal IT project planning in most cases. Assessing the infrastructure, however, is a bit out of the ordinary for IT planning, which is why we expanded our model to specifically include it. In BC/DR planning, the infrastructure, which includes the building and facilities of the company, the utilities to the building, and the external infrastructure such as transportation and utilities, must also be assessed. Let's look at these four components using the earlier power outage example.

People

If the power goes out in the building, how likely is it that people will be able to get anything done? Many offices have no windows, so they'd be working with emergency lights that illuminate only exits and major hallways, for example. People will be distracted, they'll be gathering around people's desks discussing the outage, not focused on work. If the outage is from a storm outside, they may be concerned about getting home safely from work that day or the status of power at their own homes. People respond in a variety of ways to small and large events. How the people in your company respond will be based on numerous factors including the kinds of work they perform, the types of people your company hires, and more. For example, if your company hires former medical personnel, they may respond well to emergencies. Looking at the employee population and understanding how they are likely to respond to small and large business disruptions will help in your planning.

Process

What about the processes? Let's assume the power to the building is out. It doesn't matter if the server room has emergency power or not, does it? Users' desktop computers aren't available and the software they use to get their jobs done is unavailable. If the company has any processes still done by people without technology, those processes can proceed but sooner or later, those processes will require computer data as input or output. In many cases, then, continuing with those few processes that are not dependent upon technology (or electricity in general) will cause systems to get out of sync. If materials are off-loaded from an incoming delivery truck and placed in inventory and paper inventory sheets keep track of materials, quantities, and locations, that may help the delivery truck get back on the road, but it causes a problem on the other end. Now, there is inventory in stock that is not included in the last computer inventory count. Will that paperwork ever end up being input or will it take a cycle count in three months to discover the problem? This is a small example of how all business processes are impacted by this single threat, a power outage, even if the process isn't directly impacted by the threat.

Technology

Clearly, technology is the heart of operations in many companies today, especially those located in industrialized nations. Without technology, most things just come to a grinding halt. It's almost hard to understand the impact of technology on our lives until you're forced to do without. If the power's ever gone out in your home, even for a few hours, you suddenly realize you can't get on the Internet to get news, you can't update your stock portfolio online or on your desktop, you can't watch TV, you can't make a pot of coffee, you can't tell what time it is, you can't recharge your cell phone, and on and on. Yes, you can fire up your laptop for as long as the battery lasts, but if you don't have a wireless Internet card, you still can't get out to the Internet. Most people find their normal lives just come to a halt. We're at a loss because we've become so accustomed to having uninterrupted power 24x7x365. Clearly, people living in areas where electrical service is less reliable are more aware of the impact, but they also have developed risk mitigation strategies—they may have generators or solar power to their homes in order to offset that risk for example, or they have simply designed their operations around these facts.

The important concept here is that technology is needed so that the *people* in the company can use the *processes* defined to conduct business. Technology, by itself, is a pervasive business tool, but it is most often useless without the context of people and process. As we continue our discussion of risk assessment, keep that fact foremost in mind. It will help as you look at risks to remember that there are four elements to be addressed with every risk: people, process, technology, and infrastructure. By incorporating this four-pronged view of risk, you can help reduce the chance that your plan will have any significant gaps.

Infrastructure

Infrastructure is sometimes included in the technology segment of "people, process, and technology" but it's useful in BC/DR planning to address it as a discrete category. Most IT professionals understand the term infrastructure from an IT standpoint, but from an organizational standpoint, it refers to things such as the building and facilities, the utilities coming into the building, and the external infrastructure such as public transportation, public utilities, communication services, and any other local, state, or national resources pertinent to your business. Corporate infrastructure typically is managed by the facilities manager or by someone assigned those duties, whether in finance, operations, or Human Resources. External infrastructure typically is managed, owned, controlled, or regulated by the local, state, or federal government. Within a BC/DR risk assessment, the risk to the company's infrastructure from various threat sources must be evaluated and assessed. Risks to the external infrastructure must also be understood, though mitigation strategies will clearly differ for resources you control or own versus those you do not. In a natural disaster or serious external event, there are usually disruptions to external infrastructure components that have far-reaching effects, which are often underestimated in the planning stages. It's difficult to realistically assess what will happen if a major freeway collapses or a nearby chemical plant explodes. Your business will have to rely upon local officials, including fire, police, and emergency medical staff, to address the external events. Your business will also need contingency plans with regard to your business operations in such an event.

IT-Specific Risk Management

Risk management across the business enterprise is a wide and varied topic, as you've seen. IT-specific risk management is a subset of overall business risk management. That said, there are some very unique risks in IT that exist nowhere else in the enterprise.

The Computer Security Division of the Information Technology Laboratory (ITL) of the National Institute of Standards and Technology issued a document outlining the steps in IT risk management, NIST Special Publication 800-30 (July 2002). Per the ITL: "ITL develops technical, physical, administrative, and management standards and guidelines for the cost-effective security and privacy of sensitive unclassified information in federal computer systems. The Special Publication 800-series reports on ITL's research, guidance, and outreach efforts in computer security, and its collaborative activities with industry, government, and academic organizations." (Source: NIST Special Publication 800-30, http://csrc.nist.gov/publications/nistpubs/800-30/sp800-30.pdf).

IT Risk Management Objectives

The ITL document provides an excellent framework for IT risk management, a complete discussion of which is outside the scope of this book. The ITL document outlines what have become industry best practices with regard to IT risk management. Clearly, the greatest risk to IT is the data stored on and traveling across IT equipment. There are three needs with regard to electronic data: *confidentiality*, *integrity*, and *availability* (we discuss these three concepts in greater detail later in this chapter).

The objectives of IT risk management are to enable the company to achieve its strategic objectives by:

- Securing IT systems more fully

- Enabling management to make well-informed decisions with regard to the purchasing and implementation of IT systems

- Enabling management to authorize (accredit) the IT systems on the basis of supporting documentation that results from the IT risk management activities

The System Development Lifecycle Model

From these three objectives, you can see that IT risk management is a subset of overall risk management because the IT systems must enable the company to achieve its objectives in a secure and cost-effective manner. IT risk management ideally is incorporated completely into a company's system development lifecycle (SDLC) activities, which have five phases (the names for the steps may vary slightly depending on whether you're focused on the lifecycle of software development or hardware and application implementation):

1. Analysis/Requirements

2. Design/Acquisition

3. Development/Implementation

4. Integration and Testing/Operations or Maintenance

5. Disposal

In some cases, a system may be in several stages simultaneously. Regardless of the phase (or the terminology), the methodology for risk management is the same. As with project management, risk management is an iterative process. As you can see from the data in Table 3.1, the phases and phase characteristics track closely with overall risk management and

BC/DR planning activities. For example, phase one is an assessment of risks and the development of requirements. Phase 2 is the development or acquisition.

Table 3.1 SDLC Phases

SDLC Phases	Phase Characteristics	Support from Risk Management Activities
Phase 1—Initiation	The need for an IT system is expressed and the purpose and scope of the IT system is documented.	Identified risks are used to support the development of the system requirements, including security requirements, and a security concept of operations (strategy).
Phase 2—Development or Acquisition	The IT system is designed, purchased, programmed, developed, or otherwise constructed.	The risks identified during this phase can be used to support the security analyses of the IT system that may lead to architecture and design tradeoffs during system development.
Phase 3—Implementation	The system security features should be configured, enabled, tested, and verified.	The risk management process supports the assessment of the system implementation against its requirements and within its modeled operational environment. Decisions regarding risks identified must be made prior to system operation.
Phase 4—Operation or Maintenance	The system performs its functions. Typically the system is being modified on an ongoing basis through the addition of hardware and software and by changes to organizational processes, policies, and procedures.	Risk management activities are performed for periodic system reauthorization (or reaccreditation) or whenever major changes are made to an IT system in its operational, production environment (e.g., new system interfaces).

Continued

Table 3.1 continued SDLC Phases

SDLC Phases	Phase Characteristics	Support from Risk Management Activities
Phase 5—Disposal	This phase may involve the disposition of information, hardware, and software. Activities may include moving, archiving, discarding, or destroying information and sanitizing the hardware and software.	Risk management activities are performed for system components that will be disposed of or replaced to ensure that the hardware and software are properly disposed of, that residual data is appropriately handled, and that system migration is conducted in a secure and systematic manner.

Source: Stoneburner, Gary; Goguen, Alice; Feringa, Alexis, NIST Special Publication 800-30, "Risk Management Guide for Information Technology Systems." Recommendations of the National Institute of Standards and Technology, July 2002, p. 5. **http://csrc.nist.gov/publications/nistpubs/800-30/sp800-30.pdf.**

There are many excellent resources on IT risk management, and rather than go into more detail on the subject here, we'll direct you to these resources:

1. The US National Institute of Standards and Technology publications are available on the Internet at http://csrc.nist.gov, including Special Publication 800-26, "Security Self-Assessment Guide for Information Technology Systems," and Special Publication 800-30, "Risk Management Guide for Information Technology Systems."

2. Operationally Critical Threat, Asset, and Vulnerability Evaluation (OCTAVE). OCTAVE provides best practices for evaluating IT security. It was developed by the Computer Emergency Response Team at Carnegie Mellon University. www.cert.org/octave.

3. Control Objectives for Information Technology (COBIT) was developed by IT auditors and provides a framework for assessing IT security, developing performance metrics, and monitoring performance over time. www.isaca.org/cobit.htm.

4. Common Criteria (International Standards Organization (ISO)) 17799. These criteria represent the international standard for testing IT security systems.

Information about the criteria can be found at www.commoncriteria.org and a copy of the criteria can be purchased from ISO at www.iso.org.

Other helpful Web sites include:

- Computer Security Institute: www.gosci.com

- SANS Institute: www.sans.org

- Center for Internet Security (CIS): www.cisecurity.org

- Computer Emergency Response Team (CERT): www.cert.org

- Critical Infrastructure Assurance Organization (CIA): www.ciao.gov

- National Institute of Standards and Technology (NIST): www.nist.gov

- Computer Security Resource Center: http://csrc.nist.gov

The IT risk assessment process intersects with our business continuity and disaster recovery planning risk assessment in that we need to assess the various risks (including, but not limited to, security) to the company and the IT systems in the larger risk arena. However, the goals are the same: to enable businesses to meet their strategic objectives. Clearly, being in business after a disaster or major business disruption is an objective of every organization.

Identifying risk for IT systems includes two major components: systems and operating environment. The systems data includes, but is not limited to:

- Hardware

- Software (OS and applications)

- System interfaces (internal, external connection points)

- People who support the IT systems

- Users who use the IT systems

- Data, information, and records

- Processes performed by the IT systems

- System's value or importance to the organization (system criticality)

- System and data sensitivity (confidential, trade secret, medical data, etc.)

The operating environment data can include:

- The functional requirements of the IT system

- The technical requirements of the system

- Users of the system

- Security policies (company policies, industry, regulatory, governmental requirements)

- Security architecture (to assess vulnerability to cyber threats)

- Level of protection needed for confidentiality, integrity, and availability (CIA)

- Current network topology, network diagrams

- System interfaces, information flow diagrams

- Data storage protection

- Technical controls (added security products, identification requirements, access requirements, audits, encryption methods, etc.)

- Physical controls (access control, monitoring, etc.)

- Organizational controls (policies and procedures defining acceptable methods and behaviors)

- Operational controls (backup policies and procedures, personnel security, system maintenance, off-site storage or computing capabilities, etc.)

- Environmental controls (power, temperature, humidity)

These items are included to give you additional insights into the areas you'll need to investigate throughout your risk assessment phase. Some items may not be relevant to you and you can delete them from your list. You and your team may have other items not listed that you want to include. It's better to be inclusive at this juncture. You can always pare down your list later, but if you trim it down too early in the process you may miss critical threats, threat sources, and vulnerabilities.

Risk Assessment Components

The risk assessment is shown in the shaded area in Figure 3.2 earlier. There are three distinct steps defined in the preceding section. So, let's begin with a discussion of *threats* and *threat sources*. If you look up the definition of threat and threat source, you'll see pretty much the same or very similar definitions. In this book, we're likely to use the two terms interchangeably as well. However, it is worth noting the distinction just to help clarify the risk assessment process.

A power outage threatens just about every business. The threat, then, is a power outage. However, the threat source is where the power outage comes from. For example, power outages can occur when ice storms break power lines, when transformers are struck by lightning or when substations or the power grid itself experience some major failure of the power infrastructure. These are all *threat sources*—where the threat actually comes from. Does this matter? Well, in general discussions of threats, it's often not very useful to discuss a threat source separate from the threat. Most of the time, we simply discuss a power outage. However, in BC/DR planning, understanding the *threats* and *threat sources* can help you uncover potential risks to your company or IT systems about which you were previously unaware. If we discuss a power outage in a general manner, you might think about power going out to the server room or even power going out to the building. If you fail to consider the possibility of power going out because a train derails and wipes out a nearby substation, have you adequately addressed all the threats? That's a judgment call in many cases. The likelihood of a train derailing and wiping out a nearby substation has got to be pretty low on the likelihood meter, unless, of course, there are 90 trains per day and the substation is adjacent to the tracks.

As you can see from this one example, there is no one set of threats that will fit every company. In the following sections, we're going to discuss threats and threat sources, but it is almost certain not to be a comprehensive list. We'll get you started in this assessment and you'll need to add details unique to your company and your geographic location(s).

Figure 3.3 shows the steps in the risk assessment segment. This is a subsection of Figure 3.1 shown earlier.

Figure 3.3 Risk Assessment Subprocess

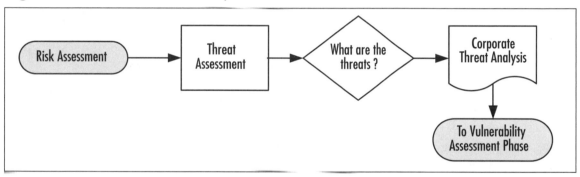

The risk assessment begins with the assessment of all potential threats and an analysis of those threats. The output from this phase is the input to the vulnerability assessment phase. Some companies may choose to look thoroughly at each threat and create a full assessment of each threat (such as threat, vulnerability and impact for power outage or flood). However, it's advisable to create a detailed threat assessment first. If you get caught up in the details of looking at a specific threat before all threats have been delineated, you may waste time and resources planning for the wrong things. You may only be able to see the relative risk of various threats once the entire list has been created. Be sure to keep this in mind and focus on generating a comprehensive threat assessment before moving into the vulnerability assessment phase. Otherwise, you risk rework, errors, and gaps in your final BC/DR plan.

Information Gathering Methods

There are numerous methods you can use to gather data about your company's risks. Some of the methods are described briefly in this section. If you have preferred methods or a process unique to your company, feel free to use those as well.

1. **Questionnaires**. Standardized questionnaires can elicit data from specific groups or individuals. Questionnaires can help limit input and feedback to those areas most useful.

2. **Interviews**. Interviews with subject matter experts can be extremely helpful in uncovering needed information. This process is particularly helpful when you have subject matter experts who cannot or should not participate on the BC/DR team but whose input is vital. Interviews can be conducted using the questionnaire instrument to help direct and focus the interview. However, sometimes a more freeform interview can yield more information. Questionnaires often contain unintentional biases and allowing an interviewee to discuss the topic without the constraints of a questionnaire can often yield data that might otherwise have been missed.

3. **Document Reviews**. Reviewing corporate and organizational documents can help identify threats, threat sources, and vulnerabilities. These documents may also be extremely helpful in understanding the company's current critical processes and functions so that systems can be properly prioritized later in the process.

4. **Research**. Internal and external research can be extremely helpful and often is needed to round out the data collected. Your team can gather reams of data on the frequency and likelihood of storms, earthquakes, or other natural events from a variety of governmental resources (many of which are referenced throughout this chapter). Your team can also gather data from local fire departments, police departments, and other local organizations. Finally, there may be a lot of data about past business disruptions or events archived within the company that may be helpful in understanding threats, threat sources, and vulnerabilities to things such as break-ins, thefts, or cyber crimes to name just a few.

These four methods should yield the data that you need to use as input to your assessment. However, it should be clear that these four methods could also yield reams of data, some (or much) of which might be useless or off-target. Before launching into your threat assessment, you should decide how to limit your data gathering efforts so that relevant data is most likely to be gathered. Review your questionnaires and interview questions to be sure they are focused and targeted on risk assessment for business continuity and disaster recovery planning. Limit your review of documents and research to those items specifically related to BC/DR. Although this might sound obvious, it is sometimes overlooked in the desire to avoid gaps or omissions in the planning process. Information overload happens quickly, so be sure to clearly define what data you're collecting so you don't end up sifting through reams of irrelevant information.

Natural and Environmental Threats

We've chosen to divide threats into natural threats—those caused by natural phenomenon and found in the environment—and human threats—those caused by humans either intentionally (including terrorism) or accidentally. Natural or environmental threats occur everywhere, but there are certain geological boundaries that determine whether you're more likely to experience a tornado or a hurricane, an electrical storm or an ice storm. We'll review some of the major threats and discuss business considerations. We'll also point you to a few resources that might be of interest or assistance to you and your team as you perform your risk assessment. You might be tempted to skip over this section or just skim the headers thinking you're already well aware of natural disasters, but resist that temptation. The information contained in the following sections is intended to expand your understanding of these threats as well as to spark thoughts on how these events may relate to your business operations.

Before we delve into the details of various natural and environmental threats, let's remember that these threats impact *people*, *process*, *technology*, and *infrastructure*. Though you'll undoubtedly take a very IT-centric view of these threats, you should also ask how these threats will impact the people, processes, technologies, and infrastructure of your business.

Fire

We've listed fire first because it is the most common disaster businesses have to deal with. Each year in the United States, fire causes thousands of deaths and injuries and costs billions of dollars in property damage. Fires are caused by a wide range of events, some of which are intentional, some accidental, and some environmental. Intentional fires come under the banner of arson and we'll discuss arson a bit later (and arson, itself, can be classified as "intentional" or "terrorism" depending on other factors).

Fires cause injury and death to people, but fires also cause damage to buildings, systems, and corporate records. If your firm does not already have a fire response plan in place, you should start your fire threat assessment by setting up a meeting with your local fire department representatives. It is important to understand what can and cannot be expected from the fire department in terms of fire and emergency response in your area. It's also helpful to have the fire department do a walk-through to help you identify and remove (or reduce) fire risks. In some cases, this type of walk-through is required by law before a business can occupy a commercial building. However, many times, it is not required or things within the facility may have changed significantly since the inspection. Be prepared for the fire inspection to yield negative results. This may impact business operations in the near term. For example, if there is a serious and imminent fire danger in your building due to improperly stored chemicals, the fire department may require the building be evacuated and locked until proper fire protections and practices are in place. If you suspect there are significant fire hazards in your facility, you should speak to senior management about it to get the situation resolved as quickly as possible. However, if you feel the danger is that serious, scheduling a meeting with a fire inspector could force the issue to be resolved before human life is lost.

Clearly, most companies do not have conflagrations waiting to happen, so you will usually get solid recommendations from your fire inspector or perhaps a few minor violations to correct. You'll learn a lot about how fires start and how they should be contained by talking with your local fire crew. You can also find out how to develop fire drills and safe fire evacuation procedures, if those are not already in place.

You may also wish to contact your insurance carrier to learn about fire prevention and protection measures that make sense for the type of business you're operating. They clearly have a vested interest in preventing fires since the fewer fires you have, the more money they make. Insurance companies will typically provide information resources and, sometimes, free training to assist you in fire prevention and containment.

Most companies develop and practice fire drills and clearly post evacuation routes throughout the building. These practices *may* reduce the company's potential liability in the event of a fire with injury or death, but more importantly, they *will* reduce the likelihood of injury or death. Evacuation maps should be posted clearly and everyone should know the shortest route outside and the closest exits. Larger facilities may also assign crew leaders responsible for making sure their area is evacuated and taking a head count once outside. If you have fire equipment such as automatic fire doors, sprinkler systems, or fire extinguishers, make sure managers and supervisors are familiar with these systems and their use. You may choose to conduct training on emergency equipment such as fire extinguishers so that people have hands-on experience prior to a real emergency. If the first time someone attempts to use a fire extinguisher is during an actual fire, they may be unable to read or process step-by-step instructions. If they've run through it a few times in nonemergency situations, they have a better chance of using the fire extinguisher effectively during the emergency. Talk with your insurance company and local fire department to learn how to prevent, contain, and manage fires in the workplace.

With regard to IT systems, you should certainly see about having chemical fire suppression systems installed in your server rooms, though most current building codes will address this issue adequately. Servers that are sprayed with water from overhead sprinklers may have a higher risk of water damage than fire damage. Those of you occupying older buildings should consult with your facilities manager or fire inspector about what type(s) of fire suppression (if any) exists in your building and more specifically, in your server room(s).

TIP

The cause of fires can be internal or external to the company. Several natural hazards can spark fires. Electrical storms, tornados, earthquakes, and drought can all cause fires to flare, so when you begin looking at fire as a potential threat to your company, also be sure to look at all potential threat sources. If you plan for an electrical fire in the building but neglect to address the potential for wildfires sparked by lightning storms or drought or the possibility that adjacent buildings may start a fire, you will have an incomplete risk assessment and leave your company vulnerable to those threat sources.

Floods

Floods are characterized by relatively high water flow that spills over the natural or artificial banks of a stream or waterway or that submerges land not normally below water level. External floods, like fires, can be caused by a variety of factors. Just to reinforce definitions used earlier, floods can be considered a threat, but the threat sources are many. Floods can be caused by:

- Heavy winter or summer rains

- Melting snow

- Swollen rivers (from rain, snow melt, broken dams, etc.)

- Broken levies or dams

- Tsunamis (which can cause flooding of large areas after the initial wave hits)

- Extremely high tides (typically caused by tsunamis, hurricanes, and powerful storms)

- Broken water mains

Depending on your location, you may be able to identify additional flood threat sources. Floods are the most common of all natural disasters, so there's a good chance your company will have to deal with flooding at some point in time. Floods can impact the building, the equipment (desks, chairs, file cabinets, computers), records (paper and electronic documents of all kinds), and people (drowning, injury, shock, etc.). Floods can also cause power outages and destabilization of the infrastructure. For example, a flood can cause landslides or the ground beneath the building to shift or sink, causing a serious failure of the building structure (or serious risk of failure). Landslides of surrounding areas can occur when the ground becomes too saturated. Landslides and other ground shifts (called *subsidence*) can not only impact your building but can disrupt the transportation infrastructure, including airports, railroads, and roadways. Flooding can also cause buildings to become uninhabitable. Doors, walls, floors, and ceilings can warp or split. Dangerous and noxious mold of various types can proliferate in the right conditions.

In the United States, the federal and state governments have various agencies that provide information about the nature, severity, and frequency of natural disasters in various geographic locations along with information on preventing (when possible) or mitigating the impact of these events. They often provide excellent emergency management resources and in some cases, free or low-cost training.

In many instances, buildings located in flood plains were built before modern flood plains were defined. In other cases, buildings are built and areas later become flood plains. Regardless, it's important to understand where the flood plains are in your area (if any) and how they might impact your business. This is most important for small and medium companies that may have moved into a leased facility. If you are not the owner of the building, you may not be fully aware of the flood risks in the area. Unfortunately, not all landlords are honest and you might have inadvertently placed your business in harm's way. You might not discover it until your insurance carrier contacts you about flood insurance requirements or until you experience a loss and discover you have no flood insurance or that your insurance was voided due to your location in a flood plain.

As with fires, it's a good idea to understand the best routes out from the building and away from the area in the event of a flood. If you're going to shut down early due to heavy storms in the area, it's also useful to let employees know in advance which roads, bridges, under- or overpasses are closed and which routes out are best.

If you are subject to flooding, you can look at standards set by the National Flood Insurance Program, part of the Federal Emergency Management Agency (FEMA), at http://www.fema.gov/business/nfip/.

The risk of flooding to your IT components follows the general flooding risks. Desktops, laptops, routers, switches, hubs, printers, and cabling are all subject to failure if exposed to water, whether the power on those devices is on or off at the time. In some cases, you might evaluate whether it makes sense to flood-proof your server room, though in most cases, the answer is likely to be no. If the rest of the building is underwater, paying to seal a server room may not be a good investment, but only you and your team can make that assessment. We'll look at business impact analysis and risk mitigation strategies (and associated costs) in upcoming chapters.

Emergency lighting systems are useful in many types of emergencies—from fire to flood to power outages, so we'll list them once here as an item "strongly suggested" for a variety of emergencies. Does emergency lighting reduce your risk of these various threats? No, but it does reduce the overall risk of injury and death of your employees. The cost of installing these systems is relatively small and if they prevent one serious injury or one death, it will have been an excellent investment. This is an example of the cost/benefit being extremely favorable—small cost, large benefit.

Severe Winter Storms

Much of the midwestern and northern states in the United States experienced severe winter storms in early 2007, and the southern and southeastern states had devastating tornadoes (discussed in a later section). Severe weather is a fact of life and if scientists are correct, it appears the globe is in for more severe weather in the coming decades. Severe winter storms include significant snow fall, strong winds, freezing rain, ice, and often freezing or subzero temperatures.

During heavy storms, people often can't get to or from work or they get stuck en route. Once at work, they may worry about their homes and families and be less productive. Driveways, sidewalks, and entries may become blocked or slippery and pose a hazard to people and vehicles. Technology can be affected by winter weather in the event that technology is exposed to the elements such as outdoor pumps, electronics, or mechanical devices that freeze and do not function.

Severe winter weather can disrupt a business by preventing it from opening for days at a time because no one can physically get to the building, employees and customers alike. In some cases, employees may be able to work from home if systems in the building are still functioning. Unlike loss of power or loss of cooling, even if the heat in the building goes out, the servers and other IT equipment will continue to function normally and may enable some portion of the business to continue. For example, if you have web servers, they may keep serving up web pages, taking online orders and tracking online orders, even if no product is leaving the warehouse.

In addition, extremely cold temperatures can cause pipes to burst, which can cause a variety of problems. The pipes that burst are typically those carrying water, but other pipes can burst potentially causing an environmental hazard (though this is more rare). Freezing pipes can cause flooding, so the threat may be flooding but the threat source is a winter storm, something you might not immediately think about.

Heavy snow and/or ice can build up on roofs and walls, causing structural damage and collapse. In cases of very heavy snow, people may be unable to remove the snow (steep roofs or large areas such as warehouses) or there may simply be nowhere to shovel the snow, as was the case in upstate New York in February 2007 when parts of the area were dealing with over 12 feet of snow.

If your company is located in a place where there is snow, ice, and cold temperatures, you may want to consider keeping supplies in the building in the event employees are unable to leave. This should include food, water, blankets, emergency medical supplies, batteries, radios, and possibly battery-powered televisions. Depending on the nature of your business, the location, and the climate, emergency power generators and portable heaters might be warranted. You should also arrange for snow and ice removal from driveways, parking lots, and sidewalks. Remember to clear snow and ice from emergency areas such as doorways, electrical boxes (main circuit panels, for example), and fire hydrants.

Common Challenges...

Winter Weather Warnings

On March 3, 2007, the St. Cloud Times (Minnesota) reported that a warehouse roof had collapsed after heavy snow, the second such collapse in that storm. It also reported 911 vehicles off the road, 411 car crashes in a 12-hour period (6:30 P.M. to 6:30 A.M.), and 99 spin outs. Airline flights in and out of the area were delayed or cancelled. And here's something you might not think about—the fire department was asking people to dig out hydrants in case of fire. When it's 20 degrees below zero, fire is usually a welcomed thought, but if your house or company is on fire, you hope the fire department can find the hydrant before the building is consumed in flames. This is a great example of a risk that you might not think about beforehand, but is obvious once it's pointed out.

One final word on winter storms—don't assume that just because you're located in a warm climate that you don't need to plan for winter storms. If you use suppliers, vendors, or contractors located in cold climates, they could be impacted by winter weather and that, in turn, could affect your business operations. Winter storms do sometimes impact warm climates, as well. In January 2007, a winter storm dropped four to six inches of snow on parts of southern Arizona, an area that gets measurable snowfall about once every 10 years. Roads and bridges were frozen; schools and some businesses were closed for the day because there is no snow removal equipment in those parts. Did anyone have to implement a disaster recovery plan as a result? Doubtful, but it does pay to avoid assumptions and look at all possibilities before dismissing them.

Electrical Storms

Electrical storms can occur any time of the year and in any climate, though they're most likely to be found during hot, humid summer months in the United States. There are an estimated 25 million lightning flashes in the United States each year. About 300 people are injured and an average of 66 people are killed each year in the United States due to lightning strikes. Lightning kills more people per year than tornados, on average, though they occur one or two at a time and are therefore less visible to the public than the death of, say, 60 people in a single tornado. Lightning can cause power outages, fires, damage to buildings, falling debris (trees, poles, etc.), and injury or death to people (and animals). High winds sometimes accompany electrical storms, and these can contribute to flying or falling debris, power outages (lines blown down or struck by falling debris), and structural damage to buildings as well.

From the Trenches...

The Positive and Negative Sides to Electrical Storms

According to the National Oceanic & Atmospheric Administration (NOAA), a department within the U.S. Department of Commerce, lightning originates in certain types of clouds in the region of snow crystals, snow, and ice pellets. The motion of the storm causes the snow crystals to become positively charged and the snow and ice pellets to become negatively charged. The positively charged particles rise within the storm while the negatively charged particles remain in the middle to lower portion of the storm. This difference of electrical potential can cause cloud-to-cloud lightning. In some cases, the negatively charged particles toward the bottom of the cloud create positive charges to form on the ground and in the immediate vicinity of the cloud. This increases the likelihood of cloud-to-ground lightning strikes. Not all lightning originates from the bottom of the cloud, however. When it originates from the top of cirrus anvil clouds, the lightning is called positive lightning.

For more on lightning, electrical storms, and other weather phenomena, visit the NOAA's Web site at http://www.noaa.gov/index.html.

Electrical storms can cause power outages, but they also can cause power spikes, surges, and dips. Power that is too high or too low can damage a wide range of electrical equipment, including all IT equipment. During electrical storms, power to buildings may fluctuate. We're all aware of the damage a power surge can do to electrical equipment, but extended low power (brownout) can also damage electrical equipment. Many companies invest in uninterrupted power supplies (UPS) to provide either battery backup to equipment, which typically just allows for an orderly shutdown of equipment. Batteries must be tested regularly and replaced and a process for doing so should be part of your BC/DR plan, if you have battery backup. UPS systems can also provide failover power, which provides ongoing power for some extended period of time when equipment loses power. Usually these systems rely on backup power generators that run on diesel or other fuel, so be sure that fuel is checked for backup power systems on a periodic basis. Checking the readiness of UPS systems should be part of normal IT operations processes and a process for ensuring disaster readiness should also be incorporated.

Many UPS systems provide power conditioning as well. This serves to keep the power from surging too high or dropping too low. If you have these systems in place, you are already familiar with them. If you do not have them in place, a bit of research will help you determine the best solutions for your firm based on the variety of unique conditions and constraints. Clearly, you can spend a little or you can spend a lot—the benefits typically do track with the costs, but you'll have to make some intelligent trade-offs based on your company's needs.

TIP

Contrary to popular belief, lightning protection systems do not prevent the building from being hit by lightning. Instead, they work on the assumption your building *will* be struck. They mitigate the damage by giving the lightning a preferred pathway into the ground, directing the lightning through the system. Also contrary to popular belief, if you're caught outside during a thunderstorm and cannot seek shelter, you should assume the "lightning crouch," going up on the balls of your feet and covering your ears in a crouched position. The old belief was that lying on the ground was best, but this actually can increase your chance of being struck. For more on lightning safety, visit the NOAA Web site's online weather school at www.srh.noaa.gov/srh/jetstream/index.htm.

Drought

You might not think drought has anything to do with you or your company, but you might be surprised by some of the statistics. For instance, drought has a greater impact than any other natural hazard and its costs in the United States alone are estimated to be between $6 billion and $8 billion annually.

According to the National Drought Mitigation Center at the University of Nebraska in Lincoln (UNL):

"Drought produces a complex web of impacts that spans many sectors of the economy and reaches well beyond the area experiencing physical drought. This complexity exists because water is integral to our ability to produce goods and provide services. Impacts are commonly referred to as direct or indirect. Reduced crop, rangeland, and forest productivity; increased fire hazard; reduced water levels; increased livestock and wildlife mortality rates; and damage to wildlife and fish habitat are a few examples of direct impacts." *Source*: http://drought.unl.edu/risk/impacts.htm

The impacts are environmental, social, and economic, and can be widespread. Droughts impact businesses in a variety of ways but many are long-term effects hard to predict or quantify. For example, areas experiencing prolonged drought may find populations shifting out of the area in search of employment or new opportunities. Natural population shifts (those not driven by drought or other natural hazards) can also impact water availability. The shift in the U.S. population toward the western and southwestern states including southern California, Nevada, Colorado, Arizona, and New Mexico, among them, has put enormous stress on water resources in those areas. So, while there may or may not be drought conditions in some areas, there are water resource problems in many areas that can have the same long-term impact as drought.

Clearly, if your company's business activities involve the use of or reliance on water as a resource, you need to look at local, state, and national plans for drought mitigation. Some of the indirect impacts of drought can affect a wider range of businesses beyond those that rely on forests, streams, and lakes, according to the Drought Mitigation Center at UNL:

> "Many economic impacts occur in agriculture and related sectors, including forestry and fisheries, because of the reliance of these sectors on surface and subsurface water supplies. In addition to obvious losses in yields in both crop and livestock production, drought is associated with increases in insect infestations, plant disease, and wind erosion. Droughts also bring increased problems with insects and diseases to forests and reduce growth. The incidence of forest and range fires increases substantially during extended droughts, which in turn places both human and wildlife populations at higher levels of risk."

TIP

Drought obviously increases the risk of wildfires, so if your company is located in an area that could be vulnerable to wildfires, you'll need to take drought into account in your planning. Wildfires sparked by electrical storms and drought are common in many areas of the United States and your BC/DR planning should address these various threat sources.

Earthquake

The U.S. Geological Survey defines earthquakes in this way: "Ground shaking caused by the sudden release of accumulated strain by an abrupt shift of rock along a fracture in the Earth or by volcanic or magmatic activity, or other sudden stress changes in the Earth." *Source*: http://www.usgs.gov/hazards/

   Most people who live in earthquake-prone regions are well aware of that fact. They feel tremors from time to time and either have lived through minor or major earthquakes or have heard stories from those who have. However, you might be surprised to know how often small earthquakes occur. Figure 3.4 shows an earthquake map from the U.S. Geological Survey Web site taken March 6, 2007. The key indicates quakes that were within the last hour, the last day, and the last week. Though the color coding is not shown (shades of gray may not adequately convey the data), it should be clear from the abundance of boxes on the map that many small earthquakes occur every day throughout the California and Nevada region.

Figure 3.4 Earthquake Map for California and Nevada

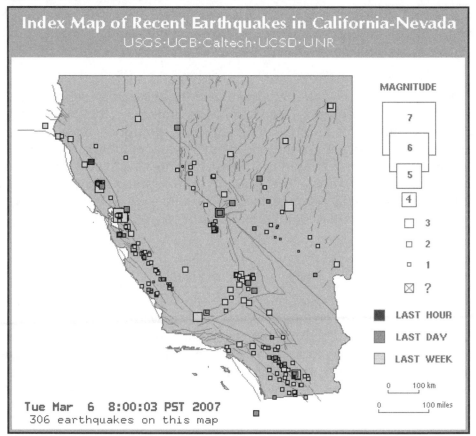

Source: http://quake.usgs.gov/recenteqs/

   Most of these quakes are below a 3.0 in magnitude, but the map certainly gives a good indication of the number of earthquakes in a given timeframe and the severity of each. Most are so small that most people won't feel them and they certainly won't cause any structure damage. However, we're all aware of the large quakes that have hit and the severe damage they caused. Damage from a single earthquake can cause in the hundreds of millions of dollars, though clearly the large earthquakes have a lower frequency of occurrence than the smaller ones that cause little or no damage. That said, scientists estimate there is a 62% chance of an earthquake with a magnitude of 6.7 or higher occurring in the San Francisco Bay Area of California between now and 2032 (*Source*: http://pubs.usgs.gov/gip/2005/15/).

Earthquakes can cause damage and have an indirect impact far from the epicenter of the quake. As witnessed in late 2005, undersea earthquakes can trigger tsunamis (though that is not the only source of tsunamis), which have a devastating effect for thousands of miles.

Although you may have earthquake preparedness plans in place if you live in an earthquake prone region, you should review your preparedness in light of the most current data available for your area. There may be new regulations regarding building safety, hazardous material handling, and others that may impact your business. Though you may be aware of many of these, a quick review of the latest information and regulations related to your area and your firm can help reduce your legal risks even if they can't change your earthquake risks. If your company is not located in an earthquake region, you should still assess the potential for a large earthquake in your area and determine whether creating a preparedness plan for such an event would make sense. Also keep in mind that your business operations can be severely impacted by an earthquake that occurs elsewhere. Key Internet infrastructure and communication sites could be disrupted, causing Internet traffic to slow or stop. Communications and power infrastructure can fail, causing disruptions far beyond the immediate impact area. Supply lines, shipping routes, roadways, manufacturers, importers/exporters, suppliers, and vendors may all be impacted by an earthquake so your preparedness needs to extend past the walls of your building and include an assessment of the risk your key business relationships face. We'll discuss this in more detail in the next chapter, but it's good to keep this in mind as you review these various threats. You'll continually need to "scan the horizon" to see how any of these threat sources might directly or indirectly impact your business operations.

TIP

Earthquakes are a good example of the need to balance the potential for a disaster with the frequency and impact of that disaster. An earthquake could be classified as a *threat*, the *threat sources* could be defined as shifting ground (subsidence), fire, gas leaks, explosions, infrastructure damage, building damage, structural failure (collapsing buildings, roadways, bridges), human injury and death, to name the most common of them. There are residual or longer-term threat sources from earthquake including disease outbreak, water shortages, water contamination, food shortages, food contamination, and even civil unrest. For more on earthquake maps, occurrence patterns and more, visit the U.S. Geological Survey's Web site at http://quake.usgs.gov/. There is an abundance of information on earthquake data from around the United States and around the world. There are useful articles on earthquake preparedness that might be useful to those of you in earthquake prone areas. Even if you've lived in an earthquake zone all your life, you will probably still learn a few helpful facts from a visit to the site.

Tornados

A tornado is a revolving column of dry air that occurs over land (unlike hurricanes, which originate over water, and can spawn tornados). Tornados have been observed on every continent on earth except for Antarctica, though the most tornados occur in the United States. Tornados are found most often in the central part of the United States, but hurricanes also can cause tornados in southern and eastern states as well. Wind speeds typically do not exceed 110 mph, though some tornados have been observed with wind speeds in excess of 300 mph, were more than one mile across, and traveled over a dozen miles on the ground.

Tornado damage usually is limited to the direct path of the storm, but the path of the storm is often extremely unpredictable. Like those that live in earthquake regions, most people know if they live in a tornado-prone area and have taken standard precautions. Tornados can disrupt businesses through physical damage to the facility, but it often has an even greater impact on the employees of a business whose homes may be destroyed or severely damaged by a storm that did not hit the business location. Employees will be worried about friends, neighbors, and family; they may lose their home or they may see their neighbor's home destroyed, causing severe emotional stress and trauma. People may be injured or killed and emergency services may not be able to reach the victims in a timely manner or at all. Tornados can also destroy needed infrastructure including telephone poles, power poles, power stations, roadways, and even emergency services (the fire house or hospital could be in the direct path of a tornado, for example). Tornados, like earthquakes, hurricanes, and many other natural disasters, can damage the infrastructure to the degree that emergency service providers are unable to reach your location for hours, days, or weeks.

Emergency preparedness information is widely available and you should be sure the information you have on how to prepare for and respond to a tornado is up to date. Again, keep an eye on the horizon and determine if any of your key business partners, suppliers, vendors, or customers are located in areas subject to tornados.

Hurricanes/Typhoons/Cyclones

Hurricanes are defined as "severe cyclones, or revolving storms, originating over the equatorial regions of the Earth, accompanied by torrential rain, lightning, and winds with a speed greater than 74 miles per hour" (*Source*: http://www.usgs.gov/hazards/). Of course, a revolving storm with winds of 65 miles per hour will also be devastating, even if it is not officially classified as a hurricane. In the United States, we've grown accustomed to hearing about Category 3 or Category 4 storms. Hurricane Katrina was alternately classified as a Category 4 and Category 5 storm because the wind speed changed over time. Regardless of the category or wind speed, hurricanes and tropical storms pose a significant threat to people and businesses (and more). Hurricanes bring destructive winds, torrential rain and flooding, storm surges of ocean water, and tornados. Hurricanes originate over warm ocean water but can travel across land and cause significant damage over coastal and inland areas. This was the case in several of the recent storms including Hurricanes Katrina, Rita, and Wilma in 2005. More than half of the U.S. population lives within 50 miles of a coast and this number is increasing every year. Many of these areas, especially the Atlantic and Gulf coast regions, are in the direct path of hurricanes.

TIP

Cyclones, typhoons, and hurricanes are all the same weather phenomenon; the name changes with the geographic location. The term cyclone is used for many types of tropical storms but most typically is used for storms that originate in the southwestern Indian Ocean. Typhoons originate in the northwestern Pacific Ocean (Pacific Rim area), and hurricanes originate in the northeast Pacific or Atlantic Oceans. However, storms can cross various boundaries, so the actual terminology used is somewhat arbitrary. A tropical storm by any other name is just as devastating.

As with many other natural hazards, there is very little you can do to avoid the risk short of relocating to an area that is not subject to those kinds of risks. Unlike tornados that can be spawned without warning, most hurricanes are tracked long before they hit land. Therefore, people often have the option of evacuating before the storm hits. As we saw with Hurricane Katrina, there were numerous problems with evacuation, including an underestimation of the strength of the storm, the inability of people to evacuate (gas shortages, traffic jams, etc.), or even the financial inability of people to make alternate arrangements (among other reasons). There is, of course, the problem with false alarms. The cost of evacuating, for individuals and businesses, is significant. Businesses have learned that they don't need to shut down for every hurricane that threatens their area, but that's a dangerous risk to take. If the speed, path, or strength of the storm is underestimated, lives can be lost. Therefore, your business leaders will need to assess the appropriate actions to take based on your company's type of business, location, and risk of hurricanes.

The cost to IT can be significant. Hurricanes bring high winds and rain, so power outages, flooding, and structural damage are the norm. If your building is destroyed by a hurricane, your IT resources are gone. How will you recover your operations in the face of total devastation? If you have a data center located elsewhere, you might recover far more quickly. If your backups were in a bank vault two miles away and the bank has been flattened, you might be out of luck. We'll look at the potential business impact of hurricanes in the next chapter.

Tsunamis

Tsunamis are defined as large destructive sea waves generated by earthquakes, volcanic eruptions, or large landslides. The tsunami that occurred in the Indian Ocean in December 2004 killed 200,000 people in 11 countries. It was a horrific example of the devastation a tsunami can cause. Though early warning systems do exist, they were not helpful in notifying people in the affected areas. Tsunamis have struck North America and the areas especially vulnerable are the five Pacific States—Hawaii, Alaska, Washington, Oregon, and California—and the U.S. Caribbean islands.

The United States has redoubled its efforts to improve early warning systems, and with the numerous communication channels now available to most people in the United States including TV, radio, the Internet, e-mail, instant messaging, land and cell phones, the likelihood of early warning is better than it was two decades ago. However, because 50% of the U.S. population lives within 50 miles of a coastline, there is also a good chance that evacuation routes would quickly become clogged if a metropolitan area such as Seattle or San Francisco needed to evacuate in advance of a tsunami warning. Therefore, it's vital for businesses in these areas to understand the potential risk and the mitigation strategies available to them. For more information on tsunamis, you can visit the U.S. Geological Survey Web site dedicated to tsunami information at http://www.usgs.gov/hazards/tsunamis/.

Volcanoes

Volcanoes are vents in the surface of the Earth through which magma and associated gases erupt. The shape of the volcano is made by the erupted material and typically forms a cone shape. Those living in volcano regions are well aware of the risks. Volcanoes, like other natural hazards, can occur with or without warning. However, unlike hurricanes or tornados, volcanoes don't change locations. Though some inactive volcanoes may be forgotten or disregarded, they are known points in the Earth from which magma and gas can erupt. Planning for a volcanic eruption typically includes evacuation because the lava flow is unpredictable in its path and moves more quickly than most people realize. It is sometimes impossible to get out of the path of lava flow; sometimes it creeps slowly forward for days and residents (and businesses) have time to collect their belongings and evacuate before watching the hot lava devour their house.

Businesses in volcanic areas should prepare evacuation plans and certainly be prepared for the possibility the building will be burned or covered by lava or that the ash that often accompanies an eruption could make the air and water hazardous. Ash particles are typically very fine and quickly clog filtration systems for water and air.

Avian Flu/Pandemics

Let's start with some definitions. An *epidemic* is an outbreak of a contagious disease that spreads rapidly to a local population. A *pandemic* is defined as an epidemic that covers a very wide geographical area including entire regions, countries, or continents. According to the U.S. Centers for Disease Control (CDC), the Avian Flu does not *currently* pose a threat as either a potential epidemic or pandemic. However, the propensity of viruses to mutate creates some future risk of epidemic or pandemic outbreaks.

The Avian Flu (also called bird flu) has gotten a lot of press in recent years. The flu naturally occurs in bird populations but in recent years has been seen in humans. Many wild birds carry the virus in their intestines but do not get sick from it. However, it is highly contagious and domesticated birds are at high risk of contracting the virus, getting sick, and dying from it if they are exposed. There are numerous variations of the virus and therefore it is difficult to assess the risk to humans. Some strains have a low chance of infecting humans, others a much higher likelihood.

The media is responsible for much of the hype and hysteria surrounding bird flu. According to the CDC,

> "Most cases have occurred in previously healthy children and young adults and have resulted from direct or close contact with H5N1-infected poultry or H5N1-contaminated surfaces. In general, H5N1 remains a very rare disease in people. The H5N1 virus does not infect humans easily, and if a person is infected, it is very difficult for the virus to spread to another person. While there has been some human-to-human spread of H5N1, it has been limited, inefficient and unsustained. For example, In 2004 in Thailand, probable human-to-human spread in a family resulting from prolonged and very close contact between an ill child and her mother was reported."

The virus is contracted through close contact with infected birds or surfaces and human-to-human spread is caused by close contact with the carrier(s). However, the danger is that all influenza viruses have the ability to change and there is a possibility a strain could occur that spreads easily among the human population. Because this virus does not naturally occur in humans, we have no natural antibodies to fight off these viruses.

The issue your company should address is not how it will handle the bird flu itself, but how it might be impacted by a pandemic flu of any sort (avian flu and pandemic flu are not one and the same). According to the CDC:

■ A pandemic may come and go in waves, each of which can last from six to eight weeks.

■ An especially severe influenza pandemic could lead to high levels of illness, death, social disruption, and economic loss. Everyday life would be disrupted because so many people in so many places become seriously ill at the same time. Impacts can range from school and business closings to the interruption of basic services such as public transportation and food delivery.

■ A substantial percentage of the world's population will require some form of medical care. Health care facilities can be overwhelmed, creating a shortage of hospital staff, beds, ventilators, and other supplies. Surge capacity at nontraditional sites such as schools may need to be created to cope with demand.

■ The need for vaccine is likely to outstrip supply and the supply of antiviral drugs is also likely to be inadequate early in a pandemic. Difficult decisions will need to be made regarding who gets antiviral drugs and vaccines.

■ Death rates are determined by four factors: the number of people who become infected, the virulence of the virus, the underlying characteristics and vulnerability of affected populations and the availability and effectiveness of preventive measures.

Source: www.pandemicflu.gov/general/index.html#impact

Not a pretty picture, but as with all disaster planning, it's better to go in well-armed with current and accurate information.

You might be arguing that if a pandemic hits, you'll have no control over the situation. Even though that may be true, you still have to provide some sort of contingency plans. Let's look at a possible scenario. Your company sells an enterprise-level software product. It's used at major corporations throughout the world. It seamlessly integrates into their messaging and communications applications and they rely heavily on this application. Your largest client, a Fortune 100 company, comes to you and asks, "What plans do you have to support this product in the event of an Avian Flu outbreak?" Here's a variation on the theme. Your company's vice president of Global Sales comes to you saying she is trying to close a deal with a Fortune 500 client that would mean tens of millions of dollars in revenue for the company. However, in order to close the deal, she needs the IT group's "Avian Flu Readiness Plan" to allay concerns by the potential client about your company's ability to respond to global service and support needs in the event of an outbreak or other pandemic.

So, you might think that if the avian flu hit, you'd shut your doors until it blew over, but you may not be able to hide your head in the sand for long. The difficulty, of course, is that IT systems are not impacted by the flu—people and processes are. So, as the IT professional in the group, you have to figure out how you can provide the services required of you in the event that tens, hundreds, or even thousands of your staff are out sick, quarantined, or unable to report to work. In pandemics, there is a fairly high mortality rate, though it usually impacts the young, old, and infirm the hardest. Still, what if key staff die or are too ill to work for an extended period of time?

It seems callous to be worried about business in the face of death and serious illness, but businesses provide important services that are needed all the time. Shutting the doors may seem like the best option and for some companies, it might be, but you can't make that determination off the top of your head. In the business impact analysis, we'll revisit this topic. As a threat, the avian flu or pandemic flu are just abstract concepts, but they are worth looking at in terms of your company's vulnerability to these events and the impact they might have on your company, your employees, your customers, your supply chain, and your community.

TIP

For more information on pandemics, you can visit the CDC Web site at www.pandemicflu.gov or the CDC's main Web site at www.cdc.gov/index.htm.

This section on natural and environmental hazards is not exhaustive, but it should give you a solid start in investigating the threats and threat sources your company might face. As you read through the various hazards, you may have thought of other threats not listed or you may have learned about threats you didn't think applied to your business. For example, not everyone would think about the risk of fire or flooding from an earthquake. If your company is located in a low-lying area and there is a dam or water containment system uphill, an earthquake can break the containment and send water downhill, flooding your building, street, or neighborhood. These kinds of examples point to the importance on doing research and being familiar with the surrounding area. You won't know that you need to plan for flooding if you don't know the reservoir is located just three miles away uphill from you or that you need to plan for wildfires because just over the crest of the hill is an open grassland. In the next section, we'll look at human-caused hazards and here too, you'll need to be aware of your surroundings. What does the company across the street or around the corner do? Do they work with hazardous materials or noxious chemicals? Could there be a biohazard or chemical spill in the area? Is there a railway that runs near your building or a major freeway? Is there a nuclear power plant in the region, is nuclear waste transported on roads near your building? These are the kinds of things to consider as you read through the remainder of this chapter, as you develop your threat list and research potential hazards in your area.

Human Threats

Human threats, like environmental or natural hazards, come in all sizes and shapes. Although we might like to distinguish between intentional and unintentional acts, that goes only to motive and intent, not impact and outcome. So, we'll look at these threats without regard to whether or not they're intentional except in a few cases. Terrorism is, by definition, intentional as is war, theft, and sabotage, to name a few. Other threats such as fire, chemical spills, or electronic data loss can be caused by intentional acts or human error. Remember the statistic from IBM cited earlier in the book —that 80% of data loss is human caused? They don't specify whether that's intentional or accidental. So, regardless of the intent, the effect is the same. Let's go through some of these human-caused hazards in some detail. You may learn new facts that help you plan better or you may simply become aware of one or more threats you didn't previously know about.

Fire

Human-caused fires can be located inside or outside as in the case of improper wiring or unattended campfires. Fires are the most common business disaster, so regardless of what other planning you do, you should certainly ensure you have a solid fire recovery plan and you practice fire drills and fire procedures regularly.

Most commercials buildings have fire suppression systems; some buildings have fire doors to prevent fire from running uncontrolled through a building. Fire extinguishers, alarms, smoke detectors, and other fire equipment should be located at strategic places throughout the building, should be well marked, and should be tested regularly to ensure proper functioning.

Arson is an intentionally set fire. The local fire department may investigate any fires in your building to determine the cause. Some insurance companies may prohibit payment of insurance claims in some circumstances so be sure to have someone from finance or legal review your company's insurance policy, especially with regard to fire.

Theft, Sabotage, Vandalism

Theft, sabotage, and vandalism are all intentional acts carried out by employees, building employees (not associated with your company), former employees, and strangers. Many of these types of problems can be effectively thwarted by having security procedures in place. These include controlling access to the grounds, the buildings, and certainly to the inner offices, labs, server rooms, and other areas within the building that contain expensive, sensitive, or strategic materials. Most IT professionals understand that security begins with controlling and monitoring physical access and the same is true for your business as it is for IT equipment.

Theft can come in many different forms, some of which might not immediately come to mind, such as:

- Software piracy (are employees stealing software from the company or installing illegal software?)

- Counterfeiting (of currency or any other commodity of value, including company checks, ID badges, software, etc.)

- Theft of proprietary information (intellectual property, trade secrets, confidential data, etc.)

- Equipment theft (from office supplies to servers and everything in between)

You might think of some of these elements, but when you view them in light of BC/DR planning, you might find you need to take a few additional steps to address these. For example, you might have a plan for how you'd get your business back up and running after a fire, but what if someone came in and stole two critical servers? First, is the data safe? Second, how would you recover? Those are the kinds of considerations to include when you think about what could be stolen or damaged within the walls of your company. Keep in mind, too, that as with most computer fraud or theft, most company fraud or theft is perpetrated by those *inside* the company, not by mysterious outsiders.

If your facility is small, be sure to have a process in place for monitoring visitors or those entering the building such as strategically locating a receptionist or someone's office near the entry way. In some companies, this falls to the HR staff to manage. In larger facilities, you may need to have a more formal method of monitoring and controlling access such as visitor sign in, presentation of identification, and the issuance of a visitor badge. Many companies require cell phones and other digital devices to be left with the front desk so that photos or recordings cannot be made while in the building. This prevents someone from stealing trade secrets or casing the property to determine the best way to burglarize the building during off-hours.

Any of these acts is intentional, and prevention is typically the best solution. However, should your business be vulnerable to theft, sabotage, or vandalism due to the location, the nature of the work you do, or the likelihood of having disgruntled employees or vendors, you should review your plans for preventing and recovering from these threats. However, since we're focusing on IT, we're going to cover theft, sabotage, and vandalism in the IT realm in just a bit.

Labor Disputes

If your company includes union workers or your company interacts with another company that includes unions and union workers, there is a risk that labor disputes will disrupt business. Remember that you have to look at your company as well as at your key suppliers, vendors, contractors, outsourcers, and even customers. A major disruption in any of those areas could temporarily or permanently disrupt your business. For example, if you're a supplier to a large manufacturing company and that company represents 75% of your sales, what happens if they shut down for six to eight months during a labor dispute? How will your company survive? Clearly, it's not desirable to have such a large portion of your revenue stream from a single source, but that's sometimes a business reality. Your executives may realize this puts your company at risk but the upside profit potential may be compelling. Therefore, as the BC/DR planning team, you need to assess this risk both internally and externally and determine both the likelihood and the impact a potential labor dispute would have on your business. Keep in mind, too, that there are various ways labor disputes can play out—from work slow downs to strikes to sabotage. Each of these scenarios may have a slightly different impact on your company and your operations, and each should be assessed as a separate threat source.

Workplace Violence

The unfortunate reality is that there is violence in the workplace. Whether from disgruntled former employees or unhappy current employees, violence can occur without warning. According to the U.S. Department of Labor's Occupational Health and Safety Administration (OSHA), homicide is the fourth leading cause of occupational injury in the United States. In 2004, the latest year for which data is currently available, there were 551 homicides out of 5,735 fatal workplace injuries.

Before you start looking suspiciously at your coworkers, keep these facts in mind: 71% of workplace homicides are robbery related and only 9% are committed by coworkers. Although any person or company could experience workplace violence, the likelihood of violence increases with these factors:

- High interaction or exposure to the public
- Exchange of money or funds
- Working very late or very early, especially alone
- Guarding valuable assets or money
- Regularly dealing with volatile situations or violent people

Cab drivers, liquor store staff, late night convenience store staff, and safety officers (police and security guards) are occupations at the top of the risk list (*Source*:

www.cdc.gov/niosh/violfs.html). If you believe your company's premises, location, or type of business may be at risk of workplace violence, you should certainly take preventive measures if you have not already. There are numerous resources available, from government Web sites to on-site training programs that can train your staff to prevent and address workplace violence effectively.

If we look at workplace violence from a business disruption point of view, a serious injury or death could result in the premises being sealed off as a crime scene for some period of time. Equipment such as computers or other needed items could be seized as part of the investigation. Employees will be impacted and productivity will suffer; good employees may choose to find employment elsewhere, and your reputation in the community, in your industry, or in the eyes of potential employees may suffer.

Tip

If you're interested in learning more about preventing workplace violence, visit the OSHA Web site focused on that topic at www.osha.gov/SLTC/workplaceviolence/evaluation.html.

The U.S. Department of Health and Human Services Centers for Disease Control and Prevention (CDC)'s National Institute for Occupational Safety and Health (NIOSH) has additional resources related to workplace safety, and you can download videos on preventing workplace violence by visiting this link: www.cdc.gov/niosh/docs/video/violence.html.

Terrorism

The very nature of terrorism is that it cannot be planned for and sometimes cannot be prevented. Therefore, your assessment should include not the threat of terrorism but the threat sources that stem from it. This goes back to many of the issues we've already discussed—what if your power goes out or there's a chemical spill or an anthrax release? (We'll cover biohazards in an upcoming section). How will you address the results of terrorism? When you look at it this way, it seems to become a slightly more manageable topic. Ultimately, each company has to assess its vulnerability based on geographic location, the nature of its business, its international business connections, its political involvement, and more. If your company is vulnerable to terrorism, there's a good chance you have a strong team in place that has already addressed this. Nuclear power plants, power stations, airports, and others have developed contingency plans because of federal or state mandates to do so. If you believe your company should have a stronger plan to prevent or address a terrorist threat, you should involve high level executives or managers in your company and discuss next steps. There may be resources at the U.S. Department of Homeland Security that can be helpful to you or you may wish to consult with a private firm that specializes in this area to address the specific threat of a direct attack on your company.

However, most companies are not the target for such an attack, but may be in the area of an attack or may experience the result of an attack. Addressing the likely threat sources in your area that potentially could be targets for terrorists will help you devise a plan that will help address the aftermath of an attack, even if you don't know whether the incident was intentional or accidental.

TIP

The U.S. Department of Homeland Security has a wide variety of resources, including information on terrorism, on their Web site at www.dhs.gov/index.shtm.

Chemical or Biological Hazards

Chemical hazards are present in a variety of manufacturing environments, whether the chemicals are the *result* of the manufacturing process or are used *within* the manufacturing process. A biological hazard, often called a *biohazard*, is defined as a danger to humans or the environment resulting from biological agents or conditions. Both chemical and biological hazards can occur as the result of an accident, sabotage, or terrorism.

If your company is not involved with chemical or biological agents, you may still face risks in this regard. You need to look in your local area and determine what other types of companies exist. If a chemical plant is located four miles away, what is the risk to your operations? What if it's next door? Fifteen miles away? Your local and state agencies may regulate these types of companies, so there may be information that's easily available to you. Some research on local companies should also reveal the nature of work that your commercial and industrial neighbors do. If you run into trouble locating this information, you might be able to contact a friendly commercial leasing (real estate) agent. Commercial leasing agents often have their fingers on the pulse of the community and can tell you who your neighbors are and what type of work they do. Your local Chamber of Commerce may also be able to provide useful information. Keep in mind that you're more likely to encounter these kinds of hazards in a heavily industrialized area, so you can look at your location and determine where the industrial areas are in relation to your operations. Areas that are heavily residential with a bit of commercial building space are less likely to pose a threat of chemical or biological hazards.

If your company is involved with chemical or biological research or manufacturing, you no doubt have safety procedures in place with regard to these agents. However, you and your team might review your safety procedures with an eye toward BC/DR planning. What types of containment procedures are in place? What sort of evacuation procedures are practiced? What are the countermeasures or remedial activities that should take place if a spill, leak, or release occur? In some cases, your company's activities (or the chemicals and biohazards) may be closely regulated by government or industry and the procedures and requirements address these questions. However, you may need to take this information as input to your business continuity and disaster recovery planning by asking how operations would be impacted by a chemical or biological hazard. Would you have to set up operations elsewhere? How long would it take to reoccupy the building, or would you have to permanently move? If the servers and other IT equipment were inaccessible due to contamination, what would you do? How would you resume operations? These are the kinds of questions you'll ask and answer as part of this process. For now, the key is to determine what, if any, risk you have to internal chemical or biological hazards.

We haven't addressed the risk of the release of these agents as part of sabotage or terrorism, but the result would be the same. If a chemical plant in the area were to be the victim of sabotage or a terror attack or just an accident by a careless employee, the net result to your firm is the same.

If a chemical or biological agent were to be released in your area as part of a terror attack, you would have to address the same types of issues such as whether the best course of action is to evacuate the building or to shelter-in-place. If you believe your company may be vulnerable to these types of threat sources, you may want to contact your local police and fire department for guidance on how to handle these types of incidents. The bottom line, however, is what data you need to develop a sound BC/DR plan that addresses the impact of this type of threat.

War

This type of threat clearly exists for many companies around the world. For U.S.-based companies, the risk is greatest for divisions of the company that may be located in areas of political or economic instability. Plans for shifting operations from areas experiencing war, civil war, or civil unrest should be made in areas that are vulnerable. BC/DR plans should examine which areas of the company may be vulnerable and how they can be protected. Remember that not only would local operations in a war zone be disrupted but vital knowledge, equipment, or assets could be stolen from the company. Therefore, the BC/DR plan must look holistically at the people, process, technology, and infrastructure of the organization in areas vulnerable to war or civil unrest and assess the vulnerability to these various threats.

Cyber Threats

We placed cyber threats last because it is a large and ever-changing topic. We'll cover some of the common threats in this section, but we recognize that they will shift and change quickly. The good news is that many, if not most, of these threat sources are ones that you and your IT team are very familiar with and as such, your BC/DR plan should be fairly easy to construct in this area. Most of these threat sources have to be assessed and addressed through normal IT operations and security assessments (and certifications), so you may already have all the data you need to include in your BC/DR plan.

The bottom line in IT is data security. Sure, a stolen server is a hassle to replace, a hacked Web site is a pain to repair, but ultimately all these threats result in compromising one (or more) of three basic areas: confidentiality, integrity, and availability (CIA).

Confidentiality refers to the protection of data from unauthorized disclosure. Unauthorized, unintended, or unanticipated disclosure can result in legal action, financial loss, loss of public confidence, or embarrassment. For example, personal medical information is protected by law and any unauthorized disclosure of such information, regardless of whether it was intentional or not, is illegal. Companies dealing with personal health information including hospitals, health clinics, doctors offices, optometrists, and prosthetics companies, to name just a few, are all subject to these kinds of regulations. However, confidentiality also can include keeping trade secrets confidential from competitors or keeping embarrassing information from spreading through unintended or unanticipated channels.

Integrity has to do with the information being protected from unauthorized or unintended modification. Hackers often have as one of their goals the modification of data—whether that data includes user permissions, network access, or business data modification such as pricing on a Web site or pay rates for employees. Integrity issues can stem from intentional acts such as wayward employees or external hackers, but it can also result from accidents and errors. Someone might make accidental changes, enter erroneous data, or corrupt a database just with a few incorrect clicks of the mouse. Inaccurate data may be from an error, it might stem from intentional fraud (such as changing payroll data or pricing on a Web site), or just bad decisions. Regardless of the intent behind it, these losses can also add up. In some cases, your company may face legal liabilities. There are almost always financial consequences including lost productivity, downtime, lost or lower sales, or untraceable losses, to name a few.

Availability pertains to critical business data being available when needed. If a database is corrupted, it is not available for use. If a Web site is hacked, it is not available for use. If a Web site is flooded with connection requests (a Denial of Service attack), it is not available for use. These kinds of availability problems can also be intentional or accidental. Lack of availability impacts productivity for the IT department (busy fixing availability issues) and for end users (unable to retrieve needed information in a timely manner).

Anyone working in IT for any length of time knows about CIA and about the various methods used to attack these three areas of data security. You are also probably painfully aware of how unintentional errors or poor decisions can impact CIA as well. One wrong setting, one incorrect click of the mouse, and users can be granted incorrect access, data can be changed or deleted, data can be exposed to unauthorized access.

Most likely as you've worked in your IT operations, you've identified the critical data for your organization and ensured it was protected against loss of CIA. These are basic operational areas that most IT departments do as part of defined security procedures. Basic steps such as creating and reviewing security logs, analyzing network traffic patterns, and other types of standard IT security operations typically are built into your everyday activities.

For your BC/DR planning, you'll need to review all your IT operations in light of the potential for these security threats occurring. For example, you might have a computer incident response team (CIRT) in place to address a breach in network security. Have you also taken it a step further and looked at what the impact on business operations would be if, say, a client database was attacked? How would your business recover from that? The likely answer is that it depends on the nature of the attack. You'd first probably need to understand whether the data was looked at, modified, or stolen and how the attack was carried out. This would dictate your next steps and would also tell you the potential impact of the incident on short- and long-term operations. So, even though you may have plans in place to prevent and address potential data security issues, you may not have business continuity and disaster recovery plans in place related to these specific threat sources.

As a starting point, you can review your IT security processes and procedures and determine whether they need to be updated or expanded to address new and evolving threats. Then, take those procedures and include the security threat sources in this BC/DR risk assessment. From there, you can treat these threat sources as you would any other threat source except that you're probably already ahead of the curve with much of this data. Look across the enterprise to see how these incidents could disrupt not just IT activities, which you probably have already determined at length, but corporate operations as well.

Looking Ahead...

Business Impact Analysis

Business impact analysis (BIA) is the next step in the risk management process and it's covered in depth in the next chapter. However, let's take a moment to define BIA so you can begin formulating ideas and thoughts about BIA while you're in the risk assessment phase. *Business impact analysis* is the process of identifying all potential impacts to your business from all identified threat sources that could disrupt the busi-

Continued

> ness. Later in this chapter, we'll analyze threat sources based on the company's vulnerability to these threats and you'll rank these in order of importance. You'll probably decide to limit your business impact analysis to those threat sources that are most likely to occur and your company's vulnerability to those threats. After you've identified and analyzed the business impact of each selected threat source, you will use that data as the input for developing mitigation strategies. It's helpful to keep the overall process in mind as you move through each phase of the assessment so that information, ideas, or suggestions relevant to upcoming phases can be captured during the process.

Cyber Crime

Cyber crime is an evolving field of study and unfortunately, there is no dearth of people interested in perpetrating cyber crimes. In the United States, there are numerous federal and state agencies that may deal with cyber crimes, depending on factors such as the suspected originating location of the crime, the scope or span of the crime, and the nature of the crime. For example, financial crimes and those related to personal identity theft for the purpose of financial gain are often handled by the U.S. Secret Service, a division of the U.S. Treasury. Other crimes are handled by the U.S. Department of Justice, and still others are handled by the FBI.

TIP

For more on these national resources, visit these links:

- U.S. Secret Service Electronic Crimes Task Force: www.secretservice.gov/ectf.shtml
- U.S. Secret Service Criminal Division: www.secretservice.gov/criminal.shtml
- U.S. Department of Justice Computer Crime and Intellectual Property Section: www.cybercrime.gov/ (*Note*: The use of the acronym IP on this Web site indicates "intellectual property" and not "Internet Protocol.")
- FBI's Cyber Investigations: www.fbi.gov/cyberinvest/cyberhome.htm
- Federal Trade Commission: www.ftc.gov/
- Securities and Exchange Commission (SEC): www.sec.gov/

For more information on state and local resources, contact your local law enforcement agency or your state's Attorney General's office.

As an added note, you can help your company's employees avoid becoming the victims of cyber crimes by providing training and education on topics such as how to avoid falling victim to phishing schemes, how to report suspicious behavior, how to avoid auction and lottery scams, and so on. This education not only helps employees avoid these problems at home, but it helps keep corporate resources safer as well.

Unfortunately, the types of crimes committed are limited only by the twisted imaginations of people intent on wreaking havoc. Following is a brief list of the headlines that appeared in mid-March 2007 on the U.S. Department of Justice's CyberCrime.gov Web site, which gives a good indication of the breadth and depth of cyber crime:

- Romanian Hacker Broadcasts eBay Customer Accounts (March 12, 2007)

- Los Angeles-Area Man Charged with Uploading Academy Award 'Screener' onto Internet (February 22, 2007)

- Software Piracy Ringleader Extradited from Australia (February 20, 2007)

- Duracell Employee Pleads Guilty to Stealing Trade Secrets (February 2, 2007)

- Man Pleads Guilty to Stealing Morgan Stanley Trade Secrets Relating to Hedge Funds (February 1, 2007)

- Former Antelope Man Sentenced to 20 Months in Prison for Fraudulently Obtaining Microsoft Software: Defendant Cracked Code Needed to Activate Software Causing More than $500,000 in Losses (January 25, 2007)

- First Conviction in Hewlett Packard Pretexting Investigation (January 12, 2007)

- Anderson Man Charged with Criminal Copyright Infringement (December 28, 2006)

- Defendant Sentenced in Online Piracy Crackdown (December 19, 2006)

- Two Michigan Residents Plead Guilty to Criminal Copyright Infringement (December 15, 2006)

As you can see from a scan of these headlines, the crimes are rather diverse—software piracy, trade secrets, copyright infringement, pretexting, online piracy—the list goes on. As an IT professional, you know how difficult it is to keep up on all the latest cyber crime methods, but you also know that if you or someone in your organization is not up-to-date, you may well fall victim to cyber crime. Rather than go into a long list of potential threats, let's just create a general list of items as a start. Remember, cyber crime is committed for three basic purposes—to make money, to earn bragging rights, or to disrupt business. Money usually is made by stealing electronic data and selling it or making unauthorized use of it. Clearly, this fits into the "confidentiality" element of the CIA model. Bragging rights often are sought by hackers, and those types of crimes often span the entire CIA framework.

Finally, disruption of business often entails the integrity or availability of data, though breaching confidentiality can also create a disruption in some cases.

The common categories of cyber crimes include:

- Identity theft (through a variety of means)

- Corporate identity theft (through a variety of means)

- Hacking corporate network or intranet (to breach confidentiality, integrity, availability)

- Hacking corporate Web site or extranet (to breach confidentiality, integrity, availability)

- Creating backdoors for unauthorized access (to breach confidentiality, integrity, availability)

- Stealing/selling confidential data (trade secrets, drawings, plans, intellectual property)

Loss of Records or Data-Theft, Sabotage, Vandalism

Although loss of records or data falls under a variety of cyber crime categories, it's worth listing as a separate category anyway. An error or an intentional act can create data loss. Even in the case of unintentional loss (error), the perpetrator is unlikely to come forward voluntarily. In some cases of error, the person may not even know they have caused a data loss. This can happen when ill-trained staff are given tasks to perform outside their skill levels. In other cases, a careless error is made and the person making it is unaware of the error or the resulting data loss. Even if asked, he or she might not understand or realize that their actions may have caused the problem. However, in most cases, the person causing the problem is aware or becomes aware of it. If your company has the type of culture that will severely punish someone making that type of error, you won't likely get people volunteering the truth. It makes it difficult, then, to ascertain whether someone intentionally or inadvertently caused the problem. If it was intentional, you have sabotage going on and you need to find the source and remove access to systems. If it was an error, it may be a one-time event or you have a training issue on your hands. In that case, you may want to restrict access to those systems until the person can demonstrate reasonable competence.

IT System Failure-Theft, Sabotage, Vandalism

IT system failure is similar to loss of data—it can be intentional or unintentional. Intentional acts that bring systems or networks down are sabotage and should be addressed as crimes. Vandalism occurs when systems have been physically broken or destroyed. Unauthorized modification of a Web site is also considered vandalism, especially if it is modified in a way to visually indicate it has been breached such as changing the text or links on a page or inserting a banner, picture, or other data in the Web site. When we talk about IT system failure and theft, we're primarily concerned with the physical theft of equipment including servers, routers, firewalls, test equipment, software, cabling, and any other IT-related asset. The theft of these items disables one or more IT systems, thus falling in the category of IT system failure.

Infrastructure Threats

Infrastructure threats are large, external environmental issues that you rarely have any control over preventing, addressing, or resolving. These issues include:

- Building-specific failures (structural damage, systems failures)
- Public transportation disruption (roads, railways, airports, seaports, waterways)
- Loss of utilities (power grid failure, gas supply failure, water supply failure)
- Petroleum or oil shortage
- Food or water contamination
- Regulatory or legal changes

Building Specific Failures

Buildings are designed and built by humans, so there's always a chance that the architect or builder could make an error that results in the part of the building becoming unstable or failing. Buildings can fail because of human error in the design and construction of the building. The materials themselves can fail as has been the case where inferior concrete was used. Buildings not built to code or that are not properly inspected can be at risk of structural failure.

Other building-specific failures include non-IT system or equipment failures (IT system failures are covered in a later section), communications equipment failures (telephone lines, communications lines, Internet connections), safety systems (fire and alarm systems), internal power failure (circuit breakers, wiring issues, circuit capacity, etc.), heating/cooling failure, and manufacturing line (production) failure. These types of systems typically are managed by the facilities manager. In smaller companies, some or all of these systems may be managed by the building's management company or directly by the building owner/landlord. If you are occupying a building you do not own or manage, you may want to set up an appointment with the manager to go over the building's systems so you understand things like how old the equipment is, the likelihood of it breaking, the estimated duration of critical repairs, and so on. For example, if you're occupying an older building and you learn that the heating system is 45 years old and that parts would be next to impossible to get, you should probably come up with a Plan B related to loss of heat. Statements like "Yeah, the next time it breaks, we're just going to replace the system" should give you a clear indication that a heat failure could take days or weeks to repair.

Public Transportation Disruption

Disruption of public transportation can have a local impact such as the inability of employees to get to work in a timely manner (if at all) or the inability of employees to evacuate due to an impending storm. On a larger scale, your suppliers, vendors, and contractors can all be impacted by transportation disruptions, so your entire supply chain should be evaluated for vulnerabilities to the same threat sources you're looking at for your company.

Loss of Utilities

Loss of utilities usually is localized to a specific region and can be caused by a number of things including weather events, sabotage, error, technology failure (a switch or transformer fails, for instance), or terrorism. In some cases, power loss can cover an entire geographic region, such as when power fails on an entire section of the U.S. power grid.

> **NOTE**
>
> An article in the Washington Post by Justin Blum points to the vulnerability of the U.S. power grid to terrorist attacks. The article, entitled, "Hackers Target U.S. Power Grid: Government Quietly Warns Utilities To Beef Up Their Computer Security" (March 11, 2005. http://www.washingtonpost.com/wp-dyn/articles/A25738-2005Mar10.html) states, "'A sophisticated hacker, which is probably a group of hackers . . . could probably get into each of the three U.S. North American power [networks] and could probably bring sections of it down if they knew how to do it,' said Richard A. Clarke, a former counterterrorism chief in the Clinton and Bush administrations." A rather unsettling thought. For more on the U.S. power grid and how it works, you can read up on it on Wikipedia at http://en.wikipedia.org/wiki/Electric_power_transmission.

Disruption to Oil or Petroleum Supplies

There has been a lot of discussion of late with regard to global oil supplies (especially as it relates to global warming issues). Regardless of your opinions on the topic, oil supplies are finite and are managed by large groups—countries or cartels—and the availability and cost of oil and petroleum products is controlled by those few. Oil and petroleum supplies can be disrupted by war, civil unrest, sabotage, weather, or political will. If supplies are disrupted, how will your business fare? Employees may not be able to get to work (long gas lines or no gas, as was evidenced in the late 1970s in the United States); suppliers may not be able to manufacture or deliver their products to you (oil is used in manufacturing and for fuel); products needed for your manufacturing may be unavailable or significantly more expensive; timelines may be pushed out due to delays in getting materials and supplies; costs may sky-rocket due to limited supplies, and the list goes on. If your company is dependent upon oil or petroleum production or supply, you clearly need to evaluate the threats and threat sources so you can create effective mitigation strategies. These tasks may fall outside your purview as an IT professional, but as with other business risks, it's important that you be as well-informed as possible so you can participate fully in developing the best BC/DR plan possible.

NOTE

Geophysicist M. King Hubbert predicted in 1956 that U.S. oil production would reach its highest level in the early 1970s. Though severely criticized by oil experts and economists, Hubbert's prediction came true in 1970. The term "Hubbert's Peak" is used to indicate the peaking of oil production in a particular area. Oil companies routinely use Hubbert's calculations to help them determine the yield of a particular oil field, so clearly Hubbert's data has been found to be accurate over the years. Kenneth Deffeyes, a geologist who worked for Shell Oil Company and later became a professor at Princeton, built on Hubbert's work and found that worldwide oil production will peak in this decade. Regardless of which side of this debate you fall on, you might find some interesting information by using the search term "Hubbert's Peak" on your favorite search engine.

Food or Water Contamination

Contamination of the food or water supply is disruptive to all life forms. Contamination can be accidental as in the case of an oil or chemical spill or it can be an intentional act of sabotage or terrorism. The impact of these events can be local, regional, or national though they usually are contained to a specific geographic region. Chances are good that if food or water is in short supply in your area, your employees will not be concerned with coming to work but with finding food or water. Your company's operations will be secondary to everyone in that event and you may not need to plan for this other than to assume you will suspend operations until the issue is resolved. Food or water contamination or shortages in other areas could impact your supply chain, so looking at your business as well as that of your business partners may turn up some risks you hadn't seen that might be worth addressing in your mitigation plan.

Regulatory or Legal Changes

Changes to regulations or legal rulings setting precedents could impact your business, but the place they're most likely to impact you is after a disaster. For example, there may be health and safety regulations that impact your ability to resume operations, especially if something has happened to your facility. Opening your doors without adhering to these regulations could result in having operations shut down or having stiff legal and/or financial penalties imposed. Changes in any of the legal areas, such as data security, could also impact your firm in the aftermath of a disaster. If a server inadvertently ends up in the wrong hands and data was not encrypted or due diligence was not used to secure the data or the computer, you may have another disaster on your hands.

The best way to address this is to have someone on your team review the current regulatory and legal requirements for your firm and do a bit of research to find proposed or impending changes. Then, determine how your company would be impacted by these changes during normal business and in the aftermath of a disaster. This is especially true for data security requirements within the IT arena. You can begin to build in or modify processes to address these changes so that they are part of your everyday operations, if appropriate. Often it's easier and less costly to scan the horizon and build in your safeguards in this manner than to try to retroactively address these kinds of issues in the aftermath of a disaster.

We've looked at a wide variety of threats and threat sources, and continually tied them back to corporate operations and, where applicable, IT. As we've mentioned, it's not exhaustive and in some cases, we did not go into tremendous detail because of changing threats or changing laws and regulations. However, this section should have given you a good idea of how to look at potential threats and threat sources as well as how to think through the potential threat and impact to your business.

Threat Checklist

The list, shown in Table 3.2, is provided for your convenience. It is a reiteration of all the threats listed in the previous sections. You may want to use this list as a starting point in your threat assessment. You can add any threats not included in the list and remove those you're confident will not impact your business. Again, be sure to avoid removing threats before you look at your company's total environment—internal, external, and extended (key suppliers, vendors, outsourcers, partners, and customers).

Examine how these threats and threat sources impact the *people*, *processes*, and *technologies* your company needs to operate as well as the *infrastructure*. In the next chapter, we'll look specifically at how these threats and threat sources can impact your immediate operations as well as how they might impact your key customers, suppliers, vendors, contractors, out-sourcers, or partners. However, if you have any thoughts on impact as you move through this portion of the risk assessment, be sure to jot them down for later use.

Table 3.2 Threat Checklist

Natural/Environmental Threats

Fire (can be human-caused)
Flood
Severe winter storm
Electrical storm
Drought
Earthquake
Tornado
Hurricane/Typhoon/Cyclone
Tsunami
Volcano
Avian Flu/Pandemics

Human-Caused Threats

Fire, Arson
Theft, Sabotage, Vandalism
Labor disputes
Workplace violence
Terrorism
Chemical and biological hazards
War, Civil unrest

Infrastructure Threats

Building-specific failures
Non-IT equipment, System failures
Heating/Cooling, Power failures
Public transportation disruption

Continued

Table 3.2 continued Threat Checklist

Infrastructure Threats

Oil, petroleum supply disruption
Food, water contamination
Regulatory, legal changes

IT-Specific Threats

Cyber threats (CIA)
Equipment or system failure
Production line equipment failure
Loss of data or records

For your convenience, in Table 3.3 we've also included a slightly different view of some of the IT specific threats you might want to consider in your planning. It's not intended to be comprehensive, but it should help get you started.

Table 3.3 IT-Specific Threats

Threat To...	Specific Threats
Hardware	Equipment failure (intentional, unintentional damage)
	Power outage
	Equipment reconfiguration (authorized, nonauthorized)
	Equipment sabotage
	Equipment theft
Software	Bugs, glitches
	Data corruption
	Data security breach (deleted, stolen, modified)
	System configuration changes (errors or sabotage)
Infrastructure	Internet connection(s)—failure, tampering, destruction
	Wireless networks—failure, tampering, destruction
	Network backbone—failure, tampering, destruction
	Cabling—failure, tampering, destruction
	Routers, infrastructure hardware—failure, tampering, destruction

Figure 3.5 shows the output from this phase of the risk assessment, which is a document listing all potential threats and threat sources you have looked at for your company. Although you may be able to skip over a few that clearly don't apply, your list should be inclusive rather than exclusive at this junction. This document will be used as the input for the vulnerability assessment phase, discussed later in this chapter. At the end of the entire risk assessment phase, you'll have a more streamlined list of threats that you'll use in your business impact analysis, discussed in Chapter 4.

Figure 3.5 Deliverable from Threat Source Assessment

Table 3.4 provides a sample of how you could organize your threat data so that you're ready to move into subsequent phases. Regardless of how you organize your data, be sure you capture it in a consistent and logical manner.

Table 3.4 Risk Assessment Table

Item No	Threat Name	Threat Source	Vulnerability Rating	Likelihood Rating	Existing Controls	Impact Rating	Overall Risk Rating
001	Fire	Internal					
002		External					
003	Flood	Internal					

This matrix shows Fire as a threat and then delineates the threat source as *Internal* or *External*. If there is a fire in the building, you may evacuate. If there is a fire in the area making leaving the area difficult or dangerous, your best solution might be to shelter-in-place. This is an example of how a single threat (fire) has two different sources and how the vulnerability, likelihood, impact, and mitigation strategies might differ for each. Therefore, to list "fire" without listing the threat sources, you might miss something in your assessments. You may choose to create additional columns or include additional details. For example, an internal source of fire could be limited to the server room or could be elsewhere in the building. Is it useful for you to make that distinction? If so, add that detail. If not, don't add unnecessary detail. Also in this table, we've included a column labeled Existing Controls, which can be used to list controls or measures that are already in place. For example, if your server room has a state-of-the-art fire suppression system, you can list that as a control. If your building has a fire suppression system and you practice fire evacuation drills, you can list that as a control. These are things that are already in place that are mitigating your risk. In some cases, these controls may be sufficient; in other cases, you'll need to add layers of control to bring the risk down to an acceptable level. By listing controls already in place, you can spend less time on risks that are already addressed and more time on those that are not addressed in an effective manner. As you go through this assessment, it can be helpful to list these kinds of items or to add/delete columns as needed. Creating the right data fields now will make your work easier later on because this matrix will help you capture information as it comes up. The goal is to find a balance between too much and too little detail.

If you have a preferred method for approaching this that will account for your threats and threat sources adequately, feel free to use it. The end result is to have a comprehensive assessment from which you can build a plan without getting bogged down in useless detail. One final word: You may start out with more detail than you need and pare down as you see how your planning is progressing. Sometimes when you have a bit of perspective on the topic, you can see more clearly what *is* and *is not* needed. Err initially on the side of inclusion, pare down later.

Threat Assessment Methodology

Before we head into the vulnerability assessment phase, we're going to discuss threat assessment methodologies that might be useful to you in evaluating various threats. In essence, there are two ways you can approach this. The first is to use a *quantitative* approach, in which you attempt to use hard numbers to represent threats, vulnerabilities, and impacts. In some companies, this may be the norm or it may be required for some reason. The second method is a *qualitative* approach, where you attempt to define the relative threats, vulnerabilities, and impacts. You use qualitative, or value-based language such as "high," "medium," and "low." We'll look at both methods and provide a few samples so you can determine which approach would be best for your BC/DR team. Keep in mind that you should pick one approach and stick with it; mixing and matching can result in unclear or meaningless data. Once you've read through this section, you should have a good idea of which approach fits best with the culture and requirements of your company.

The reason we're discussing these two different approaches is because if you're fighting an uphill battle in your company with regard to the cost or benefit of this BC/DR plan, you may need to come up with quantitative assessments to sway decision-makers. They may not be convinced by statements like "more" or "extremely high"; they might be convinced by "20% chance" or "$250,000 loss." If you have support for your BC/DR plan, you may opt for the qualitative approach, which is less precise but faster and easier to derive.

Quantitative Threat Assessment

A quantitative assessment can be defined as observations that involve measurements and numbers. They are specific and measurable. If you say, "The server costs $850 more than the desktop system," you are making a quantitative statement. In contrast, if you say, "The server is more expensive than the desktop system," you are making a qualitative assessment since "more" is not specific and measurable.

Let's start with the threat of a power outage. We can look at the possible threat sources—what could cause a power outage? As you learned in the previous sections, there are numerous potential sources of a power outage. Let's begin with an electrical storm with lightning that causes a localized power outage. If your building is susceptible to power outages in storms and those outages do not affect neighboring buildings, you may start with a power outage that impacts just your building. Again, this is where a thorough threat assessment is helpful. You can gather actual data from the National Weather Service or other reliable sources on the number of storms per year in your area, on average. If you want to know more about local weather conditions and history, you might be able to get some valuable information from your local news station's meteorologist. Sometimes they are more than willing to share information with you and they might be able to point you to resources you'd otherwise miss. You can also gather data from the power company on the number of times power has gone out in your building or your area over the past five years or on average. Once you have that data, you have very specific, quantifiable information. When looking for quantitative data on any of your threats and threat sources, you may need to be creative.

From the Trenches…

Finding Your Data Weasels

In every company, there are always a few people (sometimes more) who relish the challenge of a good data search. At one company, there was a young man whom people had affectionately nicknamed "the data weasel" because he could "ferret" out just about anything—there was not a piece of information he could not eventually retrieve (though his typical turnaround time was less than 30 minutes). There may be people in your company who participate in "Google races" to see who can find a specific piece of information the fastest (think of the TV show *Who Wants to Be a Millionaire?*—who in your company would make a great "Phone-A-Friend"?). If you need data on the number of storms in your area that have dumped more than 10 inches of snow in the past 10 years, or number, frequency, and duration of local power outages, turn to your company's data weasels. Give them specific data to search for

Continued

and a deadline for completion and leave them to it. Be sure to ask them to capture the source of their data so you can be sure it's credible before relying upon it. By leveraging the natural skills and interests of people in your company or on your team, you can effectively delegate tasks to others who will enjoy the challenge and will produce great results.

What if you can't easily find that data? You can gather anecdotal evidence, though this by its very nature is much more qualitative than quantitative. You can probably talk to your facilities person or staff who have worked at the company for several years and get their input. Will they know exactly how many storms with lightning have come through the area? No. Will they have an idea of how many times power has gone out in the building or the area? Probably. They may not have an exact number but you may well get a response like, "It seems to happen every year or two" or "I can't remember the last time that happened and I've been here five years."

To create a quantitative assessment, we need to make sure we're comparing "apples to apples," so all numbers will be converted to annual numbers. For example, if you have a power outage every other year, the annual power outage would be 0.5 chance of an outage. If an outage occurs once every four years, you have a 25% or 0.25 chance per year because the risk is only 1 in 4 that you'll have a power outage in any given year. The numbers should all be annualized so that comparisons are accurate.

Let's look at a risk diagram, shown in Figure 3.6. Remember that we'll quantify some of these other numbers through our assessments later in this chapter and in upcoming chapters. We'll quantify these in this section so you can see how the process works and then you can develop the remainder of the needed input values later as you develop the data. In other words, you can do your likelihood assessment later and input the values later but we'll review the entire model here so you can see the road ahead.

Figure 3.6 Risk Assessment Methodology—Quantitative

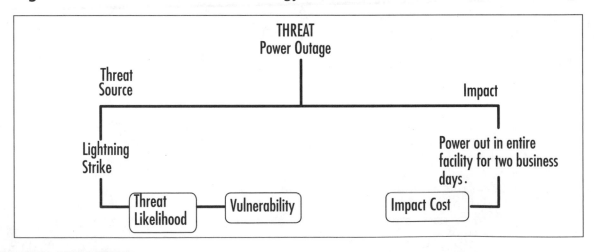

On the left side of the diagram, you can see one threat source listed. Ideally, each threat source for a power outage should be addressed in this manner. In this case, the threat source we're looking at is a lightning strike. First, we assess the likelihood of occurrence. If your data indicates this happens once every four years, then you would enter the value of 0.25 under the threat likelihood since we want to know the likelihood in any given year. Next, you would assess your vulnerability. In this case, let's say that every time there's a serious lightning strike (once every four years), your power goes out. That means that when lightning strikes, your power will go out. That's a 1:1 ratio, or 100%, so we'll insert the number 1. Now, on the threat source of this equation, we have:

0.25 x 1 = 0.25

This is sometimes referred to as the *risk value*. We're keeping the example simple so we can walk through the logic of it. In English, this equation says that although there's only a 25% chance the threat will occur, there's a 100% chance it will affect us when it occurs.

Next, we move to the other side of the diagram and look at the impact. In this example, we're assuming that when the power goes out, it's out in the entire facility and it's out for two days. (Of course, if your building loses power every four years for two days, you should be having a chat with your power company.) What is the cost of not being able to work for two days? Your servers are down, your employees cannot access desktops, printers, or IT resources, there are no lights, no heating/cooling—no productivity. How many sales do you lose? What impact does this have on your customers? Your inventory? Your order backlog? This is where having your team comes into play as members from different organizations can help you make these assessments. This is some of what we'll cover in the next chapter, but for now, let's assume the following:

- Lost sales per day: $18,000, total cost $36,000 for the outage

- Fixed costs per day: $4,200, total cost $8,400 for the outage

- Damage to reputation: unspecified, arbitrary value set at $2,000, total cost: $4,000

Note that you may decide to set a value for damage to reputation at a daily rate just to give you some measurement that can be used consistently. If you have a method for calculating this to a more exact figure, feel free to use it. Otherwise, a daily rate for these unspecified costs may help by providing an "order of magnitude" estimate. In this case, we're using $2,000 per day and this amount will be used for any "damage to reputation" suffered from any threat source. Therefore, we'll be able to understand the impact of a one-day event versus a month-long event. Though there is a multiplier effect that occurs with an extended period of downtime or outage and you may choose to address this, we're using a single value. We also recognize that the use of an arbitrarily set value, such as $2,000 per day for "damage to reputation," is qualitative in nature. Still, assigning a dollar value to it and using it consistently across all threats will mitigate that to some extent. Now, let's calculate our impact costs:

$36,000 + \$8,400 + \$4,000 = \$48,400$

So now we know that if this threat occurs, it will cost the company \$48,400. Now, let's input that into our earlier equation as shown here and reflected in Figure 3.7.

$0.25\% \times \$48,400 = \$12,100$

Figure 3.7 Total Risk Cost per Year of Power Outage from Lighting Strike

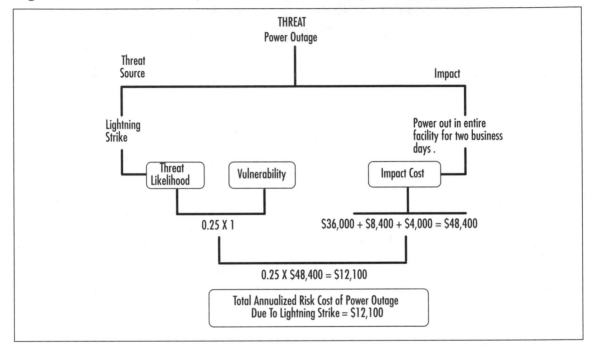

If you know that your annualized risk cost is \$12,100 for a power outage from a lightning strike, it's much easier to determine whether a \$10,000 backup power generator makes sense. You may also decide that it's worth investing \$5,000 to have the power company install equipment that will make your power system less likely to fail. These are your risk mitigation strategies that we'll develop in Chapter 5. You can see that having an annualized value can certainly help you and your team come up with a number of reasonable risk mitigation strategies. Clearly, a solution that costs \$100,000 is probably not a good investment because it would take you about eight years to recoup your investment (\$100,000 / \$12,100 = 8.26). How cost effective is additional equipment for \$5,000 to make the problem go away? Probably very effective and as such, an excellent investment. We use terms like "probably" in this case because there may be mitigating circumstances we don't know about. For example, suppose that \$5,000 solution costs \$1,000 annually to maintain. Is it still a good deal? What if it lasts for only four years? Is it still a good solution? Without knowing all the details, it's hard to make a solid assessment. This is part of what's covered later in Chapter 5. For now, we'll use straight numbers and assume that a $5,000 solution is just that—and as such, it would make a lot of sense.

The second thing to consider when looking at potential mitigation strategies is that there may be additional *benefits* to a particular solution. If, for example, the $100,000 solution would also extend the serviceable life of all computers in the building by 1.37 years, it might be a better solution. You'd need to calculate the value of extending all computers' serviceable life by 1.37 years and comparing that to the cost of the solution. Another possibility is that the $100,000 solution meets industry requirements that will be enforced starting in three years; the $5,000 solution does not meet requirements. Now what's your best option? This is certainly a place where getting your finance folks into the loop will help; they can develop what-if scenarios for your different options and you, as the IT expert, can help everyone understand the additional benefits (or risks) that come with various solutions. Some solutions you'll consider may inject a new risk into the mix; some solutions will mitigate risks in areas you hadn't expected. Keeping your eyes open for possibilities will help you maximize your results.

Will calculating your values be this easy? Probably not. You most likely have far more factors to consider when calculating the cost of an outage, for example. However, you can also decide as a team what degree of accuracy you require in order to create an effective plan. If exact numbers are not required, you can use a qualitative model.

Qualitative Threat Assessment

Qualitative assessments use words or relative values to express risk, cost, and impact. The first step in using a qualitative system is to define the scale you want to use and then use it consistently. You can use systems like those shown in Table 3.5 or Table 3.6, or you can develop a customized scale to fit your needs.

Table 3.5 Qualitative Scale Examples

Numeric	Frequency	Impact
6	Constant	Extremely high
5	Very frequently	Very high
4	Frequently	High
3	Infrequently	Low
2	Very infrequently	Very low
1	Never	Extremely low

One suggestion is that you use a scale with an even number of variables; the one we used has six. This forces a choice between two options, "frequently" or "infrequently" or "high" or "low," and can prevent someone from selecting the middle value (present when there are an odd number of choices) to be safe. Whatever scale you use or whatever number of variables you opt for, be sure to define these elements to everyone's satisfaction. It's important to have a shared understanding of what these values mean so that when you're using them for the risk assessment, you're all using them in the same manner.

When assessing likelihood, you can define a scale that works for your organization. Table 3.6 shows the likelihood matrix developed by the National Institute of Standards and Technology. This matrix is specific to security risk vulnerabilities but provides a good example of how to define these types of qualitative assessments.

Table 3.6 NIST Likelihood Matrix

Likelihood Level	Description
High	The threat-source is highly motivated and sufficiently capable, and controls to prevent the vulnerability from being exercised are ineffective.
Medium	The threat-source is motivated and capable, but controls are in place that may impede successful exercise of the vulnerability.
Low	The threat-source lacks motivation or capability, or controls are in place to prevent, or at least significantly impede, the vulnerability from being exercised.

Source: National Institute of Standards and Technology, "Risk Management Guide for Information Technology Systems," Special Publication 800-30, July 2002, p. 21.

Now, let's look at the same example we looked at previously only this time, let's use the qualitative method. First, we map out the threat, as shown in Figure 3.6 earlier and repeated here in Figure 3.8 for your convenience.

Figure 3.8 Power Outage Threat Assessment—Qualitative

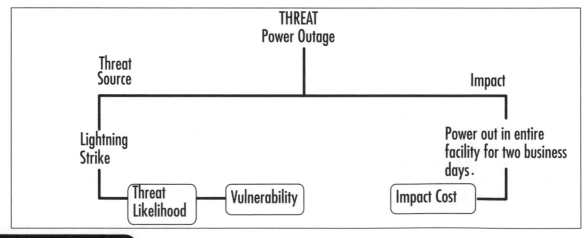

Now let's assign values. Let's say we know that these outages happen once every four years. We might determine that deserves a rating of "infrequently" and we can assign it the value of 3. Using the same system, we can say that the vulnerability when the storm hits is 100% (per our quantitative assessment), which would place it on the scale as "extremely high" and give it a rating of 6. So, the left side of our equation = 2, 6.

On the right side, we want to assess the impact cost, but we're not using exact dollar amounts. We could say, well the cost of being down two days would be about average because we can catch up later without too much trouble and our fixed costs aren't through the roof. Therefore, you might assess your impact cost as being "low" or a level 3. If you take the average of these, you have 2 + 6 + 3 = 11 / 3 = 3.66. This puts it on the scale at "high" if we round up (any number above 3.5 would be 4, any number 3.5 and below would be 3). This is depicted in Figure 3.9.

Figure 3.9 Total Risk Value per Year of Power Outage from Lighting Strike

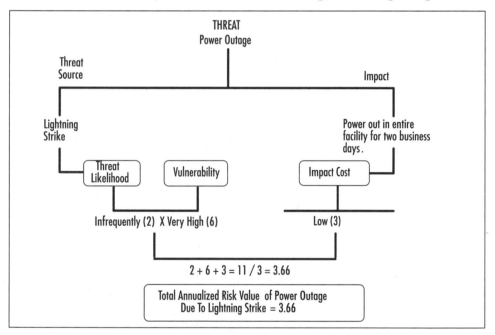

You might decide you don't like converting these assessments to numbers—that's fine. You might also decide you want a scale with a few more options, say a 10 item scale—that's fine, too. The point here is that you can make assessments without hard dollar figures and still come up with a meaningful assessment. In the case of the power outage, you might argue that the value of 6 for "very high" under vulnerability skews this data in a way you don't like because it's not weighted, for example. However, when you do this assessment using this scale for a number of threat sources, you may find that your data shakes out as expected. For example, you might perform this same assessment on a power outage from an internal failure and decide its total risk value is 3.5. You can then look at these two sources and ask, "Do we really have a slightly greater risk value if we experience a two-day power outage every four years versus our internal power failure that could take us down for a week but only happens once every eight years?" If the answer is no, you may want to go back and better define your scale or reassess the values you used in one or the other assessment. However, in most cases what you'll find is that after a few of these, you get the feel for the scale and you begin to see that your data tracks with the reality of the situation. Once you're confident your scale is working, you can tackle the more difficult or more intangible threat sources.

Another rating scale could range from 1 to 100 to give you a bit more fine-tuned result. An example of this is shown in Figure 3.10. If you really want to keep it simple, you can use a five-element, single-rating system and come up with something similar to that shown in Figure 3.11. In Figure 3.10 and Figure 3.11, the costs are delineated in terms of the relative impact cost of 1) loss of revenue, 2) damage to servers, 3) damage to the database, and 4) damage to user computers. These two examples assume that the servers were able to shut down without incident but that there was damage to a database as a result of the sudden loss of power. This is just an example to show you how you might assess your IT components. You might also choose to delineate things like firewalls, routers, and cabling in your list, if it's helpful in making a qualitative assessment.

Figure 3.10 More Refined Qualitative Scale

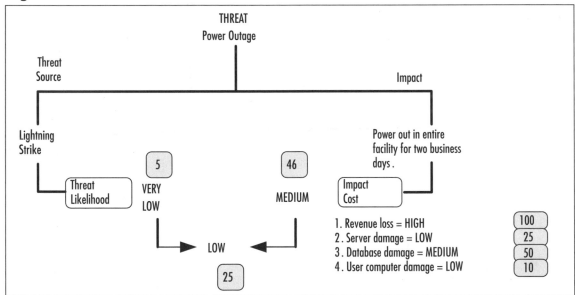

Figure 3.11 Simple Qualitative Scale

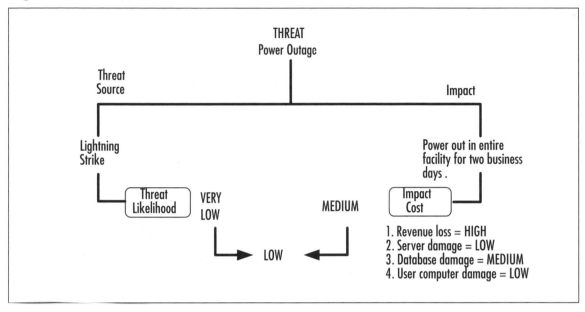

Whether you choose to use a quantitative system or a qualitative system, be sure everything is clearly defined and that you apply these ratings consistently. What you'll end up with at the end of your risk assessment phase is a chart, table, or document delineating each threat, the likelihood of that threat, the vulnerability to that threat, and the impact should that threat occur. From there, you'll develop your risk mitigation strategies because you'll be able to see the big picture and create optimal solutions for your firm.

Vulnerability Assessment

A *vulnerability* is defined as the weakness, susceptibility, or exposure to hazards or threats. A vulnerability in a software program, for example, is a weakness that poses a problem if discovered and exploited. Vulnerabilities in the case of business continuity and disaster recovery are the various areas of the business and IT systems that are exposed or susceptible to the threats defined in the previous assessment phase. Vulnerabilities can be exploited intentionally or triggered unintentionally. As you know, a change to a security setting in one area of the operating system can create a vulnerability elsewhere in the system without the IT administrator even being aware of it. The analysis of vulnerabilities in BC/DR planning must include IT systems, but it should not be limited to IT systems. Clearly, people, processes, technology, and infrastructure are vulnerable to the threats delineated earlier. Therefore, while our focus will continue to be IT-related data, we have to cast a wider net so that the BC/DR plan is complete.

The result of the threat assessment becomes the input to the vulnerability assessment, as shown in Figure 3.12, which is the second section of the larger image presented at the beginning of the chapter in Figure 3.2.

Figure 3.12 Vulnerability Assessment Phase of Overall Risk Assessment

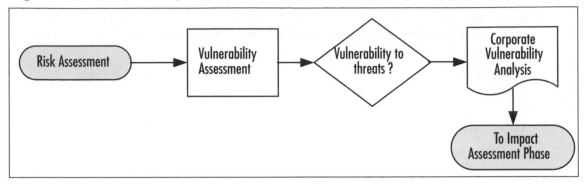

Some people like to break out this assessment into likelihood of occurrence and vulnerability to the threat based on the likelihood. Others prefer to keep it simple and use a straightforward vulnerability assessment. The value in breaking it out, as shown in the previous section, is that the likelihood of something occurring may be high but the vulnerability to that threat source may be low. Conversely, you could say something is unlikely to occur but if it does, you are very vulnerable, as would be the case with a major earthquake or hurricane. Thus, the value in breaking it out into likelihood and vulnerability might be that you can address those issues unlikely to occur but to which you're very vulnerable as a separate element of your BC/DR planning. Another approach is to simply look at the vulnerability as the likelihood of occurrence and then assess the potential impact through the business impact analysis.

Once you have your threat sources listed in detail, you may choose to subdivide your list and assign segments to appropriate subject matter experts. For example, you could give the entire list to four subteams, each specializing in a particular area such as IT, facilities, finance, HR, and/or operations. You could also create subteams to look at each threat source from the people, process, technology, and infrastructure framework. You could divide your team in two and have half address the internal threat sources and the other half address the external threat sources. However you subdivide the work, you should hold a final full team review to ensure there are no gaps. If time and resources allow, you could have each group's deliverable be handed to another subgroup so that each group's work is reviewed by another group (i.e., Group A's results are reviewed by Group B; Group B's results are reviewed by Group C; Group C's results are reviewed by Group A). The point is not to chastise another group for errors or omissions, but simply to make sure every angle has been considered and reviewed appropriately.

> **NOTE**
>
> The Business Continuity Institute (www.thebci.org) provides certification in business continuity planning. In conjunction with the Disaster Recovery Institute (www.drii.org), they have developed and published a set of guidelines that describe best practices. You can download the PDF file using this link: www.thebci.org/10Standards.pdf.

People, Process, Technology, and Infrastructure

We've continually pointed out that BC/DR planning activities require research into the impact on people, processes, and technology. In the case of BC/DR plans, infrastructure might be a fourth category that could be included (some might include it in technology). Infrastructure in this case includes the building itself, heating/cooling systems, power to the building, among others. When you're assessing your business risks, these four areas come into play. When you begin your vulnerability assessment, these four areas should also be considered. If you've done a thorough job in your threat assessment, these areas will be covered, but a quick reminder to look at your plan from this perspective at this point will help ensure there are no gaps.

People

When performing a vulnerability assessment, you need to ask and answer the question: How vulnerable are our staff and the people in our community to these threats? Some threats may not impact people beyond their ability to be productive at work (a one-hour power outage, for example). Other threats may not only impact your staff but the surrounding community, as is the case in major natural disasters. If you look at how vulnerable people are to the various threats, you can determine your overall risk with each threat source. For example, people are particularly vulnerable to phishing and social engineering. This is not a systems vulnerability—there is little a system can do to stop a person from deciding to respond to a phishing or social engineering ruse. So, if someone willingly hands over a "power user" account name and password, the system is vulnerable, but only because a person was vulnerable first. Looking at threats and vulnerabilities in this light will help not only in determining the overall risk value of each threat source but in developing effective mitigation strategies later.

Process

How vulnerable are your business and IT processes to these various threat sources? In some cases, your processes may not be very vulnerable at all, as might be the case in a brief power or server outage. You already have processes in place for normal IT operations and minor outages and equipment failures are probably covered in your standard operating procedures. Other processes might be very vulnerable, such as is the case when a natural disaster occurs. In those cases, it's typical that all business and IT processes are vulnerable because there is nothing about a disaster that is "business as usual." For example, how would you handle, process, and fulfill customer orders after a disaster that made your building uninhabitable for weeks? What could you do to get back up and running? How would your processes have to flex or change? As you review the vulnerability of each of your critical business processes to the various threat sources, you'll begin to see which processes need to be reviewed, revised, or reinvented for use in an emergency. We'll discuss this in the business impact analysis as well as in the mitigation strategy development chapters.

Technology

Clearly, technology is vulnerable to numerous threat sources and as an IT professional, you're aware of most (if not all) of them. How vulnerable is your server to an internal or external attack? How vulnerable is your web server? These are questions you've probably already addressed through standard IT security assessments and operating procedures. As you go through this particular risk assessment process, you also need to broaden your outlook a bit and ask how vulnerable your systems are to the disaster threat sources such as floods, hurricanes, and fires. Since your perspective is an IT-centric perspective, you probably have the most detailed information available on this subject. As you go through the vulnerability assessment, this data should be captured. Don't assume that your standard operating procedures have addressed the vulnerabilities and don't assume that your current "emergency plans" will be adequate for all threat sources. Approach this topic with fresh eyes to see what else you can add to the process.

Infrastructure

Clearly, infrastructure is vulnerable to some threat sources and not to others. A building is vulnerable to flooding if it's in a low lying area or in a location that could flood. If the building is at the top of a hill overlooking the town, there's a good chance it is not vulnerable to external flooding, but any building could potentially be vulnerable to internal flooding (broken plumbing within the facility). As you review your threat sources, your facilities expert will likely be in the best position to understand the vulnerability of the company's infrastructure to threat sources. As a team, you may all want to think about the vulnerability of external infrastructure in your area to threat sources since these will clearly impact your business. For example, is there a seaport, power plant, airport, or dam nearby that could be vulnerable to the threat sources? If so, what impact would these have on your business?

Vulnerability Assessment

A vulnerability assessment can be qualitative or quantitative, but in many cases a qualitative assessment is used. It's difficult to put a hard number on a vulnerability, so using a rating scale such as those shown in Table 3.5 is usually most effective. The key is to get an accurate picture of the vulnerability to each threat source. When viewed in total, you'll be able to make any needed adjustments to individual vulnerability ratings. For example, you might use a scale of 1 to 100, with 100 being most vulnerable and 1 being least vulnerable. When you view your final list of threat sources and vulnerabilities, you might see that some of the vulnerability ratings are out of sync with the rest—those can be modified so that your overall vulnerability picture is accurate. As with other rating systems, be sure that you define it and then use it consistently. Also, because you may choose to subdivide the work, it's vital that everyone have the same understanding of the ratings and apply them in the same manner.

A vulnerability assessment typically uses various data sources as input. These include prior risk assessments, security requirements, security test results, regulatory requirements (HIPAA, GLBA, etc.), and prior problems. According to the National Institute of Standards and Technology's "Risk Assessment Guide for Information Technology Systems," the types of vulnerabilities that exist and the methodologies needed to determine vulnerabilities will vary depending on the nature of the system and in particular, the phase of the SDLC it is in. Accordingly,

- If the system has not been designed yet, the vulnerabilities assessment should focus on the organization's security policies, planned security procedures, and system requirements definitions, and vendor's (or developer's) security product analyses (white papers, etc.).

- If the system is in the process of being implemented, the vulnerabilities assessment should focus on more specific information such as the planned security features, security and design documentation, and the results of system certification, testing, staging, and evaluation.

- If the system has been implemented and is operational, the vulnerabilities assessment should include the analysis of the system's security features, security controls (technical, operational, and environmental), and standard IT operating procedures.

However, we are looking beyond just IT systems to the larger organization, so we need to expand our view of vulnerabilities just a bit. Even though these other areas may fall outside your direct line of authority, you should be familiar with them so you can participate fully on the BC/DR planning team or head it up effectively, whichever the case.

The vulnerability assessment can be accomplished using the same methods described earlier in the threat assessment: questionnaires, interviews, document reviews, and research. In addition, you can develop scenario questions based on the identified threat sources to help you assess vulnerability. For example, you might ask your subject matter experts to respond to a set of questions similar to those shown in Table 3.7.

Table 3.7 Vulnerability Assessment Questions

Statement	High	Medium	Low
1. If the plumbing pipes in the building were to burst, what is the vulnerability of our IT systems to water damage?			
2. If the building were to catch on fire, what is the vulnerability of our IT systems to water damage from fire suppression systems?			
3. If the building were to become flooded by heavy rains, what is the vulnerability of our IT systems to water damage?			

As you can see from these sample statements, we've identified three threat sources for water damage/flooding: internal flooding, water damage from fire suppression systems (which might not be in the server room but adjacent to or above the server room), and external flooding. We have also not asked (here) what the likelihood is of these threats occurring. We are simply assessing how vulnerable we believe these systems would be should these events take place.

Once the vulnerability assessment is complete, you can develop a risk value for each of the threat sources. This risk value can be derived numerically if you've used either a quantitative system or a qualitative system that uses numbers for the scale. Once the vulnerability assessment is complete, you should analyze the data. This analysis should include a thorough review of all the threat sources, likelihood, and vulnerabilities ratings. A final assessment of the data should allow you to adjust ratings that seem out of balance with other data, which is sometimes the case with qualitative assessments. The interim "risk value" for each threat source should be reviewed and verified. The value is considered interim at this point because we have not yet conducted the impact analysis. Therefore, you will have a "risk value subtotal" in a sense, which can be reviewed at this juncture. If you recall, the risk equation can be stated as:

Risk = Threat + (Likelihood + Vulnerability) + Impact

Rather than repeat the material presented in the previous section, we'll leave it to you to go back through your threat source list and perform the vulnerability assessment. The result of this phase is a document that lists, at minimum:

1. All potential threat sources (except those purposely excluded).
2. The likelihood of each threat source occurring.
3. The vulnerability of your company and IT systems to those threat sources.
4. Interim risk value for each threat source.

The deliverable from this phase is the vulnerability assessment and analysis, as shown in Figure 3.13. This data is used as the input to business impact analysis phase, covered in detail in the next chapter. You might have a team meeting to review the final data and present a report to your project sponsor and your corporate executives at this juncture. This can help bring visibility to your efforts and should underscore the need for the BC/DR planning project itself. The document can also be provided to various subject matter experts within the firm for one final review to ensure there are no gaps or errors at this point in the process. If you have a formal sign-off procedure in place, you may want to obtain formal approval for this document before moving onto the next phase of your project plan.

Figure 3.13 Deliverable from Vulnerability Assessment

Looking Ahead...

Business Impact Analysis

The next step in the risk assessment is to perform the business impact analysis (BIA). It is in this phase that you will look at the company's business processes (including those associated with IT functions) and develop a rating or assessment of the criticality of those systems. Then, you can determine which business functions must be restored and in what order. Clearly, in the aftermath of a disaster or business disruption, functionality must be restored in a methodical and logical manner and the BIA will provide that roadmap. The input to the BIA is the output from this phase of the assessment, so don't launch your BIA until you've completed this phase to your satisfaction.

Summary

Business continuity and disaster recovery planning begins with a thorough risk assessment. Risk assessment is part of a larger risk management process found in most businesses. The four major components of the BC/DR risk assessment are threat assessment, vulnerability assessment, impact assessment, and risk mitigation strategy development. In this chapter, we focused on threat and vulnerability assessment. In order to perform a thorough threat assessment, you need to look at threats and threat sources both internal and external to the company. It is often helpful to assess risk based on the potential risks to people, process, technology, and infrastructure. People are not only the company's employees but its vendors, partners, customers, and the larger community in which it operates. Processes are all the business and IT processes used in the business. Processes are used to generate revenue, track expenses, and manage operations from facilities management to human resources and beyond.

From an IT-centric viewpoint, the key components to address in the risk assessment include hardware, software (OS and applications), system interfaces (internal, external connection points), people who support the IT systems, users who use the IT systems, data, information and records, processes performed by the IT systems, system's value or importance to the organization (system criticality), and system and data sensitivity (confidential, trade secret, medical data, etc.). The operating environment in which the IT systems function include a wide variety of elements. Among them are the functional and technical requirements of the systems; security policies, procedures, and controls; network topology and information flow diagrams, data storage protection policies, procedures, and controls; encryption, physical, and environmental controls.

The methods used to gather data for any of the assessment phases typically includes questionnaires, interviews, document reviews, and research. Questionnaires can be helpful in structuring desired input but also can have the downside of containing built-in biases, often unintentionally. Interviews can be conducted with subject matter experts and yield more useful information than questionnaires but may also generate a lot of tangential or unneeded data. Reviewing documents and performing research can supplement the questionnaire and interview process.

Once you've defined the methods you'll use to gather the necessary data, you can begin your review of various threats. We discussed many different types of threats that fall into three primary categories: natural and environmental threats, human-caused threats, and infrastructure threats. Infrastructure threats are caused either by natural or human causes, and it's important to delineate these because they involve people, processes, and technologies outside of the company and the company's control. As such, they sometimes can be overlooked in BC/DR planning. Natural threats include those we might commonly think of such as fire, flood, or earthquake, but we also discussed other less obvious threats including volcanoes, droughts, and pandemics. Human-caused threats can be intentional as in the case of terrorism, labor disputes, or workplace violence, or they can be unintentional as can be the case with fire, flood, or a security breach. Infrastructure threats include those to the building as well as external to the building and the company. Public transportation including roads, rails, seaports, and airports are all external infrastructure elements that need to be assessed. Other external elements include threats to water and food supplies, biological and chemical hazards, and public utilities such as the power grid, petroleum and fuel supplies, or telecommunications.

The threat assessment methodology begins with a list of all potential threats and threat sources. Each threat source is then evaluated. Some people like to assess likelihood of occurrence and vulnerability to the threat; others prefer to include both likelihood and vulnerability in a single assessment. Regardless of whether you choose to break them into two distinct ratings or one rating, the likelihood of occurrence and vulnerability rating(s) should be assessed for each threat source. The argument for making two separate assessments is that a threat may have a high likelihood of occurring but your company and its people, processes, technology, and infrastructure may not be vulnerable to those threat sources. Others would argue that if there is a low vulnerability, the likelihood of occurrence doesn't come into play and should therefore not be assessed separately. Either method is acceptable as long as you make a conscious decision as to how to proceed and use the same process throughout your risk assessment cycle.

You can perform a quantitative assessment in which actual values such as dollars or frequency are known. The benefit to this type of assessment is that you can generate hard data that can be used in a standard cost/benefit analysis. The downside is that not all values are easy (or possible) to derive and an unacceptable amount of time or money may be required to generate that data. You can also perform a qualitative assessment in which values are relative. These types of assessments use labels such as high, medium, and low or an arbitrary numbering system such as 1 to 100 where 1 = no chance or extremely low, 50 = medium chance or about average, and 100 = will occur or extremely high chance. These kinds of systems are much easier to implement but typically generate less specific data that is unsuitable for a standard cost/benefit analysis. In analyzing threat data, qualitative measurements are often sufficient to generate a clear picture of the threats facing the organization.

The vulnerability assessment may include the likelihood of occurrence or it may be a separate rating. However, the same processes can be used to evaluate vulnerability as were used to assess threats. Questionnaires, interviews, document reviews, and research can help in generating data needed to assess the actual or relative vulnerability to a threat. This rating is compiled with the threat assessment data and is used as the input to the business impact analysis phase, discussed in the next chapter.

The bottom line is that your risk assessment activities will end up generating a list of threats and threat sources that you'll be able to evaluate. You can sort the list and decide which risks you need to address, which can be accepted, and which should be transferred.

You'll have the data to make this decision once you complete the business impact analysis, the third major step in the risk assessment process.

Solutions Fast Track

Risk Management Basics

☑ Risk management is a business process used to manage all kinds of risks facing business today.

☑ A standard risk management process includes four phases: threat assessment, vulnerability assessment, impact assessment, and risk mitigation strategy development.

☑ When assessing threats for a business continuity and disaster recovery plan, you can use the framework of people, process, technology, and infrastructure to ensure you're looking at all aspects of your business.

☑ IT-specific risk management is related to three objectives: securing systems more fully, enabling management to make sound IT purchasing decisions, and enabling management to authorize/accredit IT systems.

☑ IT risk management often uses the framework of the system development lifecycle (SDLC) model. As you perform your BC/DR assessments, systems being considered, developed, and implemented must be assessed. Some BC/DR risk controls may already be in place, others can be incorporated in the SDLC process.

☑ Risk can be expressed as an equation: Risk = Threat + (Likelihood + Vulnerability) + Impact.

Risk Assessment Components

☑ Risk assessment components include threat, vulnerability, impact, and mitigation. In this chapter, we focused on threat and vulnerability assessments.

☑ Threats typically are categorized as natural/environmental in nature or human-caused. Threats to the infrastructure, caused by both natural and human actions, are delineated separately in order to ensure they are adequately addressed.

☑ Natural and environmental threats to the business must be assessed not only in terms of how they directly impact the company, but also how they indirectly impact the company. Disasters or business disruptions to your business customers, partners, and vendors can have a major impact on your business and must be assessed.

☑ Human-caused events include acts of terror, theft, and sabotage, but they also include things you might not consider such as labor disputes or workplace violence.

☑ Threats to infrastructure are typically outside your direct control, but they often have a direct (or indirect) impact your business. These include damage to airports, seaports, highways, and rail stations as well as problems with the delivery of utilities.

☑ Though IT-specific threats are caused either by natural events or human actions, we listed them separately in order to delineate the various threats to be considered in an IT-centric plan.

Threat Assessment Methodology

☑ Four major types of tools can be used to assess threats: questionnaires, interviews, document reviews, and research. Each will yield specific data. A comprehensive review will use all four methods and combine data.

☑ Questionnaires can have built-in biases that may limit the information gathered.

☑ Interviews can avoid the built-in bias but can generate tangential data that either gets you off track or simply is not helpful in your BC/DR planning.

☑ Document reviews and research should be a part of the threat assessment process to review what is already known and to gather statistical and factual data pertinent to your BC/DR plan.

☑ Threats can be assessed using quantitative or qualitative assessments. A quantitative assessment uses hard numbers that can be used in a cost/benefit analysis. A qualitative assessment uses arbitrary numeric values or labels such as high, medium, and low to assign relative value. Although this cannot as easily be used in a cost/benefit analysis (if at all), it is often easier to derive these types of values.

☑ Threats should be assessed with regard to the various threat sources, the likelihood of occurrence and the vulnerability of an asset to the threat source.

☑ Some people may prefer to assess threat with likelihood and vulnerability assessed as two separate values; others may prefer to assess these two traits as one value. Either method is acceptable as long as you consider both the likelihood of an occurrence and the vulnerability of an asset to that occurrence.

Vulnerability Assessment

☑ The vulnerability assessment uses the output of the threat assessment phase as its input. The complete list of threats and threat sources is evaluated with an eye toward the likelihood of such an occurrence and the vulnerability of corporate assets to those threats.

☑ Corporate assets and operations may be highly vulnerable to an event that may be unlikely to occur, such as an earthquake or volcanic eruption. As these factors are evaluated, you can create an overall risk value for each threat source.

☑ The data from the vulnerability assessment is analyzed and is used as input to the business impact assessment phase, discussed in the next chapter.

☑ The key at this juncture is to be inclusive and not rule out anything until the assessments are complete. This helps prevent inadvertently creating gaps in your BC/DR plan.

Frequently Asked Questions

The following Frequently Asked Questions, answered by the authors of this book, are designed to both measure your understanding of the concepts presented in this chapter and to assist you with real-life implementation of these concepts. To have your questions about this chapter answered by the author, browse to **www.syngress.com/solutions** and click on the **"Ask the Author"** form.

Q: Some events like volcanoes and earthquakes are so enormous, it doesn't seem to make sense for us to evaluate and plan for these kinds of events. Any comments on that?

A: Major events are overwhelming for everyone, even trained emergency response teams. However, in order to survive such events, some amount of planning is needed. If you live in an area prone to earthquakes, you are pretty likely to have an "earthquake readiness kit" on hand—battery operated radio, extra batteries, water, food, a blanket, and a flashlight, for example. Even though there's a small chance of an earthquake, you've prepared the best you can. The same holds true for business. If an earthquake decimates your place of business, your employees' first concerns will be for their own welfare and the welfare of family and friends. Once those concerns are addressed, some employees will be able to focus on trying to get the business back up and running. If you have a remote data center and you can dial in on existing phone lines (if they're working) or by using wireless cards in your laptops (if the wireless network is working), you may be able to start recovery efforts. Whether it takes days, a week, or a month, at some point the emergency of the situation will subside and someone will have to turn their attention to getting the business back up and running. The livelihood of your employees depends on it, the financial health of your company and community depend on it. Having a plan in place is better than responding off-the-cuff after a disaster, even if the plan doesn't perfectly fit the aftermath of the disaster event.

Q: I understand the threat and threat source assessment part of the process, but assessing the likelihood and the vulnerability seem like the same thing to me. Plus, doesn't that all roll into the impact analysis? Why break it out into all these layers?

A: Everyone approaches assessments a bit differently. For example, some companies might perform their assessments solely through questionnaires sent to various subject matter experts whereas others might conduct focus groups (group interviews, in essence). The key is to gather the right data and assess it thoroughly. In some ways breaking these elements apart is a bit contrived because they are all so closely interrelated. However, by delineating them very distinctly, you can select the method that works best for your team while ensuring that you are covering your bases. Ultimately, you need to look at all potential threat sources and evaluate them based on your company. This includes how likely the event is to occur, how vulnerable your company is to that event, as well as what would happen to your company if that event occurs. These elements can be thought of as different facets of the same element—each view gives you slightly different information that, when viewed as a whole, gives you an accurate risk picture for your company.

Q: Our company uses a fairly formalized SDLC process. How does that relate to this BC/DR risk assessment process?

A: SDLC, as you know, defines major steps in system development. These steps typically are defined as initiation, development/acquisition, implementation, operation/maintenance, and disposal. This fits into the BC/DR risk management process because at each SDLC phase, you can evaluate the various threats and your IT systems' vulnerabilities to those threats. In addition, many of the SDLC activities are designed specifically to reduce a variety of risks and rather than replicating that work, you can build upon it. For example, during the development/acquisition phase, you can use your threat source assessment data to determine which threats might impact the IT project (or equipment) as well as to determine what elements of the IT project (or equipment) provide various levels of risk mitigation. For example, let's say you're looking at an enterprise-wide data storage solution and you have created a requirements list in the initiation phase of the SDLC framework. As part of the BC/DR risk assessment, you can also look at this project and determine which potential solutions that meet the technical and functional requirements under SDLC could also mitigate risks in your BC/DR framework. A remote storage solution, for example, might fit both needs and cost the same as another solution that meets only the SDLC needs. It's possible a solution might fit both the SDLC and the BC/DR needs and cost more but you can justify the cost by showing how it not only mitigates IT-specific risk but provides excellent risk mitigation from the BC/DR process as well.

Q: It seems a quantitative assessment is far better than a qualitative assessment where you can pretty much "make up" the answers you want. Why would anyone use a qualitative assessment?

A: Every company runs differently and the BC/DR process needs to match the company's standard operating culture to some degree. It's hard enough in many companies to get approval for a BC/DR project without forcing the company to operate outside its normal manner. For example, some companies run on very little quantitative data; decisions are made based on qualitative data, gut feelings, or instincts. Whether you agree with that approach or not, it is how many companies operate. So, using qualitative data may be the only way you'll be able to get the project through the starting gates. In addition, qualitative data can be quite accurate and valid if a scale is well-defined and used consistently. For example, if you devise a scale of 1 to 100 and define 1, 25, 50, 75, and 100 very clearly, there's a good chance the decisions and evaluations made against that scale will fall into line, meaning the data can be extremely accurate. The bottom line in risk management and assessment is to determine which risks you need to address and which risks you're willing to accept or transfer. If you come to that using an artifical scale or through hard calculations, the results are likely to be quite similar, assuming you don't make up the answers you want but stick to the qualitative scales you've all agreed to use.

Chapter 4

Business Impact Analysis

Solutions in this chapter:

- **Business Impact Analysis Overview**
- **Understanding Impact Criticality**
- **Identifying Business Functions and Processes**
- **Gathering Data for the Business Impact Analysis**
- **Determining the Impact**
- **Business Impact Analysis Data Points**
- **Preparing the Business Impact Analysis Report**

- ☑ **Summary**
- ☑ **Solutions Fast Track**
- ☑ **Frequently Asked Questions**

Introduction

In Chapter 3, you learned about risk management and the process for assessing risks. In this chapter, we turn our attention to the process of business impact analysis. Risk assessment looks at the various threats your company faces; business impact analysis looks at the critical business functions and the impact of not having those functions available to the firm. These two assessments look at the company from two different angles. The risk assessment starts from the threat side, and the business impact analysis starts from the business process side. When you're managing general business risk, you might actually start with the business impact analysis. However, in planning for business continuity as an outgrowth of disaster recovery, it makes more sense to understand the full picture regarding risks and threats and then look at business impact. However, if you have a methodology you use that starts with business impact analysis, that's fine. Both outputs—from the risk assessment and the business impact analysis phases—are used as input to the mitigation strategy development. As long as you have those ready before you start the mitigation phase, which we'll discuss in Chapter 5, you should be all set. Figure 4.1 depicts where we are in the planning process thus far.

Figure 4.1 Business Continuity and Disaster Recovery Planning Process

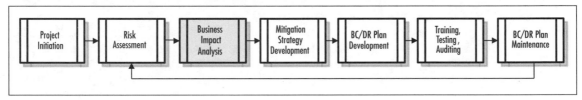

You can see, in Figure 4.2, that we'll be focusing on the third and final segment of the risk assessment phase introduced in Chapter 3 (refer to Figure 3.2 in Chapter 3 for the full diagram). In this chapter, we're going to concentrate on the impact of various business functions on your operations. We'll begin with discussing the general framework of performing a business impact analysis and conclude with the specifics of performing an impact analysis for your business continuity and disaster recovery (BC/DR) plan.

Figure 4.2 Impact Assessment Process

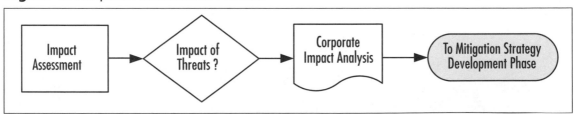

Business Impact Analysis Overview

The fundamental task in business impact analysis (BIA) is understanding which processes in your business are vital to your ongoing operations and to understand the impact the disruption of these processes would have on your business. From an IT perspective, as the National Institute of Standards and Technology (NIST) views it: "The BIA purpose is to correlate specific system components with the critical services that they provide, and based on that information, to characterize the consequences of a disruption to the system components." (*Source*: NIST "Contingency Planning Guide for Information Technology Systems, NIST Special Publication 800-34, p. 16). So, there are two parts to the BIA: the first is to understand mission-critical business processes and the second is to correlate those to IT systems.

As an IT professional, you certainly understand the importance of various IT systems, but you may not be fully aware of the critical business functions performed in your company. Even if your role in this project is limited to managing the IT elements in this BC/DR plan, you should still pay close attention to the material in this chapter for two main reasons. First, understanding the critical business functions is important in terms of understanding how to recover IT systems in the event of a significant business disruption. You might think that System A is most critical, based on a number of assumptions you're making. However, through this process, you might find that System B or C is really what keeps the company up and running on a day-to-day basis or that without System D, System A doesn't really matter. Second, if you have any aspirations at all of moving up the corporate ladder toward that CIO job, your understanding of the overall business will certainly help you achieve those goals. Today's CIO needs to have a solid background in technology *and* business, so understanding the critical business functions in your company will pay off in many ways for you.

According to the Business Continuity Institute (www.thebci.org), a recognized leader in business continuity management and certification, there are four primary purposes of the business impact analysis:

- Obtain an understanding of the organization's most critical objectives, the priority of each, and the timeframe for resumption of these following an unscheduled interruption.

- Inform a management decision on Maximum Tolerable Outage (MTO) for each function.

- Provide the resource information from which an appropriate recovery strategy can be determined/recommended.

- Outline dependencies that exist both internally and externally to achieve critical objectives.

Source: The Business Continuity Institute, Good Practices Guidelines, 2005, p. 21.

Business impact analysis is the process of figuring out which processes are critical to the company's ongoing success, and understanding the impact of a disruption to those processes. Various criteria are used including customer service, internal operations, legal or regulatory, and financial. From an IT perspective, the goal is to understand the critical business functions and tie those to the various IT systems. As part of this assessment, the interdependencies need to be fully understood. Understanding these interdependencies is critical to both disaster recovery and business continuity, especially from an IT perspective. Would it make sense for your IT staff to spend three days trying to recover System D if System A is still out of commission? Until you perform the BIA, there may be no real way to know.

Business impact analysis includes the steps listed earlier, but we can break them out into a few more discrete activities or steps:

1. Identify key business processes and functions.
2. Establish requirements for business recovery.
3. Determine resource interdependencies.
4. Determine impact on operations.
5. Develop priorities and classification of business processes and functions.
6. Develop recovery time requirements.
7. Determine financial, operational, and legal impact of disruption.

The result of performing these seven steps is a formal business impact analysis, which is used in conjunction with the risk assessment analysis to develop mitigation strategies (discussed in Chapter 5).

The two primary impact points of any business disruption are the operational impact and the financial impact. The operational impact addresses the nonmonetary impact including how people, processes, and technology are impacted by a business disruption and how best to address that impact. The financial impact addresses the monetary impacts and how a business disruption will impact the company's revenues.

Upstream and Downstream Losses

In addition to the direct impact of a business disruption such as an earthquake or flood, there are also indirect impacts you should consider. These can be viewed as upstream and downstream losses. *Upstream losses* are those you will suffer if one of your key suppliers is affected by a disaster. If your company relies on regular deliveries of products or services by another company, you could experience upstream losses if that company cannot deliver. If you run a manufacturing company that relies on raw materials arriving on a set or regular schedule, any disruption to that schedule will impact your company's ability to make and sell its products. This is how a disaster elsewhere can impact you, even if your company is unharmed. *Downstream losses* occur when key customers or the lives in your community are affected. If your business supplies parts to a major manufacturer that is shut down due to a hurricane or earthquake, your sales will certainly suffer. Similarly, if your company provides any type of noncritical service to your community and there is a flood or landslide, your sales could take a hit while residents of the community deal with the disaster. If you operate a chain of restaurants or movie theaters or golf courses, residents will be more focused on dealing with the disaster than on entertainment and leisure pursuits. These are considered downstream losses even if your business, itself, has not taken the direct impact of a disaster.

Keep in mind, too, that people, businesses, and communities are interrelated; very few (if any) companies exist in isolation. A natural disaster or serious disruption can create a chain reaction that ripples through the business community and impacts the local or regional economy.

From the Trenches…

Protecting Your Assets

Business continuity and disaster recovery planning can certainly help you mitigate some of your risks. In Chapter 5, we'll develop specific strategies for doing so. However, keep in mind that various types of insurance can help as well. This is considered *risk transference* and is a well-accepted business practice. Consider looking into business income interruption and extra expense insurance. If a business disruption occurs, you could have both an immediate and long-term impact to your company's revenues. Not only will it not be business-as-usual, you'll have the added expenses of lost productivity, lost customers, and higher costs. Some of your out-of-pocket expenses might ultimately be covered by insurance, such as the loss of equipment from a storm or building collapse. Other expenses, however, won't be covered. When revenues decrease and expenses increase, it can create a devastating financial picture for your company. Some basic business insurance policies cover expenses and loss of net business income, but it may not cover business interruptions that occur away from your business, such as to your key supplier, vendor, customer, or even your utility company. This type of insurance can typically be purchased as additional coverage to an existing policy. We're not suggesting you purchase additional insurance (and we have no connections to the insurance industry), but we do suggest you look at your financial exposure and your current insurance policy and decide if you're properly protected. Of course, insurance alone will not protect your business from failing in the face of a serious disruption or event—that's where a solid BC/DR plan comes in.

Understanding the Human Impact

Although this chapter is focused on recovering business systems, it's clear that people are a major factor in business continuity efforts—not only from a planning and implementation perspective but from the impact perspective as well. If a natural disaster strikes, it's possible that some or all of your company's employees will be impacted. It's possible that some may die or be seriously injured. Although no one likes to think about these possibilities, they cannot be ignored in a BC/DR plan. As you assess business functions and business processes, you'll also need to identify key positions, key knowledge, and key skills needed for business continuity. In some sense, this begins to cross over into what is traditionally called *succession planning*. In publicly traded companies or high profile start ups, the company often purchases what's called *key man insurance*. This insurance covers the cost of losing a high ranking executive in the company, the assumption being that if someone at that level were suddenly unavailable to carry out that function, the business would suffer financial losses.

Key Positions

Succession planning in companies covers many areas, but typically it's discussed in terms of replacing key employees as well as how to transfer the reins of the company from one leader to the next. Succession planning can include training employees to move up the corporate ladder and assume leadership positions. From a risk management perspective, it can also address who will replace key employees in the event of a planned or unplanned departure. For example, if a company was started by a couple of business partners, at some point before their retirement, they should spend time identifying their successors—whether family members or trusted employees—and identifying the path to hand over the leadership of the company. When done in a thoughtful and predetermined manner, this can help smooth the transition. In terms of BC/DR, this plan can help identify who should step up should something happen to the company's founders or executives.

Beyond key man succession and planning, the BC/DR plan needs to look at key positions within the company and understand the role of each in the business continuity realm. For example, if you have complex database applications, you may identify a database administrator (DBA) as a key role in the business recovery process. Ideally, your existing database administrator would take care of this, but what if she was unable to respond to the business disruption because she was injured or unable to get to the site (or worse)? Rather than identifying specific people, you should identify roles, responsibilities, skills, and knowledge needed. Even though you'd prefer your own DBA to recover the system, if she was unavailable for any reason, you would know that you need a DBA to recover your systems and you could go to external sources to locate a temporary or permanent DBA replacement.

Human Needs

Beyond replacing needed skills and positions, it's important to keep the human impact in mind throughout your planning. As mentioned earlier in the book, everyone responds to disasters differently. If a portion of the building catches on fire and burns, it's likely that those employees in the area at the time the fire breaks out will experience the event in a variety of ways. Some people will evacuate and stand in the parking lot laughing about the close call, even as the fire engines pull in. Others probably will be frightened by the experience and may become shaky, disoriented, or panicky. Still others might seem fine immediately afterward but days or weeks later, they begin to display odd behavior that might be the result of a delayed onset of stress from the event. Clearly, the bigger the event (earthquake, tornado, hurricane), the bigger the human toll in terms of death, injury, and emotional distress.

A good business continuity plan will address the human factors for two reasons. First, addressing employee needs is simply the right thing to do. Although there are companies that may demand that employees report to work following a serious business disruption or face termination, most companies understand that everyone will have different needs. Some may report back to work, some may need to deal with family problems, some may be physically or emotionally unable to return to work immediately. The company's policies with regard to employee needs and requirements in the aftermath of a business disruption or natural disaster should be developed by your Human Resources department; however your BC/DR plan must take these varied responses into consideration. If your IT systems recovery effort hinges on two experienced network administrators, you need to address these as risks in your plan and develop mitigation strategies along with them.

The second reason for addressing employee needs in your BC/DR plan is because it makes good business sense. The ideal scenario might be that everyone is fine and shows up to work, but reality is often far different from that. You can demand that people show up all you want, but if faced with a choice between work and family, between work and health, people will usually choose family and health first. In some cases, insisting people return to work before they are ready can make things worse—they may not be able to concentrate and therefore may make recovery efforts worse instead of better. Incorporating this reality into your plan will mean that you and your team come up with appropriate alternatives that can address the lack of key staff in the aftermath of a business disruption. This helps the employees who may be unable to come back immediately and also helps the company recover in the fastest, most efficient manner possible.

We won't dwell on the human element in this chapter, but we will mention it again in key places to keep it foremost in your mind so that as you determine the impact of various risks, you can also keep the human factor in mind.

Understanding Impact Criticality

As you're thinking about your company and its critical functions, which we'll review following this section, you should keep a rating scale in mind. Later, after you've compiled your list, you can assign a "criticality rating" to each business function. It's important to have an idea of your rating system in mind before you review your business functions so you can spend the appropriate amount of time and energy on mission-critical functions and less time on minor functions. For example, when you sit down with the finance group, you want to keep them focused on defining the mission-critical business functions while listing all business functions that would be needed for business continuation.

Criticality Categories

You can develop any category system that works for you but as with all rating systems, be sure the categories are clearly defined and that there is a shared understanding of the proper use and scope of each. Here is one commonly used rating system for assessing criticality:

- Category 1: Critical Functions–Mission-Critical
- Category 2: Essential Functions–Vital
- Category 3: Necessary Functions–Important
- Category 4: Desirable Functions–Minor

Obviously, your business continuity plan will focus the most time and resources on analyzing the critical functions first, essential functions second. It's possible you will delay dealing with necessary and desirable functions until later stages of your business recovery. Many companies identify these four areas and set timelines for when each of these categories will be functional following a business disruption. Let's look at each category in more detail. You can use these category descriptions as-is or you can tweak them to meet your company's unique needs.

Mission-Critical

Mission-critical business processes and functions are those that have the greatest impact on your company's operations and potential for recovery. Almost everyone working in a company has an innate understanding of the mission-critical operations within their department. The key is to gather all that data and develop a comprehensive look at your mission-critical processes and functions from an organizational perspective. What are the processes that must be present for your company to do business? These are the mission-critical functions. One way to get people to focus on the mission-critical functions is to ask (whether through questionnaire, interview, or workshops) what the first three to five things people would do in their department following a business disruption once the emergency or imminent threat of a business disruption subsides. This often gives you the clearest view of the mission-critical business functions in each department.

From an IT perspective, the network, system, or application outage that is mission-critical would cause extreme disruption to the business. Such an outage often has serious legal and financial ramifications. This type of outage may threaten the health, well-being, and safety of individuals (hospital systems come to mind). These systems may require significant efforts to restore and these efforts are almost always disruptive to the rest of the business (in the case that any other parts of the business are actually able to function during such an outage). The tolerance for such an outage, whether from the IT system or the function/process it provides, is very low and the recovery time requirement is often described in terms of hours, not days.

Vital

Some business functions may fall somewhere between mission-critical and important, so you may choose to use a middle category that we've labeled "vital" or "essential." How can you distinguish between mission-critical and vital? If you can't, you may not need to use this category. However, you might decide that certain functions are absolutely mission-critical and others are extremely important but should be addressed immediately after the mission critical functions. Vital functions might include things like payroll, which on the face of it might not be mission-critical in terms of being able to get the business back up and running immediately but which can be vital to the company's ability to function beyond the disaster recovery stage.

From an IT perspective, vital systems might include those that interface with mission-critical systems. Again, this distinction may not be helpful for you. If not, don't try to force your systems into this framework; simply don't use this category. You'll end up with just three categories—mission-critical, important, and minor. If that works for you, that's fine. If you use this category, your recovery time requirement might be measured in terms of hours or a day or two.

Important

Important business functions and processes won't stop the business from operating in the near-term but they usually have a longer-term impact if they're missing or disabled. When missing, these kinds of functions and processes cause some disruption to the business. They may have some legal or financial ramifications and they may also be related to access across functional units and across business systems.

From an IT perspective, these systems may include e-mail, Internet access, databases, and other business tools that are used in a support function, whether to support business functions or IT functions. If disabled, these systems take a moderate amount of time and effort (as compared to mission-critical) to restore to a fully functioning state. The recovery time requirement for important business processes often is measured in days or weeks.

Minor

Minor business processes are often those that have been developed over time to deal with small, recurring issues or functions. They will not be missed in the near-term and certainly not while business operations are being recovered. They will need to be recovered over the longer-term. Some minor business processes may be lost after a significant disruption and in some cases, that's just fine. Many companies develop numerous processes that should at some point be reviewed, revised, and often discarded, but that rarely occurs during normal business operations due to more demanding work. In some sense, a business disruption can be good for those small business functions and processes as they may be reworked or revised or simply pared down after a disruption. You may use the process of performing your BIA to recommend paring down these minor business functions as well, though your time is better spent focusing on the mission-critical and vital elements. You may make notes about which functions and processes could be pared down outside of the BC/DR planning process and hand this off to the appropriate SMEs for later action.

From an IT perspective, these types of system outages cause minor disruptions to the business and they can be easily restored. The recovery time requirement for these types of processes often is measured in weeks or perhaps even months.

TIP

Be sure to prompt participants to think about all business processes throughout the year. Some functions and processes occur only during certain times of the year, such as tax season, year end, holidays, and such, and these might be missed during the process. If they're important enough processes, there's a good chance they'll be included, but project management best practices don't rely on luck—they rely on process. Be sure you to ask about any special processes that occur throughout the calendar year that might not immediately come to mind for participants.

Recovery Time Requirements

Related to impact criticality are recovery time requirements. Let's define a few terms here that will make it easier throughout the rest of the analysis to talk in terms of recovery times. As you read through these definitions, you can refer to Figure 4.3 for a representation of the relationship of these elements.

Maximum Tolerable Downtime (MTD). This is just as it sounds—the maximum time a business can tolerate the absence or unavailability of a particular business function. (*Note*: The BCI in the UK uses the phrase Maximum Tolerable Outage (MTO) instead.) Different business functions will have different MTDs. If a business function is categorized as mission-critical, or Category 1, it will likely have the shortest MTD. There is a correlation between the criticality of a business function and its maximum downtime. The higher the criticality, the shorter the maximum tolerable downtime is likely to be. Downtime consists of two elements, the *systems recovery time* and the *work recovery time*. Therefore, MTD = RTO + WRT.

Recovery Time Objective (RTO). The time available to recover disrupted systems and resources (systems recovery time). It is typically one segment of the MTD. For example, if a critical business process has a three-day MTD, the RTO might be one day (Day 1). This is the time you will have to get systems back up and running. The remaining two days will be used for work recovery (see Work Recovery Time).

Work Recovery Time (WRT). The second segment that comprises the maximum tolerable downtime (MTD). If your MTD is three days, Day 1 might be your RTO and Days 2 to 3 might be your WRT. It takes time to get critical business functions back up and running once the systems (hardware, software, and configuration) are restored. This is an area that some planners overlook, especially from IT. If the systems are back up and running, they're all set from an IT perspective. From a business function perspective, there are additional steps that must be undertaken before it's back to business. These are critical steps and that time must be built into the MTD. Otherwise, you'll miss your MTD requirements and potentially put your entire business at risk.

Recovery Point Objective (RPO). The amount or extent of data loss that can be tolerated by your critical business systems. For example, some companies perform real-time data backup, some perform hourly or daily backups, some perform weekly backups. If you perform weekly backups, someone made a decision that your company could tolerate the loss of a week's worth of data. If backups are performed on Saturday evenings and a system fails on Saturday afternoon, you've lost the entire week's worth of data. This is the recovery point objective. In this case, the RPO is one week. If this is not acceptable, your current backup processes must be reviewed and revised. The RPO is based both on current operating procedures and your estimates of what might happen in the event of a business disruption. For example, if a tornado touches down in your town and your data center is without power, you may implement your BC/DR plan. If you have an alternate computing location, you may transfer operations to that location. Your next step would be to determine the status of the data. Are you attempting to update systems using backups or were these alternate locations kept up to date? When was the last data backup performed relative to business operations? What do you need to bring systems up to date? These are the questions you'd need to answer after a business disruption. Therefore, it's important to define your RPO beforehand and ensure your recovery processes address these timelines.

Let's look at how these elements interact. Figure 4.3 graphically depicts the interplay between MTD, RTO, WRT, and RPO. If your company has mission-critical and vital business processes that do not interact with computer systems of any kind, you still need to perform a business impact analysis in order to understand how these manual systems may be impacted by a business disruption, especially natural disasters. At the end of this chapter, we'll walk through an example to help illustrate these concepts. Most companies use technology and computer systems to some extent and the graphic in Figure 4.3 shows how the recovery time is impacted by a business disruption.

Figure 4.3 Critical Recovery Timeframes

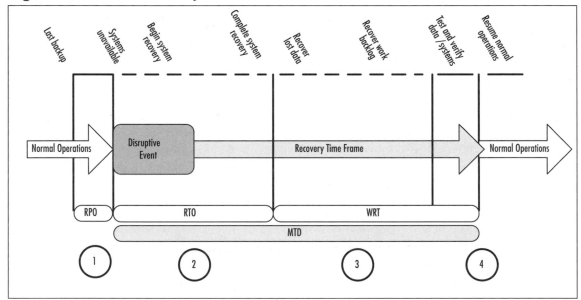

- **Point 1**: Recovery Point Objective—The maximum sustainable data loss based on backup schedules and data needs

- **Point 2**: Recovery Time Objective—The duration of time required to bring critical systems back online

- **Point 3**: Work Recovery Time—The duration of time needed to recover lost data (based on RPO) and to enter data resulting from work backlogs (manual data generated during system outage that must be entered)

- **Points 2 and 3**: Maximum Tolerable Downtime—The duration of the RTO plus the WRT.

- **Point 4**: Test, verify, and resume normal operations

During normal operations, there is usually some gap between the last backup performed and the current state of the data. In some operations, this may be minutes or hours; in most organizations it is hours or days. This timeframe is the recovery point objective. In most organizations, this is the same as the period of time between backups. We see at circle 1 that there is a gap showing the point of the last backup and the state of current data, just before the disruption occurs. That's the point at which one or more critical systems becomes unavailable and business continuity and disaster recovery planning activities are initiated. The first phase of the Maximum Tolerable Downtime (MTD) is the recovery time objective. This is the timeframe during which systems are assessed, repaired, replaced, and reconfigured. The RTO ends when systems are back online and data is recovered to the last good backup. The second phase of the MTD then begins.

This is the phase when data is recovered through automated and manual data collection processes. There are two elements of work recovery time. The first is the manual collection and entry of data lost, typically because systems went down between backups. The second phase addresses the backlog of work that may have built up while systems were down. Most companies try to recover the data up to the disruptive event to bring the systems current and then address the backlog, but your business processes may dictate a different recovery order. The key is to understand that there is a delay between the time the systems are back online and the time when normal operations can resume. During the periods indicated by circles 2 and 3, emergency workarounds and manual processes are being used. These are processes that will be developed later in your BC/DR planning process. For example, if a CRM system is down, what processes will your sales, marketing, and customer sales service teams use to interface with and manage customer service delivery? You'll define that in the planning process. Circle 4 indicates the transition from diaster recovery and business continuity back to normal operations. There may be some overlap as manual processes are turned back over to automated processes and you may choose to do it in a rolling fashion—perhaps by department or geographic region.

As you collect your impact data, you'll also need to begin determining the recovery time objectives. You may choose to create a rating system so you can quickly determine recovery time objectives. For example, you might determine that mission-critical business systems or functions should have recovery windows as follows:

- **Category 1**: Mission-Critical—0–12 hours

- **Category 2**: Vital—13–24 hours

- **Category 3**: Important—1–3 days

- **Category 4**: Minor—more than 3 days

You and your team, with input from the subject matter experts, can determine the appropriate maximum tolerable downtime (MTD) requirements. For some companies, a mission-critical business function could have an MTD of a week. For others, it might be 0 to 2 hours. There is an inverse correlation between the amount of time you can tolerate an outage and the cost of setting up systems that allow you to recover in that time frame. If you can't afford much downtime, you'll clearly have to invest more in preventing downtime and in having systems in place that allow fast recovery times. If you're a small company and can afford a longer MTD, you can spend less on preventing or recovering from outages.

Let's look at an example. In a small company, you may very well be able to do without even mission-critical systems for a couple of days or a week if you really had to. It's possible that you contract with an outside IT service provider to maintain, troubleshoot, and repair your computer systems. If you want a guaranteed two-hour response time, your monthly maintenance costs will be significantly higher than if you sign up for a guaranteed next business day response. So, if you really can't afford to be without that mission-critical business function for more than about eight hours (two-hour response time, six-hour repair time), you'll have to pay more to your service company and you'll probably also have to purchase additional computer equipment to provide some redundancy to prevent extended downtime. These costs add up and the less disruption your business can afford, the more it will cost you to prevent or mitigate those risks. We'll discuss this in more detail in Chapter 5, but it's within the business impact analysis segment where you have to begin making these kinds of assessments.

It's important to note during your impact analysis and subsequent mitigation planning phases that there is an optimal recovery point. Figure 4.4 shows the inverse relationship between the cost of disruption and the cost of recovery. Earlier in this book, we discussed the fact that any business continuity and disaster recovery plan had to be tailored to the unique needs and constraints of the organization. This is particularly true when it comes to the financial costs involved with disruption and recovery.

You can see that the longer you allow a disruption to go on, the more expensive it becomes to the business. Conversely, the longer you have to recover, the less expensive recovery itself becomes. This makes sense when you understand that the longer a business disruption goes on, the more lost revenues, lost sales, and lost customers you accumulate. At the same time, if you need to recover your systems immediately, it's going to cost more to implement things such as zero downtime solutions and hot sites. If you can afford to take a bit more time to recover you have more options, and these options are typically less expensive. If you start plotting these points, you will find an optimal point between these two costs, shown in Figure 4.4 by point A. Each company's intersecting points (point A) will be different based on your company's financial constraints and operating requirements.

Figure 4.4 Optimal Balance between Cost of Disruption and Cost of Recovery

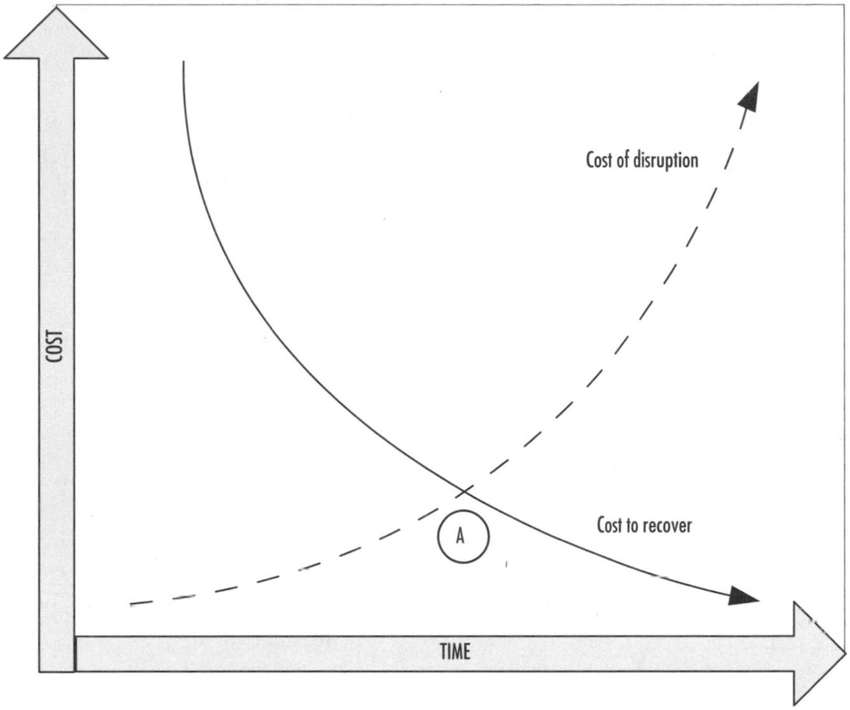

Looking Ahead...

Making the Business Case Makes Your Life Easier

During the assessment and implementation of IT systems over the course of the past few years, you may already have addressed (and invested in) some of the elements needed to reduce the time to recover or to reduce the cost of a disruption. If so, be sure to make note of these systems or investments and be sure to include them in your planning. One way to help make the business case for continued investment is to show how the systems already implemented have made an impact or have contributed to your BC/DR plan. For example, suppose you implemented a mirrored site to allow users to gain access to key data more quickly. That mirrored site also serves as a backup and reduces the cost of disruption to a single site. It also reduces the amount of time it takes to recover, thereby pulling your point A down and to the left (toward lower cost, less time). This investment, then, has contributed to optimizing your balance between cost of disruption and cost to recover while also improving user productivity. Being able to establish and articulate these kinds of IT benefits within your organization may not only win support for your BC/DR plan, it might also help you move up the corporate ladder.

Next, let's look at what the entire analysis process looks like, as shown in Figure 4.5. After we explore this, we'll take a look at the specific data required for inputs and outputs to this process.

Figure 4.5 BIA Inputs and Outputs

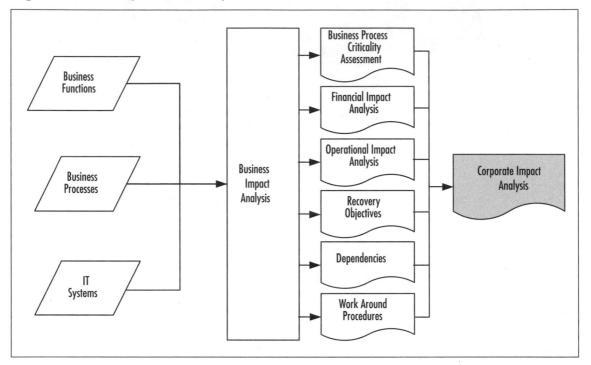

In this segment of BC/DR planning, we're looking at business functions, processes, and IT systems to determine criticality. Business functions can be defined as activities such as sales, marketing, or manufacturing. Business processes can be defined as how those activities occur. Are your sales conducted via a Web site, via telephone, via sales calls? How are orders processed? How are employees hired? These are business processes, they describe how the functions get done. By first identifying business functions, you then can focus on the key processes in each function to develop a comprehensive view of your company. The third input area, shown in Figure 4.5, is IT systems. In most companies, the business processes are carried out in part through computer systems, applications, and other automated systems. Identifying mission-critical business functions and processes and how they intersect with IT systems will help you map out your business continuity and disaster recovery strategies.

Once you have compiled that data, you'll perform the analysis to generate the needed outputs, including the criticality assessment, the impact assessments (financial and operational), required recovery objectives, dependencies, and work-around procedures. The work-around procedures will enable you to get critical business functions back up and running as quickly as possible. These work-around procedures may be used during the RTO and WRT periods discussed earlier and shown in Figure 4.3. As you can see, the output is a comprehensive corporate impact analysis. This is the same output shown in Figure 4.2 and is the end of the larger risk assessment phase in our overall BC/DR planning process. The impact analysis will be used as input to the risk mitigation planning segment of the BC/DR project and we'll discuss that in Chapter 5.

Identifying Business Functions

In this section, we're going to walk through some of the more common business functions found in business today. It's not a comprehensive list but it's intended to do two things. First, you can include these in your BIA and you'll know you've got the major items covered. Second, you can use this to spur your thinking to include other areas that might be related to the items listed. You should begin by listing all the business functions that come to mind unless it's clear they should *not* be included. As with your risk assessment, it's best to begin by scanning the wide horizon and narrowing your focus later on. It's always easier to cut than to try to find gaps later.

When possible, it's advisable to create a list of all the functional areas of the business and gather SMEs from each area to discuss the critical business functions. Although it's more time consuming to get everyone in a room together, you will more quickly discover interdependencies in this manner. If SMEs sit quietly by themselves and come up with the critical business functions alone, they might miss the elements that are vital to other areas. An alternate method of gathering this data is to have the SMEs generate a list of questions to ask others in their area and compile the results. When the compiled results are ready, the subject matter experts from all areas of the company can meet to go over the results with the specific mission of finding interdependencies. How you manage this aspect of the project will have everything to do with how your company runs on a day-to-day basis.

The common business functions include those shown here. They're listed in alphabetical order, not necessarily in the order in which you would review these areas. The order in which these are reviewed will be dictated by the project management processes you've defined, the data gathering methods you choose, and the structure of your company. Following this section, we'll discuss the specific data points you need to gather from each of these areas.

1. Facilities and Security
2. Finance
3. Human Resources
4. Information Technology
5. Legal/Compliance

6.   Manufacturing (Assembly)

7.   Marketing and Sales

8.   Operations

9.   Research and Development

10.   Warehouse (Inventory, Order Fulfillment, Shipping, Receiving)

As we look at these business functions, keep your business in mind and think about the key processes that occur in each functional area. After you've documented your key business processes, you will assign a criticality rating to them similar to the ones discussed earlier. As a reminder, you may also want to document key positions, skills, and knowledge in these functional areas. For example, what would the impact be if your head of facilities was injured in a building collapse and your company needed to operate from an alternate location? Who would head that up? What skills or knowledge would be needed in order to temporarily (or permanently) replace your facilities manager in the aftermath of a business disruption? These human factors should be assessed in conjunction with the major business functions.

Facilities and Security

Your company may be located in a single office in a small office building or it may span several continents. Regardless of how many physical locations your company operates, you need to understand the critical processes performed by facilities and security management with regard to your business operations. If a business disruption were to occur, what processes and procedures would be needed in order to get your business back up and running? For example, if the building is damaged or destroyed, physical security of the building will be disrupted. Employees won't be able to just swipe their badge at the front door. Is this a critical business function or not? It depends. If the building is destroyed, it doesn't matter that they can't get into the building. You don't just need an alternate process, you need an alternate location. Once an alternate location is established, you need facilities support. So, the critical business function, in this example, is having a place of business ("facilities"). Security and access are secondary. Notice how it helped to think of a specific scenario—it focused our thinking so we could see the key areas. Is having a place of business a critical business function? Not in the formal definition of a business *process*, but it's certainly important. Security usually involves a process—adding employees to access lists, providing employees with badges, IDs, or other identification, and granting them appropriate access to company resources. This might be highly important during normal business functioning, but does it impact the company's mission-critical operations? It depends on your business. If you work in a secure research environment, facilities and security may be mission-critical. If you work in a software development firm where employees could check code out of an online library and work from home, facilities and security may not be mission-critical at all. Facilities and security, though, may have some critical business functions beyond these macro-level functions just mentioned. For example, is facilities involved with the receiving or shipping of products, inventory, or other tangible goods? If so, these may be critical business functions to be included.

Finance

By definition, the financial workings of the company are critical business functions, but not all financial functions are mission-critical functions. For example, tracking receivables and payables are critical business functions because without the ability to keep track of what others owe you and what you owe others, you have no idea about the financial status of the company. Employee payroll is another critical business function (which is a financial transaction that might fall under the purview of the Human Resources department). If employees are not paid, if appropriate withholding and other taxes and deductions are not taken, your company faces serious problems, with employees and with state and federal authorities.

If your company has legal obligations to pay back a loan from a bank or make payments or reports to investors, these also might be critical business functions to be included in your analysis. In some cases, you may have some leeway with regard to repayment if you experience a natural disaster, but don't count on it. Your financiers don't care, they just want payments on time and in full. Therefore, keeping track of these kinds of financial and legal obligations may be considered critical business functions, depending on the nature of your company and its financing structure.

Accounting, finance, and reporting functions within finance should be reviewed and analyzed. There are many interdependencies in financial functions that cross over into HR, marketing, sales, IT, and operations. If key IT systems were to go down, which business processes would be impacted? Which processes and functions would have to get back up and running first in order to keep the business going?

Human Resources

If your firm experiences some sort of natural disaster, your Human Resources staff will be busy trying to fulfill a number of roles. Employees will usually contact HR for information on the status of the building, the status of the company, whether they should report to work, where they should report to work, and so on. Employees may also use HR as a clearing house for information about the well-being of other employees or information on the broader community. Finally, employees will be looking to HR for information on how, when, and where they'll get paid. In fact, this will likely be the first question many employees ask, especially if the business disruption happens just prior to or on payday. The staff in HR will be in the best position to provide guidance on the kinds of issues for which employees come to them. From there, you can compile a list of critical business functions. Remember, create a list of all business functions, then prioritize them later. If IT systems were to go down, which HR functions and processes are mission-critical? How would they be accomplished in the absence of IT systems? How would this impact other areas of the company?

IT

Critical business functions for IT? It seems like almost all of them are critical most of the time, especially if you judge by the phone calls, hallways pleas, and e-mails begging for assistance when one of the applications, servers, or hardware goes down. However, ultimately, the hardware and software should support the critical business functions, so the IT functions, in large part, will be driven by all the other departments. HR might say "we have to have our payroll application"; marketing might say "without our CRM system, we can't sell any products"; manufacturing might say "without our automated inventory management system, we can't even begin to make anything." Therefore, the IT department's critical business functions are driven externally, to a large degree. However, there are also business functions that occur within the IT department critical to the company's ability to recover and continue doing business after a disaster. For example, the IT department needs to create backups of all data that changes after a disaster. If a disaster happens on a Tuesday and you're able to get some systems back up and running by the following Monday, backups need to start on Monday, as soon as data begins being generated, saved, or changed. Therefore, backup processes can be viewed as critical business functions from the IT perspective. Managing security is another critical aspect. As you look at these functions, you'll find addtional critical IT functions.

Legal/Compliance

There are numerous mission-critical business functions related to legal and compliance areas of your company. If your firm is subject to legal or regulatory statutes and requirements, you're already well aware of these constraints. You need to view these constraints and requirements in light of a potential business outage to determine which of these are mission-critical, which are vital or important, and which are minor in nature. For example, if your firm deals with private or confidential personal data, it must be protected at all times, even if you move to a manual system for the duration of a system outage. Which systems, then, should be recovered first? Which business processes are mission-critical? Those related to remaining in compliance, both in terms of business process and business data, should be ranked very high on your list. The legal and financial consequences, as discussed in the case study earlier in this book (see Case Study 1, "Legal Obligations Regarding Data Security") can be enormous.

Manufacturing (Assembly)

If your company is involved with the manufacturing, assembly, or production of tangible products, you obviously need to scour this area for mission-critical functions since your ability to produce your products is the engine that drives your company. There may be some systems that can come online later, but there are likely to be certain systems that must be up and running in order for any manufacturing, assembly, or production to occur. Identify these business processes and systems by understanding what would happen if the production equipment were to be damaged or destroyed. Next, understand what would happen if the production equipment was left in tact but upstream or downstream events impacted your customers or vendors. The impact analysis needs to include both internal and external elements. What business processes should you put in place to deal with the potential loss of a key supplier? We'll look at risk mitigation strategies in detail in Chapter 5. For now, you should be identifying the potential impact of various business disruptions to your manufacturing operations, keeping both internal and external (upstream/downstream) disruptions in mind.

It's also important to understand the interaction between any manufacturing/assembly automation equipment and IT systems. If IT systems go down, how are automation systems impacted? If automation systems go down, how are IT systems impacted? What manual processes can be implemented in the absence of either automation systems or associated IT systems?

Marketing and Sales

Marketing activities help create demand for the company's products and services by establishing or expanding knowledge of the company and its products/services. Sales activities are those actions that actually create a sales transaction and bring revenue into the company. Some companies may determine that marketing activities in the aftermath of a business disruption can be put on hold while sales activities should be a top priority. Other companies may see marketing activities as mission-critical in the aftermath of a business disruption because they are businesses that need to stay in touch with customers, keep their products/services in front of customers, and cannot afford to let rumors and erroneous information about the company's status float around, especially in today's world of instant, on-demand news. How you approach marketing and sales functions in your firm from a business continuity and disaster recovery standpoint will depend largely on the size of your company, its market visibility and other internal factors. Clearly activities that support the company's ability to perform sales transactions will most often be considered either vital or mission-critical activities and systems.

Operations

If your company doesn't manufacture, assemble, or produce tangible products, it probably develops and sells intangible products such as service, software development, research, analysis, and others. Whatever it is your company does, it sells something in order to generate revenue. Therefore, your operations are what end up generating those goods and services that are sold to customers. As with manufacturing and assembly, operations are what generate sales and therefore are almost always part of the most urgent mission-critical business functions. Although "operations" is a rather broad and vague term, each company knows exactly what its operations are and how these operations contribute to revenue generation. It is within that scope of knowledge that these activities should be assessed for criticality.

Research and Development

Some companies or organizations are funded through investors, through grants, or operate as nonprofits. They may be dedicated solely to research and development and may not generate revenue in the traditional sense of the word. However, every organization needs funding and that funding almost always comes with some sort of expectations and requirements about what is to be achieved with that funding. Therefore, you can view activities that bring in funding as your sales activities and can assess their criticality in that light. For example, if your organization does biochemical research and you're funded by federal or state programs, you still have business functions related to deliverables to consider. Is the next round of funding predicated upon the successful delivery of the results of current development or testing? If so, you have several mission-critical systems to consider along with assessing the impact of a business disruption to your research. Do you have live cultures growing in a lab that need to be tested and assessed? If so, what would happen if the research building was destroyed by fire or by an earthquake or tornado? How would your research be impacted and how would you recover? Though these are a bit different from traditional business functions and are not related directly to IT systems, these are questions that should be asked and answered if you're in this business.

Warehouse (Inventory, Order Fulfillment, Shipping, Receiving)

If your company deals in tangible goods of any kind, you have processes for handling inventory, order fulfillment, returns, shipping, and receiving. In some companies, these functions are handled by outside firms. For example, you may manufacture or assemble a product that is sent out daily on trucks to some other company that handles the remaining inventory processes. Nonetheless, your company has to keep track of what it makes and what it ships out at minimum. So, there are two elements here, the actual manufacturing or assembly (covered earlier) and the tracking, storing, and moving of these products. These two functional areas are closely tied together and the interdependencies in these areas should be given special attention. If IT systems go down, how are these activities impacted? If the building is ravaged by fire or flood, how are these activities impacted?

Other Areas

There may be other functional areas not listed here that exist in your company. If so, be sure to explore each functional area and determine the various business processes used in each area along with their relationship to the business's IT systems.

Looking Ahead...

Flaws Exposed

It's important to understand that a business impact analysis is a thorough business assessment that involves an unbiased study of the entire organization. When you start looking at the workings of the company in a very close and detailed manner, things may start to look less than stellar, like when you shine a very bright light on something and you suddenly see all its flaws quite clearly. Your corporate executives might take one of two positions. In the best case, they will appreciate the opportunity to closely examine the company's operations and find ways to improve it along the way. In the worst case, they will hesitate, stonewall, or misdirect you in order to prevent you from uncovering business processes that are broken, inefficient, or worse, illegal. So, be prepared for a variety of reactions from the top to the bottom of your organization. Also, if you're so inclined, you might begin preparing your organization for this level of scrutiny, being sure to communicate the positive aspects of this process.

Ideally, you can double your mileage from this project by using it as an opportunity to perform your BIA and to streamline business operations. Just be prepared for a few bumps in this road, especially if you suspect that the business processes are not too pretty in some areas of the company. Remember, too, that a well-executed BIA can help you garner *more* support for your BC/DR planning project as people in the organization begin to understand the undesirable effects a disaster or disruption would have on the business. Sometimes seeing the flaws is motivation enough to fix them.

Gathering Data for the Business Impact Analysis

As we discussed in Chapter 3, there are four primary ways of gathering information: questionnaires, interviews, documents, and research. This holds true for the BIA as well. Before you can develop questionnaires or interviews, however, you have to know what you're looking for. You may choose to gather subject matter experts who then create questionnaires or interview questions. As a project team, you may create a number of very specific questions or scenarios to be presented to subject matter experts (SME) in the form of questionnaires or interviews. The additional information will come from either the project team or SMEs reviewing documents or performing targeted research.

Where to start this sometimes daunting process? One of the best places to start is with your company's organizational chart. Lacking that, try the company's phone directory—electronic or paper. In many cases, the functional areas of the company are clearly spelled out.

This can be a good place to determine sources for subject matter experts as well. You can begin by creating a list of each functional area such as each division or each major work area such as manufacturing, warehouse, operations, development, among others. List subdepartments or subdivisions under each of the major headings, as appropriate. Now, you should have a comprehensive list of the major and minor departments, which are often the functional areas, in your company. Check for duplication and remove any areas that are repeated or that clearly should not be included. The key at this juncture is to generate a comprehensive list of business functions that can later be prioritized. Also remember there may be internal or external dependencies that raise the criticality of particular business functions.

As previously discussed, asking questions and providing scenarios to consider can help people focus on specific business issues and generate better responses. Some questions you might ask of your subject matter experts to help them focus on the key aspects of the impact analysis include these:

1. How would the department function if desktops, laptops, servers, e-mail, and Internet access were not available?

2. What single points of failure exist? What, if any, risk controls or risk management systems are currently in place?

3. What are the critical outsourced relationships and dependencies? What are the upstream and downstream risks to your business function?

4. If a business disruption occurred, what workarounds would you use for your key business processes?

5. What is the minimum number of staff you would need and what functions would they need to carry out?

6. What are the key skills, knowledge, or expertise needed to recover? What are the key roles that must be present for the business to operate?

7. What critical security or operational controls are needed if systems are down?

8. How would this business function in a backup recovery site? What would be needed in terms of staff, equipment, supplies, communications, processes, and procedures? (This crosses into the disaster recovery element, which we'll discuss more in a later chapter.)

Data Collection Methodologies

For the business impact analysis, it is advisable to collect data through questionnaires, interviews, or workshops, which are in many ways group interviews. Additional data can be gathered using documents and research, but this data should be gathered only to support or supplement data gathered through direct contact with business subject matter experts. The reason for this is fairly obvious. Only those who actually perform various business functions can assess the criticality of those business functions. You could sit down and read documents all day long and never get a clear picture of what's really mission-critical and what's just important. Therefore, you should rely primarily on questionnaires, interviews, and workshops for this segment of your data gathering. Let's look at methodologies you can use for these three data gathering methods.

Questionnaires

Questionnaires can be used to gather data from subject matter experts (SME) in a fairly efficient manner. Though it takes time to develop a highly useful questionnaire, SME's responses will be consistent, focused, and concise. They can fill out the questionnaires regarding their business units, business functions, and business processes at a time that is convenient for them (within a specified timeframe), thereby increasing the likelihood of participation. On the downside, questionnaires that are sent out may be ignored, pushed aside, or forgotten. In order to generate a timely and meaningful response to your team's questionnaire, you can create a methodology that will increase your response rate.

First, it's important to appropriately design the questionnaire. If it's full of useless questions, if it's visually confusing or overwhelming, you'll decrease your response rate. The questionnaire should be clear, concise, easy to understand, and fast to fill out. If you want to use a Web-based questionnaire that records data in a database, so much the better. You can send out reminders with a link to the questionnaire as frequently as needed. With a paper-based questionnaire, there's a lot of moving of paper and the increased likelihood that the paper will be misplaced, lost in a pile, or simply thrown out.

It's also important to explain the purpose of the questionnaire to the participants in a manner that helps them buy into the process. Focus on what's in it for them, not for you. They probably don't care that *you* need this data, but they will care that this data could help prevent some problem in *their* jobs. Ideally, you should hold a kick off meeting where the questionnaire is introduced and explained, the purpose of it is clearly articulated, and the process for completing the questionnaire is explained. For example, you might let people know that the questionnaire is available at a particular location, that it takes a total of three hours to complete per department, but that it can be completed in segments and the questionnaire-in-progress can be saved for later completion. You should let people know who the contact person is if they run into problems and when the questionnaire must be completed.

If your company is the type of company that likes to have a bit of fun in these kinds of meetings, you can also announce small prizes that will be awarded to departments or individuals who complete theirs correctly first, who are most thorough, and so forth. Be careful, though, you don't want to leave the impression that this is a race to the finish (where important details can be lost) or that "cute" answers are appropriate. You can, however, announce that for any SME that submits a complete and thorough questionnaire by the deadline will be entered into a hat for the chance to win some prize such as a portable music player, a new cell phone, dinner for two at a nice restaurant, among others. Sometimes small incentives to do the right thing can go a long way in getting people to participate in the manner expected and needed. Considering how vital this particular data is to your entire BC/DR plan, it's usually worth a small investment to get people to participate appropriately, if this type of activity fits in with your corporate culture. Be sure to provide information on how respondents can get assistance with the questionnaire—either from a technical stand-point (if it's an electronic or Web-based questionnaire) or an administrative standpoint. If they don't understand exactly what a question means, who should they contact? How should they contact them? What is the contact person's e-mail, location, phone number, and work hours? Be sure to provide this information so you don't inadvertently create road-blocks for yourself.

Finally, let the team know how they'll learn about the results of the questionnaire. Most people dislike spending time filling out a form only to never hear about it again. If they are willing to take the time needed to provide this data, there should be some reciprocity. For example, if this data is all pumped into a database, a report on each respondent's data could be provided back to them for verification. Once the data is reviewed by your team, there may be additional questions. Respondents should be told, in advance, about the process for following up with them regarding their responses to the questionnaire.

Once questionnaires are completed, you and your team should review them to ensure they are complete. In some cases, you may choose to create a process whereby certain ques-tionnaires are followed up by an interview. This might be in the case of the most critical business functions or where questionnaire data indicates there may be confusion, conflict, or incomplete data. Any follow-up interviews should follow a specific format as well so that targeted data can be collected.

Interviews

If your team has decided that data will be gathered through interviews, you'll still need to create a questionnaire type of document that will provide the interviewers with a set of questions to which they gather responses. Free form or informal interviews will yield incon-sistent data across the organization and you'll have a wide array of meaningless data. Develop a questionnaire and use it as the basis of the interview process. Each interview should follow a predefined format and the questions asked of each respondent should be the same. Develop a questionnaire, interview, or question sheet from which the interviewer will work and also develop a corresponding data sheet onto which the interviewer can record responses. Look to find methods to speed up the interview process. For example, don't use a rating system of ten elements that use 1 as NEVER and 10 as ALWAYS with eight other word/number combinations. This will be cumbersome for the interviewer to describe and will be almost impossible for the interviewee to remember. If you choose, you might say, "On a scale of 1 to 10 with 1 being never and 10 being always, how often would you say you access the CRM database on a telephone sales call?" This sort of sliding scale can be used because the respondent does not have to remember 10 different descriptions—what does three mean again? However, the danger is that each respondent is going to give you a different sliding scale number if the range is 10. Instead, you might use a three-element scale without numbers. "How often do you use this system during a telephone sales call? Never, sometimes, or always?" That's much easier for the respondent to remember and evaluate and it's also more likely to generate a more consistent response across all respondents.

Our goal is not to go into the pros and cons of various data gathering methods, but to point out that there are unintentional problems you can build into a questionnaire or survey that can skew your results. If your organization has a group that develops market surveys or questionnaires, you may ask them to review your questionnaire before rolling it out. They might spot something you missed and help you gather better data. We all know the output is only as good as the input, so making sure your data gathering methods are clean will help on the other side of this assessment process.

Once an interview is conducted, the data needs to be reviewed and verified by the interviewee. Due to the nature of an interview, it's possible one of the people (interviewer, interviewee) misunderstood the question or response. Therefore, once the data is prepared, it should be reviewed by the interviewee before being finalized. You want to avoid having the interviewee rehash their previous responses, but you do want to provide an opportunity for additional insights and information that clarify previous responses. Follow-up interviews, if needed for clarification, should be scheduled as quickly after the initial interview as possible so that the data, response, and topic are still fresh in the interviewee's mind.

Workshops

Data collection workshops can be an effective method of gathering needed data. If you choose this method of gathering data, you might still choose to create a questionnaire so that you can be sure you cover all the required data points. Identify the appropriate level of participating personnel and gain agreement as to participants. Choose an appropriate time and place for the workshop, ensure the appropriate amenities will be available (white boards, refreshments, etc.). Develop a clear agenda for the meeting and distribute this, in advance, to meeting participants. Identify the workshop facilitator and clearly define his or her role in the process. Identify workshop completion criteria so the facilitator and participants are clear about what is expected, what the required outcomes are, and how the workshop will conclude. The facilitator's job is to ensure the workshop objectives are met, so these objectives must be clearly articulated prior to the start of the workshop. Develop or utilize an appropriate process for dealing with issues during the workshop so that participants stay on topic and focused on the key objectives. Some companies use the concept of a "parking lot," where issues are written up on note cards and collected or written on sticky notes and posted on a white board or an empty wall. Use an issue tracking methodology that allows you to stay on topic but make note of issues. Also identify the method you'll use for addressing those issues that cannot be (or should not be) resolved during the course of the workshop. Finally, ensure that the results of the workshop are written and well documented and that participants have the opportunity to review the results for errors and omissions before they are finalized.

TIP

Select the format for data gathering that is least intrusive on people's time and that is most aligned with how you normally work. Business continuity and disaster recovery planning are often very low on people's priorities and anything you can do to reduce the effort it takes to provide the data you need will pay off.

Determining the Impact

We've delineated some of the more common business functions. Now, let's turn our attention to some of the specific impacts to a business. As with other lists, this one is extensive but not necessarily exhaustive. Be sure to review this list and remove any items that do not pertain to your business and add any elements that are not included that do relate to your business. Remember, too, that a business disruption can run that gamut from a hard drive failure to an earthquake that levels your building to a pandemic that impacts an entire region or nation. Once you've looked at all the potential impact points, we'll discuss specific data points to collect and analyze as well as how to put those together with your risk assessment data. The impact of any business disruption may include:

1. **Financial**. Loss of revenues, higher costs, potential legal liabilities with financial penalties.

2. **Customers and suppliers**. You may lose customers and suppliers due to your company's problems or you may lose customers or suppliers if they experience a business disruption or disaster.

3. **Employees and staff**. You may lose staff from death, injury, stress, or a decision to leave the firm in the aftermath of a significant business disruption or natural disaster. What are the key roles, positions, knowledge, skills, and expertise needed?

4. **Public relations and credibility**. Companies that experience business disruptions due to IT systems failures (lost or stolen data, modified data, inability to operate due to missing or corrupt data, etc.) have a serious public relations challenge in front of them. These kinds of failures require a well-thought-out PR plan to help support business credibility. What impact would system outages or data losses have on your public image?

5. **Legal**. Regulations regarding worker health and safety, data privacy and security, and other legal constraints need to be assessed.

6. **Regulatory requirements**. You may be unable to meet minimum regulatory requirements in the event of certain business disruptions. You need to fully understand these regulations and their requirements related to business disruptions, both natural and man-made.

7. **Environmental**. Some companies may face environmental challenges if they experience failures of certain systems. Understanding the environmental impact of system and business failures is part of the business impact analysis phase.

8. **Operational**. Clearly operations are impacted by any business disruptions. These must be identified and ranked in terms of criticality.

9. **Human Resources**. How will staff be impacted by minor and major business disruptions? What is the impact of personnel responses to business operations? What are the qualitative issues to be addressed (morale, confidence, etc.)?

10. **Loss Exposure**. What types of losses will your company face? These include property loss, revenue loss, fines, cash flow, accounts receivable, accounts payable.

11. **Social and corporate image** (strongly tied to public relations). How will employees, customers, suppliers, partners, and the community view your company? How will its image be altered by a minor or major business disruption?

12. **Financial community credibility**. How will banks, investors, or other creditors respond to a minor or major business disruption? If the cause is a natural disaster, the challenges are different than if the cause is man-made. If the company failed to secure or protect data or resources, there are additional consequences both to the corporate image and to the company's credibility in the marketplace.

(Adapted from the Disaster Recovery Institute)

After you've compiled a list of your business functions and processes, you should assign a criticality rating to them. Payroll, accounts payable, and accounts receivable usually qualify as mission-critical business processes. Furniture requisitions for new employees usually fall to the bottom of the list as minor. Rate all your identified business processes and sort them in order of criticality. You might end up with a table or matrix that looks something like that shown in Table 4.1.

Table 4.1 Business Function and Criticality Matrix

Business Function	Business Process	Criticality
Human Resources	Payroll	Mission-critical
	Employee background checks	Important
Finance	Debt payments/loan servicing	Vital
	Accounts receivable	Mission-critical
	Accounts payable	Mission-critical
	Quarterly tax filings	Mission-critical
Marketing and Sales	Customer sales calls	Mission-critical
	Customer purchase history analysis	Vital

Business Impact Analysis Data Points

The number and type of data points you collect in your business impact analysis is largely a function of the size and type of company in which you work. Smaller companies will have fewer data points, larger companies will have more. However, you can also inundate yourself with too many data points if you don't take a focused approach. Some companies are extremely slow moving, analytical types of companies in which all data must be collected and assessed. Other companies move at the speed of light (typical in start ups) and want to grab just the high points and move on. The plan you devise needs to find a balance between information overload and superficial data. Be sure to include enough detail so that you can actually develop strategies that will help your company survive a serious business disruption, but don't allow the information floodgates to open and overwhelm you with minutiae.

Table 4.2 shows various data points you can consider collecting along with a brief description of the purpose or focus of that data point. Feel free to modify this to suit your unique needs.

Table 4.2 Business Impact Analysis Data Points

Data Point	Description	IT Dependencies
Business function or process	Short description of the business function or process (we'll use "function" from here on).	Describe primary IT systems used for this business function.

Continued

Table 4.2 continued Business Impact Analysis Data Points

Data Point	Description	IT Dependencies
Dependencies	Description of the dependencies to this function. What are the input and output points to this function? What has to happen or be available in order for this function to occur? What input is received, either from internal or external sources, that is required to perform this business function? How would the disruption of this business function impact other parts of the business? How and when would this disruption to other functions occur?	Describe IT systems that impact or are impacted by this business function. Are there any internal or external IT dependencies?
Resource dependencies	Is this business function dependent upon any key job functions? If so, which and to what extent? Is this business function dependent upon any unique resources? If so, what and to what extent (contractors, special equipment, etc.)?	Describe secondary/support computer/IT systems required for this business function to occur.
Personnel dependencies	Is this function dependent on specialized skill, knowledge or expertise? What are the key positions or roles associated with this function? What would happen if people in these role were unavailable?	Describe key roles, positions, knowledge, expertise, experience, certification needed to work with this particular IT system or IT/business function.
Impact profile	When does this function occur? Is it hourly, daily, quarterly, seasonally? Is there a specific time of day/week/year that this function is more at risk? If there a specific time at which the business is more at risk if this function does not occur (tax time, payroll periods, year end inventory, etc.)?	Describe the critical timeline related to this function/process and related IT systems, if any.

Continued

Table 4.2 continued Business Impact Analysis Data Points

Data Point	Description	IT Dependencies
Operational	If this function did not occur, when and how would it impact the business? Would the impact be on time or recurring? Describe the operational impact of this function not occurring.	Describe the impact on IT if this business function does not occur. Describe the impact on operations if this business function does not occur.
Financial	If this function did not occur, what would be the financial impact to the business? When would the financial impact be felt or noticed? Would it be one time or recurring? Describe the financial impact of this function not occurring.	
Backlog	At what point would work become backlogged?	Describe how a backlog would impact IT systems and other related or support systems.
Recovery	What types of resources would be needed to support the function? How many resources would be needed and in what timeframe (phones, desks, computers, printers, etc.)?	What resources, skills, and knowledge would be required to recover IT systems related to this business function?
Time to recover	What is the minimum time needed to recover this business function if disrupted? What is the maximum time this business function could be unavailable?	How long would it take to recover, restore, replace, or reconfigure IT systems related to this business function?
Service Level Agreements	Are there any service level agreements in place related to this business function? What are the requirements and metrics associated with these SLAs? How will SLAs be impacted by the disruption of this business function?	How would IT service levels be impacted by the disruption or lack of availability of this business function? How do external SLAs impact IT systems?

Continued

Table 4.2 continued Business Impact Analysis Data Points

Data Point	Description	IT Dependencies
Technology	What hardware, software, applications, or other technological components are needed to support this function? What would happen if some of these components were not available? What would be the impact? How severely would the business function be impacted?	What IT assets are required to support/maintain this business function?
Desktops, laptops, workstations	Does this business function require the use of "user" computer equipment?	What is the configuration data for required computer equipment?
Servers, networks, Internet	Does this business function require the use of back-end computer equipment? Does it require connection to the network? Does it require access to or use of the Internet or other communications?	What is the configuration data for required servers and infrastructure equipment?
Work-arounds	Are there any manual work-around procedures that have been developed and tested? Would these enable the business function to be performed in the event of IT or systems failures? How long could these functions operate in manual or work-around mode? If no procedures have been developed, does it seem feasible to develop such procedures?	Are there any IT-related work-arounds related to this business function? If so, what are they and how could they be implemented?
Remote work	Can this business function be performed remotely, either from another business location or by employees working from home or other off-site locations?	Can this business function be performed remotely from an IT perspective? If so, what would it take to enable remote access or the ability to remotely perform this business function?

Continued

Table 4.2 continued Business Impact Analysis Data Points

Data Point	Description	IT Dependencies
Workload shifting	Is it possible to shift this business function to another business unit that might not be impacted by the disruption? If so, what processes and procedures are in place or are needed to enable that function?	Are there other IT systems or resources that could pick up the load should a serious disruption occur?
Business/data records	Where are the business records related to this function stored or archived? Are they currently backed up? If so, how, with what frequency, where?	How and where are backups stored? Based on data provided, is the current backup strategy optimal based on the risks and impact?
Reporting	Are there legal or regulatory reporting requirements of this business function? If so, what is the impact of a disruption of this business function to reporting requirements? Are there reporting work-arounds in place or could they be developed and implemented?	Are there other ways reporting data could be generated, stored, or reported if key business functions or systems were dis-abled?
Business disruption experience	Has this business function ever been disrupted before? If so, what was the disruption and what was the outcome? What was learned from this event that can be incorporated into this planning effort?	Has IT ever experi-enced the disruption of this business function in the past? If so, what was the nature and duration of the disruption? How was it addressed and what was learned from the event?
Competitive impact	What, if any, is the competitive impact to the company if this business function is disrupted? What would the impact be, when would the impact occur, when would the potential loss of customers or suppliers occur?	

Continued

Table 4.2 continued Business Impact Analysis Data Points

Data Point	Description	IT Dependencies
Other issues	What other issues might be relevant when discussing this particular business function?	Are there other IT issues related to this specific business function that should be included or discussed?

Once you've collected all these data points for all your business functions and processes, you have a comprehensive understanding of your business, its key functions, and what would happen if those functions were disrupted. In the next chapter, we'll discuss how to develop risk mitigation strategies based both on the various risks your company faces and on the criticality of the various business functions as defined in this phase of the assessment.

Common Challenges...

Data Overload

The difficulty with the business impact analysis is that it can generate huge volumes of data that need to be sorted, assessed, and analyzed. There is no shortcut to getting this done, but it might help to keep the outcome in mind. The result you're looking for is an analysis of the critical functions and processes used in your company to conduct your company's business. Using the scenario approach can really help you focus in on the end result. If servers go down, if power goes out, if fire rages, if tornados strike, what are the most important things your company needs to accomplish to get business going again? We'll address the disaster recovery elements in an upcoming chapter—the things you need to do to stop the impact of the disruption or emergency before business can resume. For now, you need to understand what is absolutely essential to keep your business running. If you can keep this in mind as you go through this process, you're likely to be able to tune out the irrelevant and extraneous data more effectively.

Understanding IT Impact

As you can see from Table 4.2, the IT functions can be correlated to the business functions and processes at each step. As you gather this data, you will need to continually correlate the business functions/processes with the IT systems used to carry out or facilitate those functions in order to avoid gaps in your planning. In most cases, the subject matter experts and participants in this analysis will discuss the relationship of the IT systems to these functions. However, it's important to continually look at the intersection of IT systems to these business functions since the SMEs and departmental representatives may not fully understand the interdependencies of data or systems across the enterprise. For example, an SME might understand that use of the CRM system is vital to her job, but she may not have a clue that the CRM system resides on a server on the fourth floor and requires data updates from three other sources. From an IT perspective, you'll see this vital CRM function as a series of servers, applications, and data flows. As you work with the BC/DR team to map out the business functions and processes, you'll need to develop a parallel map of how that information intersects with IT equipment and functions.

In addition, you'll need to develop an understanding of how long it would take to replace or repair IT equipment based on the assessment of criticality. When you move into the risk mitigation phase, you might decide that the most optimal solution is to implement a fully redundant system for three key functions because the replacement or repair time for these systems exceeds the maximum tolerable downtime. The analysis of the data gathered in this phase must include IT-specific data so that you can optimize your risk mitigation strategies (coming up in Chapter 5).

The impact of IT on business functions (and the impact of business functions on IT) is usually already pretty well understood by the IT department through normal IT activities. However, the information gathered in this business impact analysis phase will bring to light new priorities, new gaps, and new challenges to be addressed through the IT department. Understanding how this data impacts IT and how IT impacts this data is key to developing a solid BIA and a comprehensive BC/DR plan.

TIP

You may want to encourage your subject matter experts to include their assessment of the impact on IT systems and the impact of IT systems on their critical business processes. By having them include this data, you can see IT from their perspective. You might learn something new about how they use IT systems or what you can do to mitigate risk to key business processes using IT technologies. At the very least, it will help flesh out your IT impact analysis.

Example of Business Impact Analysis For Small Business

Let's look at an example to help make this entire process a bit more tangible. A company of about 125 employees works out of a single location. They're situated in a light industrial area surrounded by warehouses and wholesalers. They sell a variety of specialty building hardware such as hard-to-find latches, fasteners, locks, and more. They purchase products from a variety of manufacturers and distributors and sell to a niche market in their region. These customers call in orders periodically. They also run a Web site that has seen sales grow significantly in the past three years, so that Web sales are now equal to non-Web sales.

The company, which we'll call ABC Hardware, does about $20 million a year in sales, about half of that online. Their facility is a large space comprised mostly of warehouse space with some office space. They ship and receive packages daily for Web operations and they ship weekly for their non-Web customer orders.

This company's risks include:

- Risk of fire in the building
- Risk of flooding in the area
- Risk of chemical spill in the area
- Risk of upstream/downstream losses by suppliers, vendors, customers

Let's focus on the risk of a fire in the building. If a fire struck the building, the damage might be contained to one of the areas, either warehouse or office. If the warehouse experienced a fire, inventory would be damaged and the ability to process inventory (receive, pick, pack, ship) would be impaired. If the office area were to have a significant fire, computer systems, including the inventory management system, would be damaged or destroyed.

So, what are the critical business functions impacted by a fire in the warehouse? First, we have the sales function because inventory would be damaged. Second, we have the inventory function because physical systems for managing inventory would be damaged.

What are the processes impacted by a fire in the warehouse? The company has processes in place for the following:

1. Picking orders.
2. Packing orders.
3. Staging orders for shipment.
4. Tracking shipments.
5. Receiving new inventory.
6. Stocking new inventory.

7.   Updating inventory systems with shipping and receiving data.

8.   Managing damaged or missing inventory.

9.   Processing returns of damaged or wrong items.

10. Inputting inventory data into inventory system.

11. Replenishing packing materials.

12. Repairing warehouse equipment.

13. Cleaning warehouse areas.

You can see from the list that items 11 through 13 are not critical processes. Other items on the list may not be mission-critical either, but we started with a full list of what goes on in the warehouse. If a fire engulfed the warehouse area, it's possible the building would be off-limits due to safety concerns, the offices might be filled with smoke and unusable, and the inventory might be smoke and water damaged by the fire suppression systems or by the water the fire department would hose in to put the fire out. Therefore, let's assume that a fire would impact all these processes listed. The company has no inventory it can ship to customers. What are the most important processes that have to get back up and running in order for the company to generate revenue and continue operations?

Remember, there are probably 14 other companies out there that are waiting for ABC Hardware to falter so they can swoop in and steal ABC's customers. ABC cannot afford to wait around for the water to dry and the smoke to clear before getting back into business. So, let's look at these first 10 items, along with criticality and comments, shown in Table 4.3.

Table 4.3 Example of Business Process and Criticality for Small Business

Business Process	Criticality	Comment
Picking orders	Mission-critical	Orders cannot be picked if inventory is damaged.
Packing orders	Mission-critical	Orders cannot be packed if they are not picked.
Staging orders for shipment	Mission-critical	Orders cannot be shipped if not picked and packed.
Tracking shipments	Mission-critical	Orders cannot be shipped if not picked and packed.
Receiving new inventory	Important	New inventory can be added to inventory system.
Stocking new inventory	Minor	New inventory cannot be stocked until damaged inventory is addressed.

Continued

Table 4.3 continued Example of Business Process and Criticality for Small Business

Business Process	Criticality	Comment
Updating inventory systems with ship/rec data	Mission-critical	No shipments going out but incoming inventory should be added so the company knows how much good inventory they have. Damaged inventory should be removed from stock as quickly as possible.
Managing damaged/ missing inventory	Mission-critical	Normally, managing damaged inventory is a minor process. In the aftermath of a fire, damaged inventory should be processed as quickly as possible to enable the company to dispose of it as quickly as possible.
Processing returns of damaged/wrong items from customers	Minor	Normally, processing damaged and returned items from customers would be a high priority. In the aftermath of a fire, this falls to a lower priority.
Inputting inventory data into inventory system	Mission-critical	In order for the company to sell its products, it needs to know, very quickly, what inventory it has that is sellable and what inventory it has that is damaged and must be discarded.

As you can see from this example, what normally might be high-priority processes shift to lower priorities in the aftermath of a fire. The key to recovery for this company is to sort out its inventory quickly so it knows what it can and cannot sell to customers. The IT systems are not damaged (though a few warehouse computers might need to be replaced) and order processing can still occur. This includes taking phone and online orders, processing orders, comparing orders to inventory levels, charging customer accounts or credit cards, and recording customer data (address, phone, etc.). Thus, the sales function for the company is relatively unharmed but the ability of the company to process and fulfill those sales is impacted.

The business impact analysis for this company now has identified the critical functions in the warehouse with regard to sales, inventory management, and shipping/receiving. The list is not exhaustive. For example, it does not include shipping supply replenishment. In the immediate aftermath of the fire, shipments cannot go out so this isn't a problem. However, it's likely that shipping supplies have been destroyed either by fire, smoke, or water, and need to be replaced before any shipments can go out. If the entire warehouse is impacted, there may be no saleable inventory and shipments will have to wait. In other cases, there may still be saleable inventory and the lack of shipping supplies would actually become a major problem. Therefore, replenishing shipping supplies as a process in the aftermath of a disruption might be mission-critical. This is how walking through scenarios helps you see the mission-critical processes more clearly.

What is the maximum tolerable downtime for these critical business functions and processes? Some of this company's customers are custom homebuilders who are working on tight timelines. They will not wait for a delayed order from ABC Hardware and will look elsewhere for these products. Therefore, ABC believes that with most of their orders, they have one week to recover operations before they begin losing serious revenue. In the risk mitigation phase of their assessment, this company's staff can devise a number of strategies to deal with this scenario either to prevent a fire from occurring or to create alternate fulfillment strategies in the event a fire does occur.

You can continue to expand this example to include other data. For example, you can include the expected financial impact, as shown in Table 4.4. The example is not complete but just shows the beginning of this process as a sample of how you might capture financial impact data. The first function, the sales function, in this example, is not immediately impacted by the fire in the warehouse. Sales are still generated through the Web site and sales people may still be able to access CRM systems and other sales tools to generate sales. The problem is not on the sales generation side but the order fulfillment side. At some point, the company's inability to process inventory and orders will affect sales. Customers whose orders are delayed may cancel, rumors may cause other customers to order from your competitors. If you can't receive new inventory or ship out existing orders, these will eventually impact sales, but not immediately. If you can forecast the delayed financial impact, that's great, but if you can't, just make a note that there is one down the line. We've also included an increased cost for customer service. If you have a fire and word gets out, customers may call about their orders, call to change or cancel their orders, or call to get assurance their order is in process. This may generate more work for customer service, which may need to bring in temporary help to staff the phones or work overtime to handle the increased volume.

Table 4.4 Financial Impact Example

Business Function	Business Process	Financial Impact
Sales	Generating new orders	Delayed impact
Warehouse	Picking orders	$2,000 per day
	Packing orders	$2,000 per day
	Shipping orders	$10,000 per day
	Receiving inventory	$4,500 per day
Customer service	Handle customer problems	$3,000

So far, we've seen little or no IT impact. The damage was contained to the warehouse and other than three computers used at the shipping and receiving stations, there was no other impact to IT. However, there are other IT tie-ins. For example, how will the company know the exact status of the inventory? When was the last inventory count performed? What is the status of the orders that were picked and packed—were they shipped or not? Which customer orders went out and which were on the dock awaiting shipment? Which returns were on the dock when the fire started and which were already processed? In this case, the company needs to quickly figure out the current status of its inventory as well as the status of customer sales and returns. It needs to know exactly what the status of everything is so that it can figure out what to do and in what order. IT may need to run special reports, print out inventory, shipment, or order lists in order to help warehouse functions get up and running again. These are disaster recovery tasks that the warehouse and IT staff will have to work together on to determine what might be needed.

You can extend this scenario and ask, what if the IT systems were located next to the warehouse and they were destroyed by fire? What if the fire started in the server room and spread to the warehouse? Now the scenario has changed significantly because not only do you have damaged inventory and uncertain status of shipments but you don't have IT system data immediately available to help sort things out. Sales data, inventory status, payables, receivables are all unavailable. The server room is charred, all systems are unusable. Now what?

Let's extend this just a bit so you can get the bigger picture. Table 4.5 shows some of the other operational impacts that might occur as a result of a warehouse fire. The impact on operations shows, for example, that customer perception is not impacted in the sales function. Customers may or may not know about the warehouse fire and if they can still place their order via the phone or Web, there is no immediate impact to customer perception. The same holds true for the customer perception of picking and packing orders. Customers usually don't know how their order shows up at their door (nor do they usually care), they care that the right products show up on time. Therefore, we begin to see a customer perception impact in the processes of "ship orders" and "receive inventory." If inventory can't be shipped, customers don't receive their orders as promised and this impacts customer perception. If inventory can't be received, it isn't available for sale and the customer sees that products are out of stock. We won't go through every cell in the grid, but you can use this to understand how various operations are impacted by a warehouse fire. The employee impact, in this case, is focused on warehouse staff, who are highly impacted by the warehouse fire. Though we did not do it in this example, you could also document the key knowledge and expertise needed to carry out these functions. For example, the key skills needed in this case are people who know how to manage inventory so that orders are properly filled and inventory levels are properly tracked. This data can be added, as appropriate. The same can be done for the IT side of the process. If IT systems were down, which processes would be impacted and how would other operations be impacted? What skills and expertise would be needed for workarounds and recovery?

Table 4.5 Operational Impact of Warehouse Fire

Business Function	Business Process	Cash Flow	Investor/ Market Confidence	Market Share	Competitive Position	Customer Perception	Employee Impact
Sales	Generate new orders	Medium	Medium	Medium	High	N/A	Low
Warehouse	Pick orders	High	Medium	Medium	High	N/A	High
	Pack orders	High	Medium	Medium	High	N/A	High
	Ship orders	High	Medium	High	High	High	High
	Receive inventory	Medium	N/A	N/A	N/A	High	High
Customer service	Handle customer problems	Low	Low	Low	Medium	High	High

As you can see, this scenario focused just on the warehouse department. The warehouse manager or someone designated by the manager should participate in this business continuity planning process. Only someone working in the warehouse is going to be familiar enough with the various day-to-day processes to generate a realistic view of the impact of various business disruptions. Once they have walked through all the risk scenarios (we mentioned fire, flood, chemical spill, and upstream/downstream impacts earlier), they can assign the criticality, the maximum tolerable downtime, the operational impact, financial impact, and the employee impact.

You may also choose to include additional columns in your impact table (or in your analysis if you choose not to use a tabular format) such as the financial impact and the legal impact. In this scenario, we also could have included the dependencies. Sales are impacted by the availability of inventory data (you can't sell inventory you don't have on hand or on order). Receivables are impacted by the ability to pick, pack, and ship inventory. Payables are impacted by the ability to receive inventory and manage missing/damaged inventory. Payroll is impacted by having to work additional hours to manage inventory damage from the fire as well as to perform work outside the normal scope of warehouse operations. Expenses go up because additional supplies must be purchased to replace the supplies lost in the fire. Sales are down because shipments cannot go out until inventory is adjusted and some customers have purchased elsewhere. The building has to be cleaned by a professional company that specializes in recovering from fire damage and that impacts operations and increases the company's expenses with an unplanned expenditure.

What you'll discover from this process is that as you walk through these scenarios, you'll begin getting ideas about how to mitigate the impact of these disruptions. In Chapter 5, when we discuss mitigation strategies, you'll find that one mitigation strategy might be helpful for three or four different risk scenarios. Thus, what would reduce your risk in the event of a fire might also be an excellent strategy for mitigating the risk of flooding or a chemical spill in the area. These economies are found only by thoroughly assessing risks and impacts so you can see the big picture and develop optimal mitigation strategies.

Now that you have identified the critical business processes for the warehouse department, you can also look at the impact a flood would have. For example, if employees cannot get to work, if trucks cannot come in to deliver inventory, if trucks cannot pick up shipments, many of these activities are impacted. If the warehouse area is flooded, you have a similar problem as you did with a fire. If the area surrounding the building is flooded but your inventory and IT systems remain in tact, you have a different set of challenges.

By identifying the critical business functions and processes, you can clearly see the impact various risk sources would have on the business. You can assign criticality and maximum tolerable downtime in preparation for developing effective strategies for addressing these risks.

If you were to continue with this example, you would define specific recovery objectives based on criticality, you would identify organizational and system dependencies, and you would define work-around procedures that could be used. This would comprise the impact analysis for the warehouse department for the risk of fire. If you expand it to include the same assessments for each threat source identified in your risk assessment, you would have a comprehensive impact analysis for your warehouse department. Each department in the company would complete this process and you'd have the risk assessment and impact analysis for the entire company. As you can see from just this small example, it's a large undertaking and may well take more time than any other part of your project. Allow enough time to get this completed but don't let it get long and drawn out. Most of this can be completed by departments in a reasonable amount of time, though the more complex the business systems, the longer it will take to perform this assessment.

Preparing the Business Impact Analysis Report

There is no standardized format for a business impact analysis report and, as with many other processes, this document will likely follow your company's standard format. At minimum, the report should include the business functions, the criticality and impact assessments (see the list is Table 4.2) and the maximum tolerable downtime (MTD) assessment for each. Dependencies, both internal and external, should be noted and the correlation to IT systems should be delineated.

This report should be prepared in draft format with initial impact findings and issues to be resolved. The participating managers, SMEs, and BC/DR team members should review the findings. Revise the report based on participant's feedback to the draft document. If needed, you can schedule a review meeting to discuss the finding in the draft. Often this is helpful (and needed) to resolve conflicts with regard to the criticality and maximum tolerable downtime ratings, since there is a correlation between these ratings and the cost of mitigating the risks and reducing downtime. Once the feedback has been gathered, revise the draft and finalize the document. This document, depicted at the outset of this chapter in Figure 3.2, is used along with the risk assessment as an input to the risk mitigation process. To assist you in preparing your final report, we've recapped the elements you may choose to include.

- Key processes and functions
- Process and resource interdependence
- IT dependencies
- Criticality and impact on operations
- Backlog information
- Key roles, positions, skills, knowledge, expertise needed
- Recovery time requirements
- Recovery resources
- Service level agreements
- Technology (IT and non-IT technology)
- Financial, legal, operations, market, staff impacts
- Work-around procedures
- Remote work, workload shifting
- Business data, key records

- Reporting
- Competitive impact
- Investor/market impact
- Customer perception impact
- Other (business-specific data not already included)

Summary

Performing the business impact analysis requires you to look at your entire organization from top to bottom. You can begin by gathering subject matter experts, whether division heads, departmental managers, or designated staff, from various parts of your company. These people should be those in the company best able to answer the questions related to critical business activities. This relates to how your company generates revenues, tracks customers and sales, and other key business processes.

Data can be gathered using questionnaires, interview, workshops, documents, and research. There are pros and cons to each approach, so be sure to select the method most appropriate to your organization. Since each company is unique, there is no "one size fits all" template you can use to delineate all critical business processes for all companies. However, throughout this chapter, we discussed a wide variety of business functions, processes, and approaches that can help you develop a comprehensive list of your company's critical processes as well as the key roles, expertise, and knowledge needed to carry out those critical processes.

Once this data is collected, each process must be assessed for criticality. In the big picture, how critical is each business process to your company's ability to continue operating? Using a three- or four-point rating system will help you look across the depth and breadth of your organization to understand which processes and functions are mission-critical, which are vital or essential, which are important, and which are minor. Your risk mitigation planning efforts will focus first on mission-critical processes and then to vital or essential processes.

You'll also need to develop your recovery time objectives (RTO) for each critical function. In some cases, you might choose to associate a recovery time with criticality ratings. For example, mission-critical functions might need to be recovered within 24 hours whereas vital or essential functions might need to be recovered within 72 hours. Alternately, you can assign criticality and then assign recovery time objectives to each process individually. This might make more sense in companies where there are numerous mission-critical processes that cannot be simultaneously addressed. Again, this is a decision you and your team have to make regarding recovery objectives. Input from division or departmental experts is key to understanding required recovery timeframes as well as key interdependencies that exist among departments, processes, and systems.

There is a relationship between the cost of recovery and the cost of downtime. Each company has to assess these costs and make decisions regarding the optimal point of intersection. The longer the company goes without a key process, the more expensive it becomes due to loss of sales and increase in costs associated with the outage. However, recovery costs go down the longer you have to recover. If you need to recover within hours, your costs to provide this type of recovery capability will be significantly higher than if you need to recover within days. The point at which downtime costs and recovery costs intersect is the optimal point for planning, though in the real world, it can be difficult to determine the exact point of intersection. Keeping this concept in mind, however, will help you find the best solutions for your company.

The business impact analysis uses business functions, business processes, and IT systems as the input points. The analysis is performed so that each process is identified and analyzed. The output for each process and function includes criticality assessment, financial impact analysis, operational impact analysis, recovery objectives, dependencies, and work-around procedures. When this is documented for each business function and key business process, you have a comprehensive look at your company and a solid business impact analysis.

Solutions Fast Track

Business Impact Analysis Overview

☑ After identifying risks and threats to the company, the business impact must be evaluated. Key business functions and processes are viewed in light of risk assessment data.

☑ The impact of disruptions not only to your business but to upstream and downstream partners needs to be considered.

☑ Consider the impact on corporate employees including physical or emotional injuries in the aftermath of a serious event or natural disaster. People respond in many ways to disasters and your plan must have the flexibility to allow for a variety of responses.

☑ For each key business process, critical objectives, timelines, dependencies, and impact must be understood and analyzed.

☑ The impact of the disruption of key business functions is assessed and prioritized so that risk mitigation strategies can be developed.

Understanding Impact Criticality

☑ Not all business functions and processes are mission-critical. Your risk mitigation strategy planning usually is limited to those functions and processes that are vital to the ongoing operations of the company.

☑ You can use a three- or four-point system of rating criticality. The four-point system ratings are mission-critical, vital (essential), important, minor. If a three-point system works better for you, you can use mission-critical, important, and minor. Define these clearly so they are used consistently across the organization.

☑ All processes should be assessed for criticality. Recovery objectives must also be assigned. Some companies assign the recovery time with the criticality. Therefore, mission-critical would have a recovery objective of 0–4 hours, for example. Other companies choose to set recovery objectives separately.

☑ The total time it takes to recover from a business disruption includes the recovery point objective, which is the lag between the time of the last good backup and the business disruption, the time it takes to recover systems, the time it takes to recover data, and the testing and verification of repaired systems. This is often called the maximum tolerable downtime (MTD) or maximum tolerable outage (MTO).

☑ There is an optimal point between the cost of downtime and the cost of recovery. The longer systems are down, the more expensive it is for your company. The shorter the required recovery time, the more expensive it is for your company. Therefore, the intersection of the cost of downtime and the cost of recovery is the optimal point. This is not always easy to determine but the concept helps in your planning efforts.

Identifying Business Functions

☑ Business functions are areas of the company that have specific roles or purposes such as sales, operations, finance, or HR. Business processes are the defined methods and actions used to achieve those purposes. Both functions and processes must be assessed in order to fully understand the company's critical work.

☑ The most common business functions include facilities, security, HR, IT, legal, compliance, manufacturing/assembly, marketing/sales, operations, research/development, and warehouse/inventory.

☑ The most common business processes include sales, invoicing, inventory management, and payroll, to name just a few.

Gathering Impact Data

☑ Gathering data for your business impact analysis is a significant undertaking. Enlisting subject matter experts (SME) from around the company is vital to your success.

☑ Using scenario-based questions, you can help SMEs understand what you're asking of them and help them envision potential problems. The more realistic your scenarios, the better data you'll gather.

☑ The data you gather should include the business function, process, criticality, time to recovery, dependencies, financial and operational impact, and other relevant data.

☑ You can use questionnaires, interviews, workshops, documents, and research to gather data. There are pros and cons to each approach; use the one that best fits your organization's way of doing business.

Determining Impact

☑ Determining the impact runs the gamut from financial to legal to operational to environmental and beyond. It's important to understand the impact to the company from these various perspectives, even if your focus is on the impact related to IT systems.

☑ The impact of a business disruption may have serious legal, financial, or regulatory consequences. These typically come from outside the organization and should be included in your planning. It's sometimes easy to miss these external elements when focusing solely on internal business impacts.

☑ The company's reputation in the community, region, or marketplace can be greatly impacted by a business disruption, especially if that disruption has to do with data security, data loss, or other sensitive areas. This should also be taken into consideration as you look at the impact analysis.

Business Impact Analysis Data Points

☑ There are numerous data points that can be collected about business processes across the organization. A comprehensive look will include these data points along with the interdependencies and impact on/with IT systems.

☑ For each critical business process, the impact to and impact from IT systems should be mapped out. In some cases, the disruption of a business process impacts IT systems. In other cases, the disruption of business processes does not impact IT but the disruption of IT systems, either primary or secondary, can impact key business processes. These interdependencies must be clearly understood and documented.

☑ External elements such as regulatory compliance, reporting, and corporate reputation must also be addressed. Again, the IT relationship must also be addressed. Often there is no leeway in meeting financial or legal obligations, regardless of the nature of the business disruption. There may be a bit of flexibility if a large natural disaster impacts the firm, but an isolated event such as localized flooding or fire will not alter regulatory, legal, or financial requirements on the firm.

Frequently Asked Questions

The following Frequently Asked Questions, answered by the authors of this book, are designed to both measure your understanding of the concepts presented in this chapter and to assist you with real-life implementation of these concepts. To have your questions about this chapter answered by the author, browse to **www.syngress.com/solutions** and click on the **"Ask the Author"** form.

Q: There seem to be far too many things to consider when doing the business impact analysis. I don't really know where to start. Any suggestions to make this process less overwhelming?

A: The business impact analysis is probably the largest data gathering aspect of this entire project and it can be overwhelming. The key to success is first to identify the various business functions then recruit experts from each function to participate. If you have to sit down and map all this out yourself, you not only will be overwhelmed, you'll also probably have lots of gaps and errors. This has to be an organizational effort, not just something the BC/DR team does off in a corner. Next, if you create a clear, concise set of questions that you want each subject matter expert to respond to, you have a much better chance of getting good data. In some companies, creating a series of workshops and working together in a less formal atmosphere may make this process a bit more interesting and productive. If you break it down by function or department and just start working your way through the data, you'll find you make it through this process a bit more easily. It's a big job but defining the segments and working systematically through it will help you get there successfully.

Q: I'm an IT analyst and a lot of this information doesn't relate to my job or role in the project. Can't I just skip over this section?

A: You could, but not if you want to have a successful project. Even if your role is limited to assessing IT functions, you need to understand how your company conducts business. Without that understanding, you won't be able to make intelligent assessments about IT systems. Sure, you know which servers are running which applications, you understand user access and security, but how does this relate to the day-to-day activities in your company? If the building were to burn to the ground with your IT systems in it, how would you prioritize your next steps? If you don't know which activities are mission-critical, you can't make intelligent assessments about which systems should be restored first. Certainly, there may be IT-related constraints with regard to the order or priority of system recovery, but you also need to consider the bigger picture. Critical business processes must resume first, regardless of where they fall in the IT world view. Therefore, participating fully in this process will make you better able to participate fully on this team and it will also help you be a more productive contributor to the overall business.

Q: You didn't spend much time talking about IT systems in this chapter. I thought this book was focusing on business continuity and disaster recovery for IT professionals. Did I miss something?

A: No you didn't miss anything. Any IT professional needs to focus on these businesswide issues, regardless of whether you're heading up the BC/DR effort or just focusing on IT needs. We didn't spend an undue amount of time on IT systems at this juncture because this section focuses specifically on the *business* impact analysis. You should include your IT systems as part of your assessment, just as you included other functions such as warehouse or marketing. However, since you know your IT systems and your IT processes intimately, we focused instead on areas that are likely to be less familiar to you. The processes and procedures discussed in this chapter, however, should be applied to your IT functions and processes as well. The interdependency of IT systems with other business functions is important and that's why we focused on that area more than strictly on IT systems. We'll look at IT systems in more detail in upcoming chapters.

Chapter 5

Mitigation Strategy Development

Solutions in this chapter:

- **Types of Risk Mitigation Strategies**
- **Risk Mitigation Process**
- **IT Risk Mitigation**
- **Backup and Recovery Considerations**

☑ **Summary**

☑ **Solutions Fast Track**

☑ **Frequently Asked Questions**

Introduction

Risk mitigation is defined as *taking steps to reduce adverse effects*. Risk mitigation is a commonly used process within traditional business risk management, but as you'll see in this chapter, there are unique aspects to risk mitigation related to business continuity and disaster recovery.

Your data gathering phase has concluded and now it's time to put all this data to work. The mitigation strategy development phase of the business continuity and disaster recovery project plan, shown in Figure 5.1, is where you develop strategies to accept, avoid, reduce, or transfer risks related to potential business disruptions.

Figure 5.1 Business Continuity and Disaster Recovery Project Plan Progress

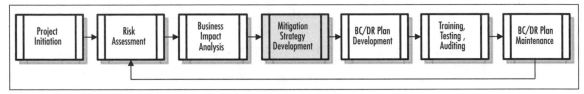

Developing the risk mitigation strategies is the last phase of risk management activities, which was shown in Figure 3.2 in Chapter 3. This last segment, depicted here in Figure 5.2, includes the inputs of the risk assessment and business impact analysis data. This information, along with risk mitigation data, is used to develop strategies for managing risks in a manner that is appropriate for your company. Once you have the risk management section completed, you can begin to draft your business continuity and disaster recovery plan.

Figure 5.2 Risk Mitigation Strategy Development Phase

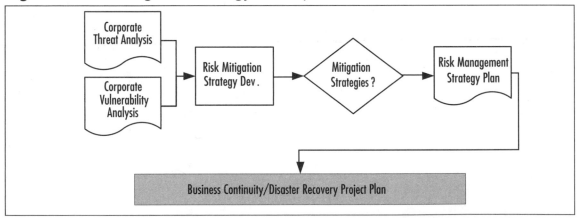

As we've mentioned before, it's important to develop risk mitigation strategies that match your company's profile. If your company is very risk averse and wants to avoid risk at

almost any cost, your strategies will be appropriate for that objective. On the other hand, if your company doesn't mind taking on a bit of risk, your BC/DR strategies will be different than a more conservative, risk-averse company's approach. There is no one-size-fits-all answer in the risk mitigation phase—you'll have to create a strategy that meets your company's financial, operational, and risk management goals.

Types of Risk Mitigation Strategies

Let's begin with a quick review of standard risk mitigation strategies. These will be useful as you develop your strategies, and a clear understanding of your options at the outset will help you and your team make better decisions. The four standard choices are acceptance, avoidance, limitation, and transference. As you read through these four options, refer to Figure 5.3, which shows the relationship between time and cost for each option and the relative cost of each option to the others over time.

Figure 5.3 Cost vs. Time for Risk Mitigation Strategies

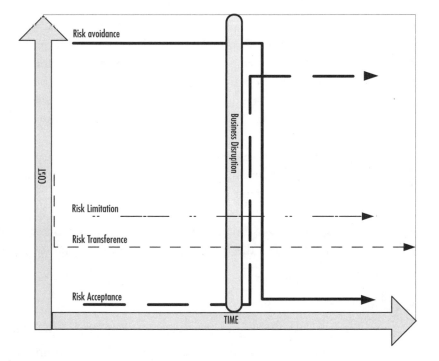

Risk Acceptance

Risk acceptance is not really a mitigation strategy because accepting a risk does not *reduce* its effect. However, risk acceptance is part of risk management. There are various reasons why companies may choose risk acceptance in certain situations. The most common reason is that the cost of other risk management options, such as avoidance or limitation, may outweigh the cost of the risk itself. Insurance companies are notorious for risk acceptance in the sense that they will allow a damaged tree limb to fall on your car and then pay for the repair of the car rather than pay to avoid the car repair expense and cut the limb down before it causes damage. In their case, they know that the limb may not fall down or that your car may not be present when the limb falls. They're taking their chances because if they paid to cut down every damaged tree limb reported to them by their customers, they'd spend a lot of money avoiding risks that were not actually going to occur. These companies spend millions of dollars each year analyzing the odds and are therefore able to make highly sophisticated risk management decisions.

Your company, on the other hand, is probably not in the odds-making business, especially when it comes to business continuity and disaster recovery planning. As you develop your strategies, you should consider the implications of "doing nothing." This can be a way of ensuring that you're taking appropriate actions because if you consider the implications of accepting the risk, you can see the potential consequences and weight them out against other options.

As you can see in Figure 5.3, the cost of risk acceptance is very low at the beginning (it may even be zero), but after a business disruption, the cost can be significantly higher than other risk management strategies. The company may be willing to save money today knowing that it will have a disproportionately large expenditure later if a business disruption occurs. That's key—you have to understand that you're betting that the business disruption will not occur or if it does, it will be far enough in the distant future that you're willing to take the financial risk.

A word of caution here: Small businesses often take the stance that they cannot afford to avoid, limit, or transfer risk and therefore, they accept risk by default. This is a mistaken and limited view and should not be the default position going into this planning. Risk acceptance should be evaluated along with the other options to determine the implications, appropriate actions, and costs of various mitigation strategies. Risk acceptance is the least expensive option in the near-term and the most expensive option in the long-term should an event occur.

Risk Avoidance

Risk avoidance is the opposite of risk acceptance because it's an all-or-nothing kind of stance. To continue with the insurance example, cutting down the tree limb would be risk avoidance. The insurance company would be avoiding the risk that the tree limb would fall on your car, on the house, or on a passerby.

In business continuity and disaster recovery plans, risk avoidance is the action that avoids any exposure to the risk whatsoever. If you want to avoid data loss, you have fully redundant data systems or you manually shut down systems and move them in advance of an oncoming hurricane. Risk avoidance is usually the most expensive of all risk mitigation strategies, but it has the result of reducing the cost of downtime and recovery significantly. Figure 5.3 shows this relationship—the cost is very high early on but the cost after a business disruption is lower than other strategies. Shutting down systems is costly in advance of a hurricane, but if they are packed and shipped to another location and fired up, the cost to recover from the business disruption is minimal. This option is not feasible for many types of risks or for many types of companies. However, it is a viable option to consider as you develop your risk mitigation strategies.

Risk Limitation

Risk limitation is the most common risk management strategy employed by businesses. You choose to limit your exposure through taking some action. For example, performing daily backups of critical business data is a risk limitation strategy. It doesn't stop a disk drive from crashing, it doesn't ignore the potential for disk failure, it accepts that drives fail and when they do, having backups helps you recover in a timely manner. In Figure 5.3, you can see that risk limitation strategies fall between acceptance and avoidance both in terms of early costs and costs after the business disruption. In a sense, it's an average of the two. Risk limitations include installing firewalls to keep networks safe, creating backups to keep data safe, practicing fire drills to keep employees safe, and more. We'll discuss various risk limitations you can take with regard to your key business processes throughout the remainder of this chapter because this is, by far, the manner in which most businesses choose to deal with their risks.

Risk Transference

Risk transference involves handing the risk off to a willing third party. Many companies outsource certain operations such as customer service, order fulfillment, or payroll services. They do this in many cases so they can focus on their core competencies, but they can also do this as part of risk management. For example, if you outsource your payroll services, you may choose to select a processing company that is not located in the same geographical region as your firm. If you're in the southeastern United States, you may choose a company in the Midwest or that has multiple processing sites around the United States so that it can process payroll regardless of weather events.

Another example of risk transference is purchasing insurance or other insurance types of services. In order to transfer risk, you usually have to pay some other company some amount of money to assume that risk, whether it's an IT shop that will manage your security or databases for you, or an insurance company that will pay for losses in the event of a business disruption. Figure 5.3 shows that, relative to other choices, your risk transference will usually cost more as some sort of up-front or ongoing fee, but that the overall cost will be somewhere in the same area as risk limitation. One important point to note, however, is that risk limitation usually has an end-point cost where risk transference can be ongoing. For example, you make insurance premium payments every month or quarter, regardless of whether or not you experience an event that requires your insurance company to step in. With risk limitation, you typically put some system in place, such as a firewall or redundant system. The cost of that implementation is finite and known and usually ends at some point in time. Thus, while the near-term costs of risk limitation and risk transference may appear to be similar, it's important to understand the *duration* of the cost with regard to these strategies.

Common Challenges...

Under the Radar

Some companies don't like to discuss risk either because they don't want to acknowledge it or because they are cavalier about the risks they face. This latter stance is most commonly found in small, entrepreneurial start-ups that have their hands full just getting the business off the ground. Often the larger a company gets, the more it is willing to discuss, plan for, and mitigate various kinds of risks. This may be, in part, due to outside pressures of financial markets or investors. If you're working in a small company that doesn't want to address risk, you may run into challenges even getting a BC/DR plan off the ground. As we discussed earlier in the book, you may be able to implement many of the BC/DR plan elements without making a big, formal process out of it. If this is the only way you can do BC/DR planning, it may be worth working in stealth mode. For example, when you look at data backup methods, you may choose to select and implement technologies and processes that not only meet your backup needs but provide an adequate level of BC/DR capabilities as well. You should certainly follow the rules, regulations, and procedures in your company, but you may find that you have a bit of leeway when it comes to implementing technology solutions that will meet the broader needs of the company, even if the company doesn't want to know about it.

The Risk Mitigation Process

In order to develop a risk mitigation strategy, you first have to know your options. In previous chapters, we looked at the various risks, threats, threat sources, vulnerabilities, and impacts. Next, we need to look at the recovery profile including the recovery requirements, options, timeframe of options (compared with maximum tolerable downtime or MTD), and cost versus the capability of options. From there, we can select appropriate options. Once these elements are known, a comprehensive strategy can be devised. The strategy will ultimately also include identifying off-site requirements and alternate facilities, and developing business unit strategies. In the following sections, we'll look at the recovery steps specifically.

Recovery Requirements

Recovery requirements typically are broken down by functional areas including facilities and work areas, IT systems and infrastructure, manufacturing and production (operations), and critical data/vital records. Your company may have other recovery requirements. If so, they should be included in this section. The recovery requirements are developed for the critical business processes identified in the business impact analysis. They help identify the resources that should be the focus of the recovery strategy since there is a cost involved with developing and implementing a mitigation or recovery strategy. If a process is not mission critical (or essential), it is likely not a good candidate for the expenditure of time and effort to develop mitigation strategies. Recovery requirements can be categorized even within the functional areas. For example, a recovery requirement category for facilities is alternate office space. Another category might be a crisis management center or a communications command center. Once you identify the recovery requirements, you can begin to review recovery options.

Recovery Options

For each critical business function or process, you have identified the impact on the organization; the dependencies to other functions; the IT dependencies, the key positions, skills, and knowledge needed; and the time requirement for recovery (among other things). Based on this data and on the recovery requirements, you can develop a variety of recovery options. Typically these options will come with varying timelines of their own as well as varying costs and capabilities. At this juncture, your primary concern is to develop a list of viable options based on the business impact analysis data you have. For example, if you have a requirement for an alternate computing facility, you have numerous options available including borrowing computer space from a local firm to setting up a colocation center outside your own geographic area and many other options in between. These options, unless absolutely outside the realm of possibility, should be listed so they can be included in the subsequent evaluation steps.

There are three basic recovery options you can consider. Each of these can also be considered part of a mitigation strategy, as you'll see. You can acquire the option *as needed*, you can *prearrange* for an option, or you can *preestablish* an option. Figure 5.4 shows the relative cost relationship of these three options.

Figure 5.4 Cost Relationship of Recovery Options

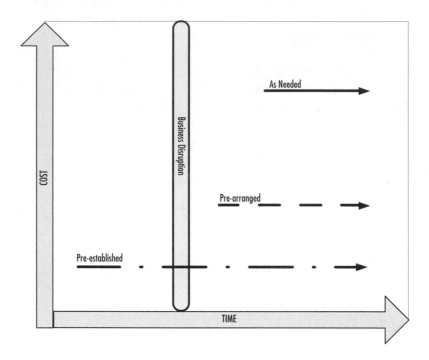

Notice that *as needed* options often take longer to implement after a business disruption and typically cost more. However, the cumulative cost may still end up being lower than *pre-arranged* or *preestablished* options, especially if one-time setup fees or recurring maintenance fees are required for those other options. *Prearranged* options are typically less expensive than *as needed* solutions, and they often can be implemented in a more timely manner since availability should be guaranteed in the arrangement. Finally, *preestablished* solutions can be implemented almost immediately, but they often have a recurring cost or a sunk cost in advance of the disruption. This can make their total cost more than other options. Clearly, *time* is one of the major factors in each of these options. Let's look at each of these options in more detail.

As Needed

Acquiring resources as needed at existing market rates and within existing market availability following the business disruption is one recovery approach. If the disruption is isolated to your company, as would be the case in a fire or building collapse, market rates and availability might be acceptable. If the disruption is broader in scope, such as an earthquake or volcano eruption, market rates may skyrocket while availability may plunge. In some cases, availability may go to zero, regardless of the price you're willing and able to pay.

Prearranged

Prearranged options involve making arrangements in advance for the quick shipment or delivery of materials, supplies, and capabilities later. These types of arrangements typically involve a contractual agreement with a vendor to supply required systems, products, or services within an agreed upon time frame following a business disruption. There is often a cost to creating these arrangements or a charge above existing market rates built into the contract. For example, if delivering new IT systems is prearranged with a computer maker, there may be an up-charge over the existing market cost of the systems for fast turnaround, expedited or custom system configurations, testing, shipping, delivery, and setup. However, these would all be specified in the contract so that costs would be contained and would be known in advance. In addition, availability requirements are included in the contractual agreement so your firm is not subject to the vagaries of the open market in the aftermath of a major event.

Preestablished

Preestablished recovery options are those that are purchased, configured, and implemented prior to a disruptive event and are used only for recovering from a disruptive event. A company-owned alternate computing site that is activated only in the aftermath of a business disruption would be considered a preestablished recovery option. Often the cost of this type of solution is lower on a per-unit basis because the expenditures can be timed and managed. However, the cost over time may be higher, depending on the cost of these preestablished options. For instance, if you purchase IT systems in the exact configuration of existing systems and have them stored at an alternate location in the event of an emergency, those systems sit idle until (and unless) there is a business disruption. Unlike working systems, these are sitting idly and are therefore nothing but nonproductive expenses. Certainly, if your company experiences a disruption, the cost of these preconfigured, preinstalled machines is suddenly a good investment. If your company never experiences a disruption, the systems become outdated and useless. Never having been in production, they must nonetheless be upgraded or replaced periodically, leading to additional costs.

TIP

For IT systems, preestablished and prearranged solutions are often best. Trying to get IT systems acquired, shipped, set up, configured, and online in the aftermath of a business disruption is a major undertaking. Anything you can do in advance, within the constraints of your organization, will be well worth it if your company faces a disruption. You'll have to balance the cost of preparing against the cost of dealing with the aftermath. In some companies, this cost can't be justified. In larger companies, it almost always makes financial and organizational sense to make arrangements in advance.

Recovery Time of Options

Once you've developed your list of recovery requirements and options, you can look at the recovery time of each option. For example, borrowing space for your computers from another local company might be prearranged and therefore it could be implemented within a matter of hours. A colocation facility, if preestablished, potentially could be online within minutes of a business disruption. Buying new computers and setting them up in a temporary work location such as a local hotel conference room or mobile office unit is another option but it typically would take days to get that set up. Having defined the maximum tolerable downtime for your critical business processes, you must now compare that data to the recovery time of the options you're considering. Any option that does not meet MTD requirements should be removed from further consideration at this juncture. In this way, only options the meet MTD requirements will be assessed in terms of cost and capability.

Cost versus Capability of Recovery Options

You should have a pared down list of recovery options based on those that meet MTD *and* recovery requirements. Next, you'll assess the cost of each of the remaining options and list the capabilities included in that cost. Some options may have various levels of cost/capability. In most cases, the higher the capability, the higher the cost. Since all mitigation strategies will ultimately have to meet the company's financial constraints, this data is critical to making the right decisions for your company. There are additional attributes that can and should be included in the cost/capability assessment. These are:

- Cost—the cost of the mitigation or recovery option.

- Capability—the capabilities of the option.

- Effort—the amount of effort it will take to implement and manage the option.

- Quality—the quality of the product, service, or data associated with the option.

- Control—the amount of control the company will retain over the critical business process.

- Safety—in cases where physical safety is a concern, this attribute rates the safety of the solution. If setting up a few braces on a faltering ceiling over the data center is among your recovery options, its safety attribute would be about zero compared to other options.

- Security—the estimates of physical and virtual (information and network access) security the option provides.

- Desirability—the assessment of the overall desirability of an option. In many cases this is a qualitative judgment based on quantitative data. If so, the quantitative data should be included. The reasons for rating desirability as high, neutral, or low should be delineated.

You can create a matrix to review these attributes and help you make sound decisions. Table 5.1 shows a grid related to various options related to acquiring critical IT systems; Table 5.2 shows options of establishing alternate computing facilities. These are two different approaches to mitigating risk and as you assess these attributes, you can make decisions as to your best options.

Table 5.1 Example: Options for Acquiring Critical IT Systems

Option	Cost	Capability	Effort	Quality	Control	Safety	Security	Desirability
As needed	High	Unknown	High	Low	Low	N/A	N/A	Low
Prearranged	Medium	Meets requirements	Medium	Medium	Medium	N/A	N/A	Medium
Preestablished	Low	Meets requirements	Low	High	High	N/A	N/A	Medium

Table 5.2 Example: Options for Establishing Alternate IT Facilities

Option	Cost	Capability	Effort	Quality	Control	Safety	Security	Desirability
Company cold site	Medium	Meets requirements	Medium	Low	High	Medium	Medium	Medium
Outsourced hot site	High	Meets requirements	Low	High	Low	High	High	Medium

Remember that these options are being considered only because they met the recovery requirements including time to recover. Therefore, they're all viable options at first glance. However, additional analysis is required to understand the particular needs of your company and the viability of these various options.

Recovery Service Level Agreements

Any agreement you enter into for recovery services should include specific metrics such as time, cost, availability, response time, throughput, bandwidth, and so on. These metrics all fall under the category of service level agreements (SLA) and can include a number of different elements including:

- Response time to initial request for services

- Technical capacities—computer equipment specifications, storage space, voice and data capacities, speeds, bandwidth availability, test equipment, among others

- Access to recovery facility and equipment

- Access to adequate work area and access for staff

- Security procedures and guarantees

- Processing controls
- Access to technical and functional support (time, response time, etc.)

Another important aspect to reviewing recovery service level agreements is to look at any existing SLAs you may have with external parties such as your clients or customers. If you have contractual agreements to process data or ship orders within a specific period of time, you will need to review your recovery options in light of those contractual agreements. Although these SLAs should have been identified in your business impact analysis as critical business functions, it's good to take the opportunity here to ensure your risk mitigation strategies address your contractual obligations and in particular, any SLAs that are currently in place.

Review Existing Controls

In some cases, you may already have all or part of these controls in place. For example, you might have a very robust data backup solution in place and by adding an additional service or two, you can meet these recovery requirements fairly easily. The reason for reviewing these controls after you've reviewed your recovery requirements is because you want to be able to look at existing solutions with fresh eyes. If you were to begin by examining existing solutions and try to fit them into your recovery options, you might have a built-in bias toward existing solutions. This is especially true if you were the one who championed or implemented the solution or if you happen to know that it was a very expensive, high-end solution. To avoid these natural biases, it's best to review your recovery options first then compare the optimal solutions to existing solutions. In some cases, you'll find that existing solutions meet requirements. In other cases, solutions might actually exceed requirements. Finally, you will undoubtedly find areas where existing solutions do not meet requirements and you'll need to address these areas.

If you find that you have solutions in place that address various recovery requirements, be sure to include these in your risk mitigation strategy document. As stated previously, this might be an opportunity to show the value of a previous investment or at least to show how an existing investment is serving dual purposes. In addition, you want to include these existing solutions in your risk mitigation strategy so that you keep these systems in mind as your systems and BC/DR plans change over time. For example, if you have a solid backup solution that is part of your risk mitigation strategy, that should be noted so that if you decide in the future to modify your backup strategy, you can evaluate the impact on your BC/DR plan. You may have a checklist (on paper or just in your mind) of things you consider when you look at technology investments—speed, compatibility, cost, security. Be sure to add BC/DR to your list so that any future investments can be evaluated in light of BC/DR requirements as well.

TIP

Leverage existing assets, processes, and procedures to the greatest extent possible, but don't be afraid to rip a solution out by the roots if it doesn't meet your immediate and long-term needs. Don't continue to support a failing (or failed) solution just because no one wants to be the one to terminate it. This BC/DR planning process can help you identify areas for improvement. It may also provide you with the financial and organizational support you need to update legacy systems that have outlived their usefulness.

Developing Your Risk Mitigation Strategy

The steps in developing your risk mitigation strategy are these:

1. Gather your recovery data.

2. Compare cost, capability, and service levels of options in each category.

3. Determine if the options remaining are risk acceptance, avoidance, limitation, or transference and which, if any, are more desirable.

4. Select the option or options that best meet your company's needs.

Now that you've gathered this data on various recovery options, you can review them in relation to cost and capabilities and service levels.

This data can now be compiled into a document in whatever format is suitable for your needs. Some people like to use a grid or matrix, others prefer an outline format. The key is to create a highly usable document that delineates the choices you've made. Let's look at a two examples. In our first sample, we look at a small segment of data that might be included with regard to backups. This uses a grid or matrix style and it should give you an idea about what data to include and how you might approach it. In our second sample, we'll use text without a grid so you can compare which method might work better for you.

Sample 1: Section from Mitigation Strategy for Critical Data

Category Selection	Option	Cost, Capability, SLAs	Risk Mitigation
Data backup—frequency	Continuous	Expensive, zero downtime, exceeds MTD	Potential solution, depending on cost to implement.
	Daily	Moderate, up to 8 hours of potential lost data, 3 hours to restore, meets MTD	Implement daily backup process to reduce likelihood of significant data loss and to reduce recovery time to meet MTD.
	Weekly	Moderate, up to five days of potential lost data, 12 hours to restore, may meet MTD	
	Monthly	Low, does not meet MTD	
Data backup—type	Full	Longest backup time, shortest recovery time, meets MTD	
	Incremental	Medium backup time, longest recovery time, exceeds MTD	
	Differential	Medium backup time, medium recovery time, meets MTD	Differential backup meets MTDs at the lowest cost.
Data backup—method	Tape backups	Longest recovery time, least expensive, may not meet MTD	
	Electronic vaulting	Long recovery time, somewhat expensive, may not meet MTD	
	Data replication	Medium recovery time, medium expense, may meet MTD	

Continued

Category Selection	Option	Cost, Capability, SLAs	Risk Mitigation
	Disk shadowing	Fast recovery time, medium expense, may meet MTD	Based on cost constraints, this option may meet MTD. This and disk mirroring will be explored in terms of cost, time, and feasibility.
	Disk mirroring	Fast recovery time, medium expense, may meet MTD	Based on cost constraints, this option may meet MTD. This and disk shadowing will be explored in terms of cost, time, and feasibility.
	Storage virtualization	Fast recovery time, high expense, removes localized failure risk, meets MTD	
	Storage area network	Fast recovery time, higher expense, removed single point of failure, may remove localized failure risk, meets MTD	
	Wide area high availability clustering	Fast recovery time, higher expense, removes single point of failure, may remove localized failure risk, meets MTD	
	Remote mirroring	Continuous availability, zero recovery time, highest expense, removes single point of failure and localized failure risk, exceeds MTD	

Sample 2: Section from Mitigation Strategy for Critical Data

Critical Data Recovery Options *(selected choice is* <u>underlined</u>*)*

1. Data backup frequency

   A. Continuous—expensive, zero downtime, exceeds MTD. Not suitable due to cost.

   B. <u>Daily—moderate, up to eight hours potential lost data, 3 hour recovery time, meets MTD. Best choice based on cost and time factors.</u>

   C. Weekly—moderate, up to five days lost data, 12 hours to restore, may meet MTD. Although cost is acceptable, the recovery time for this option just barely meets MTD and does not provide any leeway. Therefore, this option is not as suitable as daily.

   D. Monthly—low cost, does not meet MTD. Not suitable due to time.

2. Data backup type

   A. Full—uses the fewest tapes, takes the most time to back up, least time to recover, exceeds MTD. Not suitable due to time to back up.

   B. Incremental—uses moderate number of tapes, takes less time to back up than full, moderate time to recover. Just barely meets MTD. Not suitable due to time to recover.

   C. <u>Differential—uses moderate number of tapes, takes less time to back up than full, takes less time to recover than incremental. Meets MTD. Suitable due to time and cost.</u>

3. Data backup method

   A. Tape backup—longest recovery time, least expensive, does not meet MTD.

   B. Electronic vaulting—longer recovery time, somewhat expensive, may not meet MTD.

   C. Data replication—medium recovery time, medium expense, may meet MTD.

   D. <u>Disk shadowing—fast recovery time, medium expense, may meet MTD.</u>

   E. <u>Disk mirroring—fast recovery time, medium expense, may meet MTD.</u>

   F. Storage virtualization—fast recovery time, high expense, removes localized failure risk, meets MTD.

   G. Storage area network—fast recovery time, higher expense, removed single point of failure, may remove localized failure risk, meets MTD.

H. Wide area high availability clustering—fast recovery time, higher expense, removes single point of failure, may remove localized failure risk, meets MTD.

I. Remote mirroring—continuous availability, zero recovery time, highest expense, removes single point of failure and localized failure risk, exceeds MTD.

As you can see from both examples, you may need to do additional research before deciding on the right backup method for critical data. It's clear that a weekly backup scheme might work, but the problems inherent in a local backup process might not be acceptable. You can also see from the data that while a weekly differential backup strategy might be acceptable, disk mirroring is also an option. In some cases, these two backup objectives might be at odds, might be redundant, or might not make sense for your organization. Once you've looked at this data, you can determine the best risk mitigation strategy for that business function and ultimately, for your entire business.

Your final strategy might be to set up disk mirroring and perform weekly backups of data that are sent to a remote data storage vault. This reduces your recovery time if something happens to a disk (mirroring) and also protects you if you have a fire in the building that destroys all disks. You should include a section to your Critical Data Recovery Options called "Selected Strategy" and delineate the exact strategy you select. When you move into writing your business continuity and disaster recovery plan, you'll have the information you need in order to begin implementing these strategies. Avoid having to review this material at length by including enough information so that the rationale behind the selected strategy is clear.

Remember, too, that when you're selecting your strategy, you should consider risk controls already in place and attempt to build on, rather that replace or circumvent, those solutions. There may be some cases where you want to completely revamp your approach, and this is the place to make those decisions. In other cases, you may simply confirm that you're covered in these areas. For example, you may already have disk mirroring and remote data backups in place. If so, you've looked at your MTD, cost, and capability requirements and determined that these solutions are acceptable. Make a note of that finding. Later, when you're looking at your BC/DR plan, you don't want to have to go back through all these steps to determine if you used due diligence in making this decision. If something goes wrong down the line, you will also have documentation to show that you used a logical and accepted methodology for making these decisions.

For each critical business process, you need to identify an associated risk mitigation strategy. Some strategies will cover more than one critical business process, so you should not end up with as many strategies as you have critical business functions and processes. For example, your data management strategies will cover many of your critical business processes. By assessing this data with the big picture in mind, you can find areas where risk mitigation choices can cover more than one critical area. If you were to look at these strategies area by area only, you might miss opportunities to generate some *economies of scale*, which come from being able to apply one solution to many problems. These solutions become less expensive when they have more than one use. As mentioned earlier, any time you can use a solution across multiple business functions, you have a stronger business case for the expenditure. If you implement a remote data storage solution that meets data availability requirements for normal day-to-day business and it also meets your business continuity and disaster recovery needs, you're going to find more support for the cost and implementation of such a solution. At the very least, you'll be able to make a stronger business case for the investment.

People, Buildings, and Infrastructure

We're including this as a separate section because depending on the nature of your business, you may not yet have addressed the elements. In looking at the business impact, for instance, you may have deal with critical business functions, but you may not have addressed the impact to people, to buildings, or to other infrastructure.

If there is a business disruption, your company may have very specific needs related to people—staff, contractors, vendors, or the community. Some of these may already be addressed in your plan. For example, in the aftermath of a natural disaster, people need ready cash so they can buy food, medical supplies, and other immediate needs. If your company is located in a rural area where access to banks and ATMs is limited, you may want to ensure that your recovery plan includes being able to cash paychecks for employees or advance them cash against future paychecks. That may already be covered in your critical business functions under payroll, but it's a good idea to think through this again to ensure you have covered all your bases. However, there may be other areas that should be addressed in risk mitigation related to people. For example, fire drills are risk mitigation strategies that are useful not only for fires but for other types of emergencies that require people to evacuate the building in a safe and orderly manner. Keep this in mind as you devise your risk mitigation strategies.

What other risks can be mitigated for people, buildings, and infrastructure? You might have a landscaping company come out and remove all trees, bushes, and grasses that are within 50 feet of the building if you are situated in a place prone to wildfires. That would be a risk mitigation strategy related to the building and infrastructure that might not show up because it's not a critical business function.

You might go back through your risk assessment and see if there are any elements related to people, buildings, and infrastructure that have not yet been addressed in terms of impact or mitigation. Add these to your assessment process here. Remember, too, that sometimes doing nothing (risk acceptance) is an acceptable solution as long as it is an active decision based on research and consideration and not just a passive default position.

IT Risk Mitigation

We've discussed business impact and risk mitigation extensively. Now, let's turn our attention to the specifics of IT risk mitigation. Although the technology you use in your company will change over time and may not be the same as that discussed here, this section should give you a good feel for how to develop a risk mitigation strategy for your IT systems.

Risks to your data include not only the natural disasters we went over earlier in this book, but data disruptions and outages due to data center outages (fire, power, etc.); hardware or software failures; network security breaches; data security breaches that can include lost, stolen, modified, or copied critical data; and disruption due to critical data not being available to legitimate users (Denial of Service attacks, etc.). Your risk and impact assessments should have covered these areas and this is a good time to check to ensure all your data risks are addressed.

Critical Data and Records

Through looking at your maximum tolerable downtime and the cost of disruptions (lost productivity, lost revenues, etc.), you have a solid understanding of the impact a loss of various critical data would have on the organization. If you don't yet have this understanding, you should go back through the risk, vulnerability, and impact assessments with an eye toward critical data and records to determine where you critical data is stored, who generates it, what they do with it, and what they would do without it.

In addition, you should assess legal and regulatory requirements related to critical data, whether this is personal medical data, personal financial data, or other data impacted by regulations, statutes, and laws. If you've been addressing this type of data for some time in your IT department, you may have the solutions in place to meet existing regulation. However, it would also be wise to consult with your legal counsel to determine if there are new or upcoming regulations likely to impact your organization in the future. These should be included in your assessment, if possible, so that you can develop a comprehensive data management plan within the scope of your BC/DR plan. Finally, you should review all existing controls as well as your proposed solutions in light of disaster recovery and business continuity. In some cases, your risk mitigation strategies might appear to be acceptable, but when you begin running through possible disaster or disruption scenarios, you discover that your strategies have a few holes in them. If you find you are covered, then you can be confident your BC/DR plan will meet your data needs. If you do discover some gaps, you can be relieved that at least you found them now and not in the aftermath of a disaster. At this juncture, you can look at potential solutions to address any gaps you discover between your existing data protection/data recovery solutions and those needed for BC/DR needs.

Critical Systems and Infrastructure

Once you understand your data management and data protection needs within the scope of the BC/DR planning process, you can begin to evaluate hardware and software solutions, vendors, and costs. There is no magic solution that will cover all your needs and if you've been working in IT for any length of time, you already know that painfully well. However, if your analysis reveals gaps in your coverage, you'll need to look at various methods of addressing those gaps from rip-out-and-replace to patching existing systems.

If you can identify hardware, software, and vendors that can meet your needs for the next three to five years, you'll be doing well on the planning horizon. Don't try to build a solution that will last for ten years, you'll waste time and money looking for the perfect solution. Instead, look for a solid solution that meets your data management, data security, and data recovery requirements now and into the next few years. Then evaluate the cost to acquire, implement, and manage the solution. You'll have to make a few compromises, as you know, but if you have your data constraints and budget as known variables, you can devise an acceptable (or even optimal) solution that fits within those parameters.

Reviewing Critical System Priorities

Through your business impact analysis, you should have developed an assessment of critical IT systems that includes a prioritization of assets. For example, you might have found through your assessment of critical business functions that these IT assets have the following priorities:

- LAN Server (user authentication, etc.)—High

- Internet access—Low

- E-mail access—Low

- CRM application server—High

- Inventory management system server—Medium

- Financial systems—Medium

Based on these assessments, you should review your risk mitigation strategies to ensure that they meet or exceed your requirements for recovery based on these priorities. Dependencies between systems, especially those deemed as high priority or mission critical, should be reviewed. There might be a preferred or required order for restoration of systems after a disruption that should be addressed in the risk mitigation strategy. For example, if it's critical to restore the LAN server before the CRM server because the CRM server requires authentication data from the LAN server, this should be part of the risk mitigation strategy. Later this will be included in your specific data recovery plans but it is in this segment (and perhaps in the business impact analysis) where these dependencies are identified.

Backup and Recovery Considerations

We're assuming that as an IT professional, you're well aware of various backup and recovery options, both those your firm has implemented and those you've learned about in the marketplace. In this section, we're going to cover some of common backup and recovery options so that you can review your risk mitigation strategies in light of these options. This may help you see options you had overlooked or forgotten about; it might bring to light new options you had not considered. We won't go into a lot of detail about these options, but we will provide a quick look at them to help ensure you have the best risk mitigation strategy possible given your current technology, organizational constraints, and budget.

Alternate Business Processes

Your risk management data should already contain your key business processes and alternate methods (workarounds) for handling these processes during a business disruption, whether that disruption is to the IT systems, the building or the surrounding area. We've covered a lot of this material, so this section is just a quick reminder in case you have overlooked any of these areas that may be relevant to your business operations.

Customer Service. During a disruption or emergency, it's vital to most companies to still have the capability to provide support or customer services. Depending on the nature of the work your company does, this may be one of your most critical business functions. IT should clearly understand which technologies are required to deliver acceptable levels of customer service during a business disruption.

Administration and Operations. We've focused on these activities in Chapter 4 in great detail, so we won't cover them again here. You should have detailed documentation on the key business administration and operations processes for your business. These details should be at the heart of your risk mitigation strategy development.

Key Business Information and Documents. Most businesses rely heavily upon electronic data of all kinds—e-mail, text documents, presentations, among others. Data essential to ongoing operations should be identified so that IT mitigation strategies can be developed. In addition, strategies for dealing with less critical data in the absence of key IT systems should be developed. You might decide, for instance, that certain data must always be available so a continuous availability solution will be implemented. Other data is essential but not mission-critical. For that data, you may develop a fast-track recovery solution.

Essential Equipment. Other equipment essential to ongoing operations should be looked at in terms of how disruption of IT and non-IT systems may impact the availability of equipment. In some companies, IT systems run manufacturing, order fulfillment, or other operations-oriented equipment. How will the disruption of business impact these systems? How will the disruption of critical IT systems impact these operational systems? What can be done to reduce the risk to these systems?

Premises. We've discussed fire drills as a way to reduce the risk of injury or death to staff in the event of a fire or other building disaster. In addition to fire drills, there may be other ways to reduce risks to the premises. Insurance is certainly part of the equation, but fire inspections, emergency lighting, and other emergency systems can be put in place to protect the premises and employees.

IT Recovery Systems

You're undoubtedly familiar with many IT recovery systems, but as part of your risk mitigation strategy development, you should scan the technological horizon to see what's available in today's market. Sometimes IT departments develop risk management strategies based on current technology and never update those strategies. Systems put in place five years ago that are not reviewed and updated can inject additional risk into your organization, and puts your BC/DR plan at risk. Clearly what was innovative five years ago may be close to being a legacy system now. What was extremely expensive three years ago has probably dropped in price significantly. Revisit your technology solutions with an eye on what's available in the marketplace today. You might decide to upgrade, replace, or supplement existing solutions. The list included in this section is not exhaustive but should spark thoughts about solutions to consider. Be sure to do some independent research to supplement this data so that you have a comprehensive and current look at your IT recovery options before developing your mitigation strategies.

Alternate Sites

The largest decision you'll need to make is whether or not to develop alternate sites. You can have a dedicated site wholly owned by your company, you can create a reciprocal agreement with another division or company, or you can go to an external vendor for a commercially leased facility. Let's look at the most common options.

Fully Mirrored Site

Mirrored sites are fully redundant sites that mirror everything going on in the live site. This is by far the most expensive and extensive IT risk mitigation strategy. For some companies, this solution might make sense. Mirrored sites provide the highest degree of availability (and therefore risk mitigation) because every transaction that happens on the live site is also processed on the mirrored site simultaneously. Sometimes a solution implemented for load balancing purposes may also serve as a risk mitigation solution. For example, if you have two mirrored sites so that users can access data quickly, it might be that this same configuration works well in the event that one or the other site goes down. Certainly user access to data will slow considerably if one of the sites goes down, but the transactions can still occur while the initial site is being repaired. Mirrored sites typically are owned and managed by the company, which can reduce the cost of implementation.

Hot Site

A hot site is usually a site leased by a commercial vendor to your company for emergency purposes. The vendor will guarantee an identical technical configuration with communications that allow you to switch your IT operations to that commercial site within a specified time frame, usually within one to four hours. These sites typically provide enough space for hardware, supporting infrastructure (racks, cables, phones, printers), and support personnel. This is sometimes less costly than a fully mirrored site, but that depends on your technology and response time needs.

Warm Site

Warm sites are partially equipped premises with some or all of the required equipment. Warm sites are often sites used during normal operations for less critical functions that are taken over for critical IT functions during a business disruption. For example, you might have a site located in your primary location and a second site in a remote office or satellite building. You might keep a server at the remote site configured with your critical business applications with Internet access to backup data. In the event of a business disruption or disaster to the primary site, the secondary site could fire up the server, restore from the most recent backup, and resume critical operations within a matter of hours.

Mobile Site

Mobile sites are self-contained units that can be transported to establish an alternate computing (or working) site. These often are contained within a mobile trailer that is delivered by truck to a specified location. Commercial vendors lease these types of units. Due to the time and expense of configuring a mobile site, these arrangements should be preestablished far in advance of anticipated demand.

Cold Site

A cold site is started up "cold" in the aftermath of a disruption. These kinds of sites are the least expensive in advance of an emergency but take the longest to bring online after a disruption. If your recovery needs are three or four days out, this might be the most cost-effective solution for you. However, your BC/DR plan should include plans for how and where you could establish a cold site should you select this option. Trying to come up with these arrangements in the aftermath of a disaster or serious disruption will be far less effective than planning in advance. That might mean identifying facilities in your area that could host a cold site, understanding how your communications needs would be met, and how you'd furnish and staff this site.

Reciprocal Site

You may be able to make arrangements with another company or another division of your company for use in the event of a significant business disruption. For example, you might make arrangements with another company in your area for reciprocal assistance in the event that one of your businesses is disrupted. However, if a natural disaster hits the area, it's possible both companies will be impacted, so you need to assess the risks of such an arrangement. If you can create an arrangement with a firm outside your geographic area, you'll reduce the localized risk. Remember, however, to create solid agreements with plenty of detail delineating how, when, where, and at what cost these reciprocal arrangements will be implemented. You don't want another company disrupting your business for minor problems, and the use of these arrangements should be very clearly defined. That said, this type of arrangement might make sense for small businesses that can't afford to contract with commercial vendors for alternate sites.

Disk Systems

Disk systems solutions continue to evolve in terms of capabilities. They also tend to become less expensive over time as well. We'll take a quick look at some of the solutions available to you today.

RAID

Redundant arrays of inexpensive disks (RAID) come in several forms. The ability to hot-swap disks from a RAID array can be an important attribute of your disk recovery strategy. We won't run through all the permutations of RAID but you should be aware that there are newer implementations of RAID including RAID10 and RAID50, among others. Also keep in mind that you can implement hardware-based or software-based RAID systems; each has pros and cons associated with them.

Remote Journaling

Remote journaling is a method in which every write and update operation is written to another device. This can be an effective *part* of a data recovery solution, but not a standalone solution. It can be helpful in cases of network intrusion or data corruption. Journaling done in real time creates a mirrored copy. This journal data can be transmitted over a communications link to another site enabling extremely fast recovery in the event of a security breach, data corruption, or other failure.

Replication

Disk replication involves copying data on to a primary and secondary server. *Shadowing* and *clustering* are two methods of accomplishing replication. *Shadowing* happens asynchronously—changes are collected and applied to the secondary server periodically. Shadowing can be part of a risk mitigation strategy, but keep in mind that any corruption or error on the primary server will be replicated to the secondary server. *Clustering* is a higher-end solution than shadowing and it provides high availability. Server clustering works in a manner similar to RAID for disk drives. With clustering, several servers are tied together and periodically synchronize with one another. If a server goes down, the workload shifts to the remaining servers. This process is transparent to users who connect to the application and have no idea which server is providing data. As you probably know, clusters provide load balancing for users and this same functionality provides a level of risk mitigation as well.

Electronic Vaulting

Electronic vaulting is the process that transmits backup data of your systems to a remote location. Backups don't need to be transported and stored off-site and in the event of a business disruption; they may be easier to access than tapes locked in a bank vault or other secure off-site location. Electronic vaulting can dramatically reduce recovery time, especially if used in conjunction with remote journaling.

Standby Operating Systems

As you well know, the operating system with associated patches and upgrades is a critical aspect to being able to get applications back online. Having standby drives with preconfigured operating systems available can reduce risks and recovery times. Every time you upgrade your production systems, you can upgrade your standby systems so that the OS is ready to go in the event of a failure, breach, or disruption.

Network-Attached Storage (NAS)

Storage with a network interface can be attached to the network in any location that provides network connectivity. Thus, a storage unit could be contained in a vault, a server room, or in the middle of a work area. These types of storage devices are easy to install and maintain.

Storage Area Network (SAN)

A storage area network is a dedicated high-speed network for data storage. Storage is independent from servers and is stored across the storage network. In most organizations, much of the LAN traffic is dedicated to backup, mirroring, "heartbeats," and disaster-recovery activities. With a SAN, these activities are restricted to the storage network and bandwidth on the LAN is freed up for more user-centric data needs.

Desktop Solutions

Your organization should already have some process in place for backing up user data. In the Microsoft Windows operating system, most users save data to the My Documents folder or to a designated network location. For enterprise applications, user data may be stored more centrally. Regardless of your configuration, it's important that critical user data be backed up periodically. Ideally, this process should be automated so it does not rely on user compliance with established backup processes. Backups of user data should also be stored securely off-site. In your business impact analysis, you may have determined that there were certain job functions that required special attention. These computers should be flagged as critical and risk mitigation strategies for key user's computers should be developed.

Creating standardized file management processes will also assist in any recovery efforts (and therefore mitigate risk). For example, requiring users to store all important documents in their named folder on a network share can help reduce the likelihood of data loss, corruption, or breach. If users travel with laptops, be sure to establish backup and security procedures for mobile users.

From the Trenches...

Lost and Stolen Laptops— It's Not Always about the Hardware

Laptops are lost and stolen everyday. Sometimes a tired traveler leaves a laptop behind, sometimes a thief wants the hardware. Other times, the thief is targeting the information on the laptop. Lost and stolen laptops have been in the headlines recently because the data on them was sensitive and unencrypted. If you have users working with sensitive data, whether that's the company's strategic direction, corporate finances, or private customer data, be sure that all data is encrypted and that the operating system requires user authentication. Even though there are ways around user access restrictions on stolen laptops, it's tough to overcome strong encryption. Although laptops will always be lost or stolen, the data on them doesn't have to fall into the wrong hands. Implement strong encryption on all laptops that deal with sensitive data and be sure users understand the importance of encryption. Ideally, the encryption system will work seamlessly in the background so the user doesn't have to take any special action to protect data. Anytime security measures can be automated, you'll end up with stronger security than if left to users to remember and employ.

In addition to implementing backup and encryption policies and procedures, risk can be reduced through standardizing hardware, software, and peripheral equipment. Reducing the number of variables not only helps in day-to-day IT activities, it can significantly reduce recovery time after a significant event. Documenting hardware, software, and configuration data along with vendor contact information can reduce the risk of serious disruption should user systems be impacted.

Software and Licensing

Software and license data must be backed up and stored in a secure off-site location along with data. It doesn't do much good to store the database information if the licensing data is lost. The licensing for each operating system, application, user, and desktop system should be captured and stored in a secure manner in the event of a partial or total disruption of business.

Web Sites

There are two primary risks related to corporate Web sites. The first is the security risk due to the nature of external (public) Web sites. As you know, Web sites are like large neon signs saying to hackers "Enter Here." Risk mitigation strategies for Web sites include implementing strong security measures along with auditing and monitoring activity on the server. Documentation on the security and configuration settings for the Web site are important in the event the web servers go down or in the event of a security breach. In addition, many corporate Web sites are used to conduct e-commerce transactions and the disruption of these transactions can have a significant impact on revenue streams and on customer perception of the company. Some companies use load balancing strategies to ensure Web sites have high availability, and these same strategies also act as excellent risk mitigation strategies. However, if a Web site is breached or data is corrupted, it's possible these problems will be replicated to all virtual sites, so additional risk mitigation strategies may be needed.

Summary

In this chapter, you learned about a process you can use to develop risk mitigation strategies for critical business and IT functions. The inputs to this process are the risk assessment data and the business impact analysis. The key steps in this process are developing recovery requirements, understanding the recovery time of the options under consideration, comparison of these times to maximum tolerable downtime requirements, a review of the cost and capabilities of each option, the service level agreements related to each option, and finally, the selection of the option to be implemented. This process is done for all critical business processes identified. An additional review should consider dependencies between processes, functions, and IT systems. As is often the case, one solution may address several key requirements. Your review of options should attempt to find the simplest, most comprehensive, and cost-effective solution that meets your company's critical business needs now and into the near future.

Solutions Fast Track

Types of Risk Mitigation Strategies

☑ Risk acceptance is a strategy in which the company accepts the potential consequences of a given risk. The company chooses to do nothing to avoid, limit, or transfer the risk. Acceptance usually has a very low cost associated with managing the risk (or zero cost), but can have a very high cost in the aftermath of a disruption.

☑ Risk avoidance is a strategy in which the risk is completely avoided. This might include shutting down critical systems and moving them in advance of a hurricane. Avoidance takes the risk to zero but often has a high cost associated with it. Therefore, the cost of managing the risk is very high but the cost of recovery is very low.

☑ Risk limitation is a strategy that falls in between acceptance and avoidance. Most companies choose a risk limitation strategy, especially for IT systems where complete acceptance or avoidance are too costly on either side of a disruption. Steps such as secure, off-site backups can go a long way in reducing various organizations risks without being too expensive in implementation or recovery phases.

☑ Risk transference is where the exposure to the risk is transferred to a third party, usually as part of a financial transaction. Purchasing insurance is the most common risk transference method, though others exist.

Risk Mitigation Process

☑ Recovery requirements are developed during the risk assessment phase and include data from the business impact analysis. You can begin by delineating the key functional areas of your company and determining the key business processes in each.

☑ Recovery requirements include the time and cost of recovery as well as any specific processes or procedures required by each functional area of the company.

☑ Recovery options are developed for each critical business process or function. Recovery options must fit within the constraints of the recovery requirement. Otherwise, they should not be considered as part of the BC/DR process.

☑ Recovery options usually fall into one of three categories: as needed, prearranged and preestablished. The cost and time to implement each type of option varies.

☑ After recovery options are delineated, each option must be reviewed in terms of the maximum tolerable downtime (MTD) for each critical business process. Any option that falls outside the MTD should be removed from further consideration.

☑ The cost and capability of remaining options should be compared. In some cases, cost will be more critical than capability; in other cases, capabilities are more important than cost.

☑ Determining the cost, capability, effort to implement, quality, control, safety, and security of each option under consideration can help you develop a comprehensive risk mitigation strategy that meets the needs of your company.

☑ Service Level Agreements are important when dealing with vendors in preestablished or prearranged contracts.

☑ Service Level Agreements may also pertain to agreements your company has with others that must be addressed in your plan. This might include customer service functions or other externally-facing functions your firm provides to others.

☑ Existing controls and risk mitigation solutions already in place should be reviewed after requirements and options are reviewed. In some cases, existing solutions meet BC/DR requirements; in other cases, existing solutions can be augmented or expanded to meet needs. In still other cases, no satisfactory controls exist and a solution must be developed.

☑ People, buildings, and infrastructure are sometimes overlooked in the BC/DR risk mitigation phase. How will risks to people, buildings, and other infrastructure be addressed through your BC/DR plan? Many of these may have been considered during the threat, vulnerability, and impact assessment phases, and they specifically should be included in the risk mitigation phase.

IT Risk Mitigation

☑ Critical data and records should be viewed in light of risk mitigation strategies under consideration. There may be additional organizational, regulatory, or legal requirements for reducing risk to critical data and records that should be addressed as part of your overall strategy.

☑ Critical systems and infrastructure should be assessed to determine optimal solutions for risk mitigation. There is a wide variety of solutions available on the market today. You can select the optimal (or acceptable) solution for your business only after assessing your businesses critical functions and specific needs.

Backup and Recovery Considerations

☑ There are numerous areas of the company that may require alternate business processes to be developed and/or available. These areas are sometimes overlooked in planning.

☑ Customer service, administration, essential equipment, and premises are four areas that require specific attention in your risk mitigation planning.

☑ IT recovery systems are numerous and include (among many others) alternate sites, disk system solutions, clustering, virtualization, and vaulting.

☑ User's desktop systems must be considered as part of the overall risk mitigation process.

☑ Software and licensing information should be stored in a secure, off-site location with backup data.

☑ Web sites are external-facing connections to the company. As such, they require special security considerations and risk mitigation strategies.

Frequently Asked Questions

The following Frequently Asked Questions, answered by the authors of this book, are designed to both measure your understanding of the concepts presented in this chapter and to assist you with real-life implementation of these concepts. To have your questions about this chapter answered by the author, browse to **www. syngress.com/solutions** and click on the **"Ask the Author"** form.

Q: I work for a small company that runs five retail stores in a single geographic region. We perform daily backups but the data is stored at each site. I've been pushing for a better strategy but have been told that we can't afford anything more. Any suggestions for getting approval for funding for more backup and recovery solutions?

A: If you can make a compelling business case for additional expenditures, you might have a chance of getting additional funding. However, in many small companies, the funding is simply not available, no matter how compelling an argument you make. In many small companies, the de facto choice for risk management is risk acceptance. Unfortunately, that decision is often made without a complete understanding of the risks and costs involved. So, one way to get your management team on board would be to delineate the potential costs of downtime, lost customers, damage to reputation, and such. Use the scenarios we've discussed throughout the book thus far as examples of what might happen. If you still run into resistance, you might be able to implement some additional risk mitigation strategies without additional expenditure. For example, what would it take to get the data backup from Retail Location 1 stored at Retail Location 2? Retail Location 2's data backup could be stored at Retail Location 3, and so forth. If one of the stores were to catch fire (remember, that's the most common business disruption), the data backup would be located at another store and could be restored relatively easily. Sometimes low-tech solutions can be good stop-gap measures, especially when the budget is tight.

Q: In our planning process, we have identified three absolutely mission-critical processes. Two are related to the use of a massive database that is stored on a server in one building. If anything happens to that server, we're dead in the water. We have management OK to implement a risk management/recovery strategy but we're in disagreement as to the best solution to implement. Any advice on how to address this impasse?

A: If you have several solutions in mind, you and your team would do well to sit down and review critical business processes against these solutions. Include attributes such as recovery point objectives, maximum tolerable downtime, cost, capability, quality, security, and effort to implement. If necessary, create a rating scale that is used for each attribute. Then, have everyone rank these attributes for each solution under consideration and come up with an aggregate score for each solution. Be sure to include external dependencies and service level agreements as well, which might tip the scales in favor of one solution or another. Look at the results and see if you can come to some agreement. In many cases when the various attributes are assessed in a rational manner, the solution that is most appropriate for the organization will bubble to the top. If not, you might have to make an executive decision based on available data.

Q: In this chapter, you talked about recovery strategies. I thought we were talking about risk mitigation strategies. What happened?

A: Recovery strategies are closely tied to risk mitigation strategies. For example, RAID systems reduce your risk of losing data due to a single disk failure. Mirrored sites reduce (or eliminate) your risk of losing the use of critical applications or data. Each of these is both a risk mitigation strategy and a recovery strategy. In many cases, these are one and the same. The key is to reduce the risk of the event occurring, which often related directly to how quickly you recover from an event. If a disk drive goes out with RAID, it's transparent to the user, and IT staff can often hot-swap a drive without users ever knowing. Is this a risk mitigation or a recovery strategy? It's a risk mitigation strategy because it reduces the risk of critical system outages including users unable to access important data or systems. It's a recovery strategy because when the outage occurs, the impact is zero and the recovery time is fast. Therefore, you have both in one solution regardless of what you call it.

Chapter 6

Business Continuity/Disaster Recovery Plan Development

Solutions in this chapter:

- **Phases of Business Continuity and Disaster Recovery**
- **Defining BC/DR Teams and Key Personnel**
- **Defining Tasks, Assigning Resources**
- **Communications Plans**
- **Event Logs, Change Control, and Appendices**

☑ **Summary**

☑ **Solutions Fast Track**

☑ **Frequently Asked Questions**

Introduction

The bulk of your work in developing your business continuity and disaster recovery plan is complete when you get to this point. Granted, you may be reading this book through from start to finish before developing your plan (recommended) and therefore you will have none of the actual work completed. However, things move quickly in the business world and there are some of you who are doing the work as you read each chapter. Either way, this is where everything comes together. The risk analysis you performed led you into your vulnerability assessment. That data helped you develop an assessment of the impact various risks would have on your business. Finally, you took all your data and identified mitigation strategies— actions you could take to avoid, reduce, transfer, or accept the various risks you found. With that, you now have to develop a plan that takes your mitigation strategies and identifies both methods for implementing those strategies, and people, resources, and tasks needed to complete these activities.

In Chapter 7, we'll go over emergency activities including disaster response and business recovery, so we'll refer only briefly to those elements in this chapter where appropriate. In Chapter 8, we'll discuss training and testing and in Chapter 9, we'll discuss maintaining the plan. All of these are elements that should be included in your BC/DR plan as well.

The plan basically needs to state the risks, the vulnerabilities, and the potential impact to each of the mission-critical business functions. For each of these, there should be associated mitigation strategies. In some cases, there will be multiple mitigation strategies; in other cases, you may have elected to simply accept the risk. However, all of this should be clearly laid out in your documentation thus far. Next, you need to determine how and when those strategies are implemented and by whom.

Your work breakdown structure will look something like this:

1. Identify risks *(complete)*.

2. Assess vulnerability to risks *(complete)*.

3. Determine potential impact on business *(complete)*

4. Identify mission-critical business functions *(complete)*.

5. Develop mitigation strategies for mission-critical functions *(complete)*.

6. Develop teams.

7. Implement mitigation strategies.

8. Develop plan activation guidelines.

9. Develop plan transition guidelines.

10. Develop plan training, testing, auditing procedures.

11. Develop plan maintenance procedures.

As you can see from this simplified list, you should already have items one through five completed. We'll discuss developing teams in this chapter as it relates to carrying out the BC/DR plan, not the planning team that you should already have in place (and who hopefully have helped you accomplish tasks one through five). We'll cover developing plan activation and transition guidelines in this chapter before heading into Chapter 7. At the end of this chapter, you'll have items one through nine complete (or will understand how to complete them when you begin project work).

As with previous chapters, we'll begin with a review of where we are in this process (see Figure 6.1). Creating the BC/DR plan entails putting together the information you've developed so far and adding a bit more detail. We'll create the BC/DR plan document in this chapter, but keep in mind we'll have to circle back later to add detail that we develop in upcoming chapters.

Figure 6.1 Project Progress

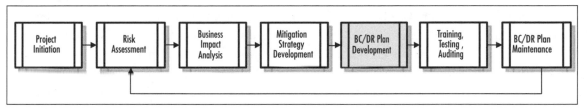

Phases of the Business Continuity and Disaster Recovery

Hopefully you'll never need to put your BC/DR plan into action, despite all the hard work you put into it. If you do need to use your plan, however, you'll need to have clear and specific guidelines for how and when to implement it. Let's begin with a quick look at the phases of the plan: activation, disaster recovery, business resumption or business continuity, and transition to normal operations.

Activation Phase

The activation phase of your BC/DR plan addresses the time during and immediately after a business disruption. In this section of your plan, you need to define when your BC/DR plan will be activated and in what manner. You don't want to activate your plan for every little glitch your business runs into, so you'll need to develop a clear set of parameters that you can use to determine if or when to activate your BC/DR plan. In addition, you will need to define how your plan is activated, including who has the authority to activate it and what steps that person (or persons) will take to initiate BC/DR activities.

Figure 6.2 Phases of Business Continuity and Disaster Recovery

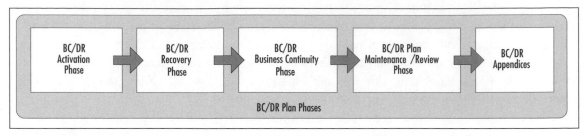

Activation includes initial response and notification, problem assessment and escalation, disaster declaration, and plan implementation. After you have begun implementing the plan, you proceed into the recovery phase, as shown in Figure 6.2.

It is in this activation phase that you should define various disaster or disruption levels so that you know when, if, and how to implement your plan. For example, if you experience a network security breach, you'll have to activate different phases of your plan than if the server room is flooded. Therefore, defining various disaster types and levels is important in understanding what should trigger the implementation of BC/DR plans. You may choose to use a three-level rating system, as described here. However, make sure that whatever system you devise, it's tailored to your specific business configuration and that it gives you the guidance you'd need to make these crucial decisions based on predetermined and agreed-upon criteria.

Major Disaster or Disruption

The possibility or likelihood of this type of disaster occurring is low but the business impact is extremely high. This event disrupts all or most of the normal business operations of the company and all or most of its critical business processes. The disruptions occur because all or a majority of systems and equipment have failed or are inaccessible. This includes destruction to the entire facility; a major portion of the facility; or entire networks, subnets, or sections of the business. Once you've defined what this level of disaster or disruption entails, you should define the process for determining which parts of your BC/DR plan should be activated and which team members should be called upon. We'll discuss triggers more in a moment; for now you should attempt to define the business systems, mission-critical functions, and major operations that when affected would cause a major disruption. This will help you develop appropriate triggers to determine when and how to activate your BC/DR plan.

Intermediate Disaster or Disruption

An intermediate disaster is likely to occur more frequently than a major disaster, but less frequently than a minor disaster (hence the "intermediate" designation). Its impact will be less than a major and more than a minor event. This type of disruption or disaster interrupts or impacts one or more mission-critical functions or business units, but not all of them. Operations will experience significant disruption, entire systems or multiple systems may fail or be unavailable, but not all of them. An intermediate event could include a fire or flood in the building that impacts IT systems and equipment, structural damage to part of the building where critical operations occur or where vital equipment is located. As with the other two levels of disruption, it's important to define not only what each tier consists of but which parts of the BC/DR plan should be activated and which team members should begin implementing BC/DR activities. As with a major disruption, clearly delineate which systems, functions, and operations would be impacted to earn an intermediate designation so you can define triggers that will address these types of situations.

Minor Disaster or Disruption

Minor disruptions occur every day in the business world and rarely, if ever, are BC/DR plans called into action. The likelihood of a minor event occurring is high, the associated disruption is low. The effects typically are isolated to one component, one system, one business function, or just one segment of a critical business function. Normal operations can often continue, almost uninterrupted, in the face of a minor disruption. Critical business functions still occur for some period of time after this type of disruption. The failure of a single system or service can typically be addressed during the normal course of business. For example, the failure of a single server, system disk, or phone system is problematic but usually does not require the activation of a BC/DR plan. There may be examples, however, where minor disruptions should be addressed by the activation of part of a BC/DR plan. If that is the case, be sure to clearly identify those disruptions along with which sections of the BC/DR plan should be implemented when and by whom.

Activating BC/DR Teams

Clearly, the BC/DR plan cannot activate itself, someone or a team of people need to make appropriate assessments of the situation and make a determination as to whether or not to activate the plan or portions thereof. Therefore, it's also important to create and maintain various BC/DR teams that handle the response to the business disruption by implementing appropriate sections of the BC/DR plan. We'll discuss the makeup of these teams later in this chapter, but for now we'll list some of the BC/DR teams you may want to define and populate as you continue in this planning process.

- Crisis management team
- Damage assessment team
- Notification team
- Emergency response team

- Business continuity coordinator or lead
- Crisis communication team
- Resource and logistics team
- Risk assessment manager

Depending on the size and nature of your company, you may or may not need some of these functions. It's also possible that one person may fill one or more roles if you're working in a small company. We'll discuss these roles in more detail in a section coming up later in this chapter.

Developing Triggers

If you're familiar with project management, you're probably familiar with triggers. Typically, risks and triggers are identified so that if a project risk occurs, a trigger defines when an alternate plan or method should be implemented. The same is true here. If you are going to implement your plan, you'll need to define how and when that should occur—those are your triggers. For example, if you use the three categories of major, intermediate, and minor, you'll need to define what actions are taken in each case. Each level of disruption should have clearly defined triggers. Let's look at a hypothetical example. You're the IT manager of a small firm and the head of the BC/DR team. You're at home one evening just sitting down to dinner when one of the data processing operators who works until 9 P.M. calls you. She reports that there was a fire in the building, it's been evacuated, and the fire department is on the scene. You ask her a series of questions and ascertain that the fire seems to have been contained relatively quickly but that some of the networking gear may have been damaged either by the fire or by the fire containment efforts. She believes the server room is in tact but she's not sure. If you have clearly defined triggers in place, you may determine that this appears to be either a minor or an intermediate disruption and that you should most likely activate a portion of your BC/DR plan. The trigger might be defined as a series of steps such as:

1. Business disruption event has occurred.

2. Disruption to business operations has occurred.

3. Initial assessment by employees on the scene indicates intermediate level damage, including the following:

   - A portion of the network is or may be out of service.

   - One or more critical servers are or may be out of service.

   - A portion of the physical facility has been impacted by the disruption.

   - It is likely employees will not be able to resume normal operations within two hours.

This is an example of a trigger you could define for intermediate types of events. As you've done previously, using scenarios helps you define these elements more clearly. By defining three statements and four attributes, you have a good understanding of whether or not to activate the BC/DR plan for intermediate outages. You also have a defined time-line—if normal business operations cannot resume within two hours. This should be tied to your overall maximum tolerable downtime (MTD) and other recovery metrics developed earlier. If your MTD is 24 hours, an intermediate disruption might be something that will disrupt normal operations for two to six hours. You and your team will need to define these various windows, but be sure to tie your triggers to your recovery metrics.

Your intermediate activation steps are related to the trigger. Once you know you should activate your plan, you should define the immediate steps to be taken. This helps remove any uncertainty about next steps and helps begin a focused response effort. An example of the first steps for an intermediate disruption is shown here.

1. If a disruption appears to be **intermediate** on initial assessment, within two hours:

   ■ Attempt to gather information from the emergency responders, if appropriate.

   ■ Activate the damage assessment team.

   ■ Notify the crisis management team to be on standby notice.

2. After two hours from event notification, gather initial evaluation from damage assessment team.

3. After three hours, notify crisis management team of next steps (stand down, fully activate).

4. Within three hours of event notification, BC/DR plan should be implemented if assessment indicates intermediate or major disruption.

Notice, though, that our description of the actual disruption levels includes trigger infor mation. How many systems are impacted? How extensive is the damage? The more clearly you can define these details, the more precise your triggers will be, and this will help you determine if and when to activate your plan. Spend time clearly defining the circumstances that will warrant plan activation at the various levels you've defined and also spend time defining initial steps to be taken in each phase so that you have checklists of next steps. We'll provide additional checklists you can use as starting points for your own lists when we go over disaster recovery steps in the next chapter as well as in the appendix materials at the end of this book.

Transition Trigger—Activation to Recovery

Another trigger to define is when to move from one phase to another. In this case, that means when to move from the activation phase to the recovery phase. This transition is one that typically occurs fairly naturally, so you don't need to over-engineer this. However, you may want to define the transition trigger like this:

1. The damage assessment team's initial evaluation indicates an intermediate disruption.

2. The crisis management team has been called in and is on scene.

3. The immediate cause of the event has stopped or been contained.

4. The intermediate section of the BC/DR plan has been activated.

You may wish to define other triggers for your transition, from activation to recovery, suitable to your organization. When defining your triggers throughout, keep your maximum tolerable downtime (MTD) and other defined metrics in mind so that you can work within those constraints. For example, if your MTD is very short, your time between activation and recovery also should be very short. In this case, you may have to err on the side of timeliness and take action with incomplete or preliminary data. You'll have to balance your need to collect information with your need to get the business back up and running as quickly as possible (and within your MTD constraints). Rarely, if ever, is there perfect data in an emergency (or any other time). Defining these triggers and constraints clearly in your plan can help you make better decisions in the stressful aftermath of a business disruption or disaster. Help the team make the best decisions possible by spending time now to define these triggers as clearly and unambiguously as possible.

Recovery Phase

The recovery phase is the first phase of work in the immediate aftermath of the disruption or disaster. This phase usually assumes that the cause of the disruption has subsided, stopped, or been contained, but not always. For example, in the case of flooding, you may decide that if it's external flooding, you will wait until waters subside to begin recovery efforts. This may be required by local officials who restrict access to flooded areas. However, in other cases, you may be able to or choose to initiate recovery efforts while flooding is still occurring. This might include placing sandbags around the entryways to the building or removing equipment that is not yet under water. As you can tell, many of your actions will be dictated by the specifics of the situation, so there's no simple rule to follow here. However, we can say that recovery efforts have to do with recovering from the immediate aftermath of the event, whether or not the event is still occurring. This phase may also include evacuating the facility, removing equipment that can be salvaged quickly, assessing the situation or damage, and determining which recovery steps are needed to get operations up and going again. The recovery phase is discussed in detail in Chapter 7.

Transition Trigger—Recovery to Continuity

You'll learn more about recovery activities in Chapter 7, so you'll need to circle back and define these triggers after you understand the information covered in that chapter. At this juncture, you can make a note that you need to develop triggers that help you know when to transition from recovery efforts to business continuity efforts. Typically, these triggers will have to do with determining that the effects of the disruption have been addressed and are not getting any worse. For example, if you experience a fire in the building, the fire is out, the assessment has been done, any equipment or supplies that can be salvaged have been, and alternate computing facilities have been activated. Those are activities that take place in the recovery phase and when these are all complete, it's time to move into the business continuity phase, which typically includes starting up systems so that business operations can resume. Defining these points should include specific events that have occurred, milestones that have been met, or time that has elapsed. Also keep your MTD in mind as you define triggers for this transition.

Business Continuity Phase

The business continuity phase kicks in after the recovery phase and defines the steps needed to get back to "business as usual." For example, if you have a fire in the building, the recovery phase might include salvaging undamaged equipment, ordering two new servers from a hardware vendor, and loading up the applications and backup data on the servers at a temporary location so that you can begin to recover your data and your business operations. The business continuity phase would address how you actually begin to resume operations from that temporary location, what work-arounds need to be implemented, what manual methods will be used in this interim period, and so forth. The final steps in the business continuity phase will address how you move from that temporary location to your repaired facility, how you reintegrate or synchronize your data, and how you transition back to your normal operations. This detail is discussed in Chapter 7. You'll also need to define triggers here that define when you end business continuity activities and when you resume normal operations. Again, as with the other triggers, you should strive to be as clear and concise as possible. You'll have enough to deal with later if you do end up activating and implementing your plan, so spend time here to save yourself a headache later on.

Although it might seem intuitive that you'll resume normal operations when everything is back to normal, things sometimes do not return to normal after a business disruption of any magnitude. Certainly, business operations will resume, but some things may change permanently as a consequence of this disruption. For example, your company may decide as a result of a major fire or flood that it wants to move to a new location and it's going to do that while operating from the alternate site. That would complicate things because it would mean moving from the alternate site to a new site, with all the concomitant challenges inherent in both resuming normal operations and moving to a new facility. Though this example may seem outside the bounds of normal business decision-making, be assured that disruptions can change the way companies see their businesses and the way they approach operations. Another example is developing a work-around that's used in the recovery phase that works so well that someone decides to use it full time. When do you transition back to normal operations if you incorporate BC/DR work-arounds? When do you officially transition back to normal operations if you decide that the new server role or network configuration actually works better than the original? It might be a simple matter of formally evaluating the change, agreeing to make it permanent, and declaring you're now running under normal operating conditions. You and your team can define these triggers in advance and you may need to modify them later but at least you won't be working with a blank slate.

Maintenance/Review Phase

The maintenance phase has to occur whether or not you ever activate your BC/DR plan. On a periodic basis, you need to review your BC/DR plan to ensure that it is still current and relevant. As operations and technology components change, as you add or change facilities or locations, you'll need to make sure that your plan is still up to date. One common problem in BC/DR planning is that companies may expend time to develop a plan but they often do not want to (or will not) expend the time and resources necessary to keep the plan current. Old plans are dangerous because they provide a false sense of security and may lead to significant gaps in coverage. If a plan is not maintained, then all the time and money invested in creating the plan is wasted as well. In addition, if you end up activating your BC/DR plan at some point, you'll want to assess the effectiveness of the plan afterward, when things settle down. You should do this relatively close to the end of the recovery and business continuity cycles so that lessons learned can be captured and applied to your BC/DR plan before memories fade and people go back to their daily routines. Reviewing the plan in the immediate aftermath of a disruption will give you valuable insights into what did and did not work. Incorporating this knowledge into your plan will help you continue to hone the plan to meet your evolving business needs. This is discussed further in Chapter 9.

Defining BC/DR Teams and Key Personnel

There are numerous people in positions that are critical to the activation, implementation, and maintenance of your BC/DR plan. Although these may not all be relevant to your organization, this will serve as a good checkpoint to determine who should be included in your various phases. You'll also need to form teams to fulfill various needs before, during, and after a business disruption or disaster. Where possible, you should specify a particular position or role that meets the need rather than specifying individuals. If your Facilities Manager should participate in the Damage Assessment Team, for example, you should specify the Facilities Manager and not Phil, who happens to be the Facilities Manager now. That will allow your plan to remain relevant whether Phil wins the lottery and leaves the company, gets hit by a bus and is out for an extended period of time, or is promoted to vice president.

Though we briefly define types of teams and their roles in the BC/DR effort, you should take time to clearly define the roles and responsibilities of each team. Having clear boundaries will help ensure that teams are not working at cross-purposes and that all aspects of the plan are covered. Gaps and omissions occur when these kinds of definitions are ill-formed. If helpful, you can create team descriptions that read like job descriptions and you can task members of your HR department on the BC/DR team to assist with or lead this activity. A good team description will identify the following attributes:

- Positions or job functions included on the team (Facilities Manager, HR Director, etc.)

- Team leader and contact information

- Team mission statement or set of objectives

- Scope of responsibilities (define what *is* and *is not* part of this team's mission)

- Delineation of responsibilities in each phase of BC/DR (i.e., when will the team be activated and deactivated?)

- Escalation path and criteria

- Other data, as needed

Crisis Management Team

In most companies, the composition of crisis management team will mirror the organizational chart. It should have representatives from across the organization and should bring together members of the company who have the expertise and authority to deal with the after-effects of a major business disruption. The crisis management team (CMT) will decide upon the immediate course of action in most cases and when necessary, they can contact senior management. They will direct the distribution and use of resources (including personnel) and will monitor the effectiveness of recovery activities. They can adjust the course of action, as needed. They should be in charge of activating, implementing, managing, and monitoring the business continuity and disaster recovery plan and should delegate tasks as appropriate.

Management

Each company has a management team or structure that oversees the business and its operations. You'll need to determine which positions from your management team should be included in your plan. Remember to review all the phases. For example, you might decide that only a member of the management team can cause the BC/DR plan to be activated.

Management might be required to decide when to transition from disaster recovery to business continuity activities or they might be the one(s) to decide how and when the BC/DR plan should be tested. Identify the positions that should participate as well as define how they should participate in each phase.

Damage Assessment Team

A damage assessment team should be comprised of people from several key areas of the company, including Facilities, IT, HR, and Operations. Your company's damage assessment team may contain other members, depending on how the company is structured and what type of business you're in. If you work in a small software development firm, you may just need the CEO, the IT manager, and the office manager to operate as the damage assessment team. In larger companies with multiple locations, you'll need to have several damage assessment teams or you may choose to create a mobile team that can fly to any site and assess damage within 24 hours of an incident. You may choose to have both a local and a mobile corporate team so that the right team can be called in. If the building floods, you may not need the mobile team to come in. However, if you have a large fire, earthquake, or other major event, you may need the support services of a mobile damage assessment team.

Operations Assessment Team

You may choose to have a separate operations assessment team comprised of individuals who can assess the immediate impact on operations. A damage assessment team may be tasked with this job, but in some types of companies, you may need a separate operations team that can assess what's going on with operations and how to proceed. The operations assessment team can also be tasked with beginning recovery phase activities, monitoring and triggering the transition from activation to recovery, recovery to business continuity, and BC to normal operations.

IT Team

Clearly, you need an IT team that can not only assess the damage to systems, but can begin the disaster recovery and business continuity tasks once the plan is activated. This IT team will work closely with the damage assessment team and/or the operations assessment team to determine the nature and extent of damage, especially to IT systems and the IT infrastructure. You may not need some of the technical specialties listed here, but this should be a good starter list for you to work from to determine exactly what expertise you'll need on your team.

- Operating system administration
- Systems software

- Server recovery (client server, Web server, application server, etc.)
- LAN/WAN recovery
- Database recovery
- Network operations recovery
- Application recovery
- Telecommunications
- Hardware salvage
- Alternate site recovery coordination
- Original site restoration/salvage coordination
- Test Team

Administrative Support Team

During a business disruption, there are a wide variety of administrative tasks that must be handled. Creating an administrative support team that can respond to the unique needs of the situation as well as provide administrative support for the company during the disruptions is important. This might include ordering emergency supplies, working with vendors arranging deliveries, tracking shipments, fielding phone calls from the media or investors, organizing paper documents used for stopgap measures, and more.

Transportation and Relocation Team

Depending on the specifics of your BC/DR plan and the type of company you work in, you may need to make transportation arrangements for critical business documents, records, or equipment. You may need to move equipment in advance of an event (like a hurricane or flood) or you may need to move equipment after the event to prevent further damage or vandalism. Relocating the company and its assets before or after a disruption requires a concerted effort by people who understand the company, its relocation needs, and transportation constraints.

Media Relations Team

You may recall that in Chapter 1, we mentioned the need to create a crisis communication plan because you will need to provide information about the business disruption/disaster to employees, vendors, the community, the media, and investors. One key area that should be well-prepped is media relations. Unlike other stakeholders mentioned, the media makes its living selling interesting stories. Since a disruption at your business may qualify as news, you might as well craft the message rather than leaving it to outsiders. Creating a team that knows how to handle the media in a positive manner and that understands the policies and procedures related to talking with the media is vital to help ensure your company's image and reputation are maintained to the greatest extent possible. Certainly, if your company is at fault, you will have to deal with a different set of questions than if your company experiences a natural disaster. Still, you'll need to manage the story either way.

Human Resources Team

The aftermath of a crisis is an incredibly stressful time for all employees. Having an HR team in place to begin handling employee issues is crucial to the well-being of the employees and the long-term health of the company. Retaining key employees, adequately addressing employee concerns, facilitating insurance and medical coverage, and addressing pay and payroll issues are part of this team's mission. This team may also be responsible for activating parts of the BC/DR team as it relates to hiring contract labor, temporary workers, or staff at alternate locations.

Legal Affairs Team

Whether your legal experts are internal or external to your company, you should identify who needs to address legal concerns in the aftermath of a business disruption or emergency. If you hire outside counsel to assist you with legal matters, you should still assign an internal resource as the liaison so that legal matters will be properly routed through the company. If you operate in a heavily regulated industry such as banking, finance, or health care, you should be well aware of the constraints you face, but having a legal affairs team can assist in making decisions that keep your company's operations within the bounds of laws and regulations. Even if you're not in a heavily regulated industry, you may need advice and assistance in understanding laws and regulations in your recovery efforts.

Physical/Personnel Security Team

In the aftermath of a serious business disruption, you will need a team of people who address the physical safety of people and the building. These might be designated Human Resource representatives, people from your facilities group, or both. If you work in a large company or in a large facility, you may have a separate security department or function that manages the physical and personnel security for the building. If this is the case, designated members of their team should be assigned to be part of the BC/DR team. If you don't have a formal security staff, be sure that the members of this ad hoc team receive training. Someone from HR or facilities might be willing to take on the role of security in the aftermath of a disaster, but they need to be trained as to the safest, most effective method of managing the situation. Training for part-time or ad hoc security teams is crucial because if a natural disaster strikes, emergency personnel such as your fire or police department will focus on helping schools, day care centers, nursing homes, and hospitals first. Your company may fall very low on the list of priorities, so having trained staff that can fill the gap in an emergency may literally mean the difference between life and death. We'll discuss training later in the book, but keep this in mind as you develop your teams.

Procurement Team (Equipment and Supplies)

Every company has some process in place for procuring equipment and supplies. In small companies, this might fall to the office manager or operations manager. In larger companies, there's usually a purchasing department that handles this function. Regardless of how your company is organized, you need to determine, in advance, how equipment and supplies will be purchased, tracked, and managed after a localized disaster such as a fire or in the after-math of a widespread disaster such as a hurricane or earthquake. This includes who has the authority to make purchases and from whom, what dollar limit the authority carries, and how that person (or persons) can get authority to make larger purchases. For example, a company might specify that three people have the authority to purchase equipment and supplies up to $2,000 per order and up to $20,000 total. Beyond that, they have to have the president or vice president sign off on purchases. This predetermined purchasing information can also be communicated to key vendors so they know the three people who are autho-rized and what the authorization limits are. In this way, if disaster strikes, the company can turn to trusted vendors who, in turn, know the rules. This can expedite the recovery process.

Keep in mind that this team needs to be large enough that there is no "single point of failure." If you authorize only one person and something happens to that one person, you'll be scrambling to obtain emergency authorization for other individuals. Instead, authorize enough people to provide flexibility but not so many as to create chaos. Also, be sure your limits are appropriate to the type of business you run. If you may need to replace computers at $1500 a piece, make sure the limits reflect that. If a purchaser has a $1,000 limit per item, that will preclude him or her from making a simple purchase needed to get the company running again.

General Team Guidelines

Though we recommend populating teams first with needed skills based on *roles* and *positions* within the company, we also recognize that ultimately *people* are assigned to the team. People should be chosen to be on teams based on their skills, knowledge, and expertise, not because someone wants to be on a team or because someone's boss placed them on a team. In a per-fect world, you could choose team members solely on competence, but we all know that in the real world, that's not always the case. Occasionally, you get the people who have the most time on their hands, who sometimes are the junior members of the team, or the least competent people in the department. You have to work within your organization's con-straints and culture, but also strive to populate your teams with the right people with the right skills. Ideally, these are the same people who perform these functions under normal conditions. It doesn't make sense to have the database administrator take on media relations duties during an emergency, just as you don't want the marketing VP managing the restoration of the CRM database, if possible. Certainly, in small companies many people are called upon to perform a variety of tasks and if that's the case, the same will be true if the BC/DR plan has to be activated. The teams also should be large enough that if one or more members of the team are unable to perform their duties, the team can still function. If you have other personnel or other parts of the organization that can wholly take up the BC/DR activities, so much the better. If not, you may also choose to designate key contractors or vendors to assist as alternates in the event of a catastrophic event. These personnel should be coordinated and trained as alternates along with internal staff.

Looking Ahead...

Specialty Vendors Help BC/DR Plans

There are numerous specialty vendors that can provide tremendous assistance to your firm in the event of a business disruption such as a fire or chemical spill. Although the numbers and types of firms in your area will vary, you should consider your specific needs in advance of any disruption and search for a firm that will meet your needs, even if that firm is located across the country. These firms provide a wide and unusual assortment of services, some of which are listed here:

- Chemical oxidation
- CO2 blasting
- Condensation drying
- Contact cleaning
- Corrosion removal
- Damp blasting
- Degreasing
- Deodorizing
- Fogging for odor removal or disinfection
- Manual hand wiping
- High pressure and ultra-high pressure jetting
- High temperature steam jetting
- Hot air drying
- Low pressure jetting
- Microwave drying

Continued

- Ozone technology
- Sanitation
- Steam blasting
- Vacuum drying
- Water displacement

As you can see, this is quite a list and it's not exhaustive. Be sure to think through the various scenarios that apply to your firm and determine which specialty services might best be outsourced to a qualified third party. You'll save yourself time and money in the long run and you'll likely get up and running much more quickly with targeted, competent help than if you try to do everything on your own.

BC/DR Contact Information

After you've developed the requirements for your teams in terms of the specific skills, knowledge, and expertise needed, you'll identify the specific people to fill those roles. Part of plan maintenance, discussed later in this book, involves ensuring that the key positions are still in the BC/DR loop and that key personnel are still aware of their BC/DR responsibilities.

Another mundane but crucial task in your planning work is to compile key contact information. Since computer systems often are impacted by various types of business disruptions—from network security breaches to floods and fires—you'll need to have contact information stored and available in electronic and hard copy. It should be readily available at alternate locations and copies should be stored in off-site locations that can be accessed if the building is not accessible. However, since this list contains contact information, it should also be treated as confidential or sensitive information and should be handled and secured as such. This information should include contact information for key personnel from the executives of the company (who will need to be notified of a business disruption) to BC/DR team members to key suppliers, contractors, and customers, among others.

Develop a list of the types of contact information you need, including:

- Management
- Key operations staff
- BC/DR team members
- Key suppliers, vendors, contractors (especially those with whom you have BC/DR contracts)
- Key customers
- Emergency numbers (fire, police, etc.)

■ Media representatives or PR firm (if appropriate)

■ Other

After you've identified the contact information you want to include, you'll need to determine where and how this information currently is maintained. In most companies, this information is stored in a multitude of locations and is not easily compiled with a few clicks of the mouse. You may need to develop a process for maintaining an up-to-date list, both electronically and on paper, of these key contacts. For example, many of your key contacts may be in a contact management application made available to everyone in the company. However, information such as executives' cell phone numbers and home phone numbers may not be included in this companywide contact database, for obvious reasons (especially if you work in a medium to large company). Therefore, you'll need to have a copy of the contact information plus information not included there. Developing a process for gathering and maintaining that data is an important part of BC/DR readiness. If a serious business disruption occurs in the middle of the night—for example, the building catches on fire—who will you contact? How will you know who to contact? Where will you find the key phone numbers you need if you can't get back into the building and you can't access computer systems? Since notification is one of the first steps in activating your BC/DR plan, you'll need to have key phone numbers available (*you* meaning the person(s) responsible for activating the plan). Develop a process for this during your BC/DR planning project and make sure that your maintenance plan includes regularly updating this information.

In addition to developing and maintaining a contact list, you should also define a contact tree. This defines who is responsible for contacting other teams, members of the company, or the management team. That way, each team member is tasked with specific calls to specific people and the notification process is streamlined.

Common Challenges

Maintaining Up-to-Date Contacts

Maintaining up-to-date contact information can be a challenge, especially since that information seems to change so frequently. If you work in a small company, you may task your office manager or other administrative support staff with maintaining this list and preparing an updated list once per month or once per quarter, storing it in designated locations and distributing it to key personnel. In larger companies, this task becomes a bit more difficult as contact information typically becomes fragmented—the contacts needed by the marketing group are not the same contacts needed by the IT group. Therefore, you may choose to have departmental responsibility for maintaining key contacts relevant to that business function. If you choose

Continued

that route, be sure you still have someone with high-level BC/DR responsibility who oversees the maintenance of BC/DR contact information, which in this example would include the departmental representatives who have the contact information for their units. The master BC/DR contact list should be maintained by someone on the BC/DR team and should include, at minimum, contact information for key executives, department heads, regional managers (other locations), and key BC/DR vendors, contractors, and suppliers. Regardless of the method you choose for managing your contact information, be sure that it includes a process for regularly updating it. Also update the contact tree once the contact list is revised.

Defining Tasks, Assigning Resources

The tasks and resources that need to be assigned have to do both with implementing the mitigation strategies you've defined as well as fleshing out the rest of the plan. First, you have to ensure your risk mitigation strategies will be properly implemented. This may mean creating project plans to address any new initiatives you need to undertake in order to meet your risk mitigation requirements. We'll assume you've got that covered as part of your risk mitigation strategy. If not, now's the time to develop your work breakdown structure, tasks, resources, and timelines for completing any risk mitigation strategies that need to be completed in advance of a disruption. This might include purchasing and installing new uninterruptible power supplies for key servers, updating your fire suppression systems, or implementing a data vaulting solution. Other mitigation strategies such as arranging for an alternate site need to be completed in advance, but activating it requires a different set of tasks that occur later. Finally, strategies that include accepting risk mean there probably are no additional tasks at this time.

Other tasks have to do with defining your BC/DR teams, roles, and responsibilities; defining plan phase transition triggers; and gathering additional data. Let's start with tasks related to some major activities including alternate sites and contracting for outside BC/DR services. Clearly, there are other tasks and resources you'll need, but this should get you started in developing your own list of tasks, budgets, timelines, dependencies, and constraints for the remaining BC/DR activities in your plan.

As you develop these tasks, keep in mind standard project management processes:

1. Identify high-level tasks, use verb/noun format when possible (i.e., "test security settings" rather than "security settings").

2. Break large tasks into smaller tasks until the work unit is manageable.

3. Define duration or deadlines.

4. Identify milestones.

5. Assign task owners.

6. Define task resources and other task requirements.

7. Identify technical and functional requirements for task, if any.

8. Define completion criteria for each task.

9. Identify internal and external dependencies.

We're not going to go through all that detail for these next two high-level tasks, but you should include this level of detail in your plan.

Alternate Site

Although this should be part of your BC/DR plan, it's worth calling it out separately due to its importance and the need for advance work. If part of your risk mitigation strategy is to develop an alternative site or off-site storage solution, you should develop a number of details before moving forward. These should be tasks (or subtasks) within the WBS just discussed, so let's look at some of the details you might include. Also keep in mind that you need to develop a trigger that helps you determine if or when you fire up the alternate site. You probably don't want to activate the alternate site if you have a minor or even an intermediate disruption, so how do you define when you should? When all systems are down or when some percentage of systems down? You have to take your MTD into consideration along with other factors such as the cost of firing up the alternate site and the cost of downtime. If your downtime is estimated to be 12 days and that cost is $500,000 but the cost of firing up the alternate site is three days and $250,000, is it worth it to activate the alternate or should you just hobble along until you can restore systems at the current location? There's no right or wrong answer, it's going to depend on your company's MTD, potential revenue losses, cost of starting up the alternate site, and so on. Have the financial folks on your BC/DR team prepare some analyses to determine metrics you can use to help determine your trigger point. As you're going through the activities listed in this section, keep these factors in mind.

Selection Criteria

Selection criteria are the factors you develop to help you determine how to select the best alternate site solution for your company. This includes cost, technical and functional requirements, timelines, quality, availability, location, and more. Be sure to consider connectivity and communications requirements in this section along with your recovery requirements such as maximum tolerable downtime.

Contractual Terms

Determine what contractual arrangements are appropriate for your company. Many vendors have predetermined service offerings and contracts are fairly standard. Other companies can accommodate a wider range of options and will work with you to develop appropriate contractual language. In either case, be sure to run these contracts past your financial staff and your legal counsel to make sure you are fully aware of the financial and legal consequences of these contracts in advance of signing them. If you're not clear what they mean operationally, be sure to talk with the vendor and add clarifying language to the contract. Do not simply take the vendor's word that a particular paragraph or section mean something. Verbal agreements are always superseded by written contracts, so make sure the contract spells it out clearly. You don't want to rely on verbal commitments made by employees no longer with the company when it comes to implementing your BC/DR solutions, so be sure to put everything in writing in advance.

Comparison Process

Be sure to specify what process you'll use to select the vendor. This might include a list of technical requirements the vendor must meet, but it might also include an assessment of the vendor's geographical location, financial history, and stability and industry expertise, among other things. Selecting the right vendor for an alternate site or off-site storage is a very important aspect to your BC/DR success and should be undertaken with the same rigor as your other planning activities.

Acquisition and Testing

Once you've selected your alternate site or off-site storage vendor and completed the contract, you will need to make whatever additional arrangements are needed for developing this solution so that it is fully ready in the timeframe you've designated. This might include purchasing additional hardware and software, setting up communications channels, and testing all solutions implemented. Create a thorough acquisition and testing plan for this phase so you can transition to it as seamlessly as possible in the event of a business disruption. During your testing phase of the BC/DR plan (see Chapter 8), you should test the process for firing up this solution on a periodic basis.

Contracts for BC/DR Services

Although we highly recommend you involve your purchasing, finance, and/or legal professionals in executing your BC/DR contracts, you should also have a general understanding of some of the elements to consider. As with alternate site considerations, keep your MTD, your costs, and potential losses in mind. Have your financial folks help you with performing financial analyses to determine what makes financial and business sense for your company. If a firm wants to charge you $50,000 for some sort of contract but your downtime estimate with associated revenue and collateral loss is only $40,000, the contract might not be worth entering. Additionally, determine your triggers for calling upon these contractual arrangements so you don't prematurely fire up these contracts or avoid using them during times when they should be activated.

Develop Clear Functional and Technical Requirements

You know from project management fundamentals that developing functional and technical requirements is often what defines the difference between success and failure. The same is true here. If you have not clearly and fully defined your functional and technical requirements, you'll get all kinds of vendor responses. The more specific you are, the more fully a vendor can address your needs. In addition, if you leave too many elements open to discussion, you'll endlessly discuss possibilities without being able to identify appropriate solutions. Have these discussions in advance, and then come to a firm agreement about the requirements. If some requirements appear to be optional or "nice to have," then list them as options and not as requirements. Pare down your requirements to the elements you absolutely must have. Remember, the more options you include, the higher the cost is likely to be. Therefore, if cost is an issue (and it almost always is an issue), be sure to list what you require and what you desire as separate items. When this information has been finalized, write up formal requirements documents that you can provide to potential vendors. Also, be sure that your requirements documents are reviewed by subject matter experts, including IT experts and those in your company who understand regulatory, legal, and compliance issues. Your requirements should meet all these needs before going out to the vendors.

Determine Required Service Levels

Service levels are typically part of technical requirements, but we've listed them separately because they are vitally important when developing Requests for Proposal (RFP) or Requests for Quote (RFQ) from vendors. You may have contractual obligations to provide certain levels of service to your customers, so you may need to specify requirements for your vendors that meet or exceed these metrics. Even if you have no externally facing service level agreements (SLA), you should still specify SLAs in your contracts with vendors. If you're contracting for Internet connectivity, you should specify bandwidth, minimum upload and download speeds, and maximum downtime per specified period, for example. These may sound like technical requirements, but let's look at how this can play out in a contract. You write up your requirements, which include bandwidth, minimum upload/download speeds, and maximum downtime. Three vendors respond to your RFQ to provide backup Internet connectivity to your company in the event of an outage from your main vendor or in the event that your company's facilities are damaged. All three companies give you quotes that indicate they can meet or exceed those three requirements (bandwidth, speed, availability). However, those are not contractual terms, those are the company saying they *can* meet or exceed those metrics. If it's not in the contract, it's just a statement of capabilities, not a commitment. A service level agreement will specify minimum bandwidth availability during a 24-hour period, 7 days a week. It would state that you will have access to [insert bandwidth metric] 24 hours a day, seven days a week until [insert termination metric or trigger]. This way, the vendor can't provide you the bandwidth you requested only from 11 P.M. to 6 A.M. on Saturdays and Sundays and short you the rest of the time. Granted, most vendors are on the level and want to provide the services to you they've agreed upon, but that's why contracts exist—to clearly define who does what, when, and at what cost. This keeps the guess work (and the finger pointing) to a minimum.

Compare Vendor Proposal/Response to Requirements

Once you receive vendor responses to your proposals, you should evaluate how closely each vendor comes to meeting the requirements of your plan. Any vendor that does not meet the requirements should not be considered further. There may be two exceptions to this. First, if your requirements are unique enough that no single vendor can meet your needs, you may have to circle back and find two or more vendors who can work together to meet your unique requirements. Second, you may discover from vendor responses that your requirements were too broad, inclusive, or vague, and that none of the vendors' responses meet your requirements exactly. In that case you may have to refine your requirements and go back out for bid. Assuming your requirements are well written, your next step is to eliminate vendors that cannot meet your needs and focus only on those vendors who addressed your requirements fully in their responses.

Identify Requirements Not Met by Vendor Proposal

If there are one or more requirements not met by any vendor, you may need to find two or more vendors to work together to provide the full range of services you need. If none of the vendors met a particular requirement, you may also choose to review that requirement and reassess it in light of vendor responses. Remember, you contract with vendors in order to leverage their specific expertise. If none of them meet a particular requirement, you may wish to talk with several of your short-list selections to find out why they did not address that aspect. It might be redundant or otherwise unneeded. In that case, you should revise your requirements to reflect this new information.

Identify Vendor Options Not Specified in Requirements

Vendors also may offer additional options not specified in your requirements. Again, based on the vendor's expertise, they may offer additional choices that can round out your requirements or plan. Utilizing their expertise can be a good way of ensuring you have the best solution in place. For example, the vendor might say (in essence), "Everyone who's asked for A, B, and C also has found that D was an extremely important option they'd overlooked. Perhaps you'd like to add D to your plan as well." They may be sharing industry expertise and best practices with you, or they may simply be trying to up-sell you. You'll have to look carefully at these options and perhaps do some independent research to determine whether these options are "must have," "nice to have," or "useless add-ons." If you have an established relationship with these vendors, they'll more than likely offer you additional options but won't put pressure on you to upgrade unless they feel it's vital to your success. However, we all know there are sales people that will try to sell you every option they can think of just to make a bigger sale, so you have to be an active participant in the transaction. Know your options, know what makes sense, do some additional research, and determine if any of the additional options would enhance your plan or fill in gaps you didn't realize existed. Don't be forced into upgrades and options you don't really need just because you have a very persuasive sales person in front of you.

From the Trenches...

Managing the Sales Process

Sometimes your purchasing department manages the purchase of goods and services, but when you're talking about the purchase of backup, storage, or alternate site services for your BC/DR plan, there's a good chance you will be directly involved. If you haven't been involved with the sales process in the past, you might find yourself being swayed by excellent sales people—the ones who can convince you that you need something you really don't. Most sales people are honest and are trying to balance their need to sell with your need for the product or service they're selling. They also realize that loyal customers are borne out of an honest sales experience, not out of strong-arming someone into purchasing more than they need. In order to be successful in the process, take time to be clear about *your* objectives *before* a sales meeting. If you intend on making a purchasing decision at that time, write down the terms or parameters you will accept. Keep these to yourself but know that this is your bottom line. If the sales person cannot or will not meet your bottom line objectives, there is no deal to be struck.

The same holds true in any negotiation—know what your bottom line is and work toward meeting (or exceeding) that bottom line. If you have developed clear requirements and you know your bottom line, you should be able to successfully navigate the sometimes tricky sales process. Negotiation skills can help you in all aspects of life and they'll certainly help you in the business world. If you're interested in learning more about the art of negotiation, there are thousands of helpful books, courses, and seminars you can turn to for more information.

Communications Plans

Earlier in this book, we discussed the need for various communications plans. In this section, we'll define various communications plans you should develop and identify some of the common elements in such a plan. If you already have communications plans in place, you can use this section as a checkpoint to ensure you've got all your bases covered. For each plan, you should define specific steps just as you would for any other process in your BC/DR plan. You should define the following:

- Name of communication team, members of team, team lead, or chain of command
- Responsibilities and deliverables for this team
- The boundaries of responsibilities (what they *should* and *should not* do)
- Timing and coordination of communication messages (dependencies, triggers)
- Escalation path
- Other information, as appropriate

Communications plans can be assigned to other, existing teams. A good example of this is that the employee communication plan may be the responsibility of the HR team. There's no need to create additional teams to execute communications plans if these activities fall within the scope of defined teams. However, in some companies, it might make sense to have most of the communications come from one dedicated communications team in order to maintain control over communications and to ensure that a single, consistent message is delivered to all stakeholders. The decision is yours and usually is based on how large the company is and how it currently operates.

Internal

The internal communication plan is really part of the BC/DR activation and implementation plan. If a business disruption occurs, you need to have a process in place for notifying BC/DR team members. This is done as part of BC/DR plan activation and is a critical aspect that should be clearly delineated. How will team members be notified and updated? What processes, tools, and technology are needed? Are these included in your plan yet? If not, add them to your WBS or in a section called Additional Resources so they are captured and addressed in advance of a business disruption.

Employee

Employee communication is also internal communication but differs because it is any communication that goes out to employees who are not part of the BC/DR implementation. If a business disruption occurs, you'll need to know how to notify all employees. You'll also need to let them know answers to the most basic questions including what happened, what is being done to address the problem, and who they should go to for more information. For example, if the building burns down overnight, employees may show up for work in the morning as scheduled. The BC/DR team may already be in action but the general employee population needs information. How this information is communicated and by whom should be identified. It often makes sense to develop an information tree so that key communicators know to whom they should go for updates and official information. For example, in a small company, you may designate the HR manager as the person who will communicate with employees on all BC/DR matters. The HR manager should know who to go to for information on the status of the BC/DR activities. This might be the Facilities manager or the BC/DR team leaders (who should be identified in the activation plans, discussed earlier in this chapter).

Customers and Vendors

Customers and vendors typically require different types of communications but the information is often similar. They may need to be notified of the business disruption, the basic steps being take to rectify the problem, the estimated time to recovery and any work-arounds needed in the meantime. If you are developing crisis communications plans for the first time, be sure to read the case study that follows this chapter, entitled "Crisis Communications 101," for more information on how to communicate in a crisis.

Shareholders

If you have shareholders of any kind (debt or equity investors, shareholders, etc.) you must communicate the nature and extent of the disruption. In most cases, they are concerned with the ongoing viability of the company and possibly the short-term financial impact of the disruption on the company. Therefore, communication with this group requires that specific issues be addressed. As you can tell, these issues are very different than, say, employee issues, so someone well-versed in investor relations should be charged with this communication. In most companies, this task falls to the CEO or a high-ranking corporate officer who can specifically address the concerns of those who have a financial stake in the company.

The Community and the Public

In addition to communicating with all the other stakeholders we've mentioned, you also will need to communicate with the general public. Local newspapers, TV, and radio stations will certainly take an interest in a localized business disaster such as a fire or flood. National and international media may also take interest if the event is unique in some way or is part of a widespread disaster. Members of the local community may also have more than just vicarious interest—they may need to understand the impact your business disruption may have on them. Businesses in communities don't exist in isolation, and what happens to one business may have a ripple effect on other businesses even if those other businesses are not customers or suppliers.

Communicating with the media is a tricky proposition and many executives at large firms go through extensive media training sessions in order to learn how to deal with the media. Although an extensive discussion of this topic is outside the scope of this book, you will learn the basics by reading the case study that follows this chapter. Additional media relations training resources are readily available online and there are hundreds of excellent books on the topic as well. As the leader of the BC/DR project plan, you may or may not be called upon to communicate with the media, but being prepared is always a good idea.

This plan should be well thought-out and you may wish to seek legal counsel with regard to what must be disclosed, to whom, and in what time frame. As you learned in the case study presented earlier in this book ("Legal Obligations Regarding Data Security" by Deanna Conn), there are numerous legal requirements regarding notification and remediation that must be met in certain circumstances. To ensure you comply with regulations and laws in your industry, be sure to seek appropriate input from subject matter experts as you craft your shareholder communication plan.

Tip

Many public relations firms specialize in crisis communications. You can work with this type of firm in advance to develop appropriate communications plans. You can also contract with these kinds of firms to assist with communications in the aftermath of a major event. In most cases, they can advise you on the best course of action, potential communications pitfalls, and provide guidance regarding certain legal issues. You may also need to get a legal opinion in certain matters, especially if death or injury occurred on your company's premises or as a result of company action. The PR firm you work with can help you understand how to communicate effectively and when to seek additional input before, during, and after your business disruption.

Event Logs, Change Control, and Appendices

In traditional IT, event logs track a variety of system and network activities. In a broader sense, you may choose to create a BC/DR event log for tracking various events and milestones. For example, your decision to activate your BC/DR plan may be based on two or three event types occurring, either simultaneously, in quick succession, or within a specified time period. These events may trigger the activation of the BC/DR plan itself, or they may signal the point in time when it's appropriate to move to the next stage in your plan. Having a chronological log of events can help clarify circumstances so appropriate decisions can be made in a timely manner.

Event Logs

As an IT professional, you're probably well versed in reviewing event logs as they pertain to systems and security events. However, in BC/DR, event logs are not necessarily logged by a computer system. In many cases, event logs are hard copies developed sequentially over time by making notes on what happens when. Event logs help you track events, in order, over time, and can help in identifying appropriate triggers for key activities.

Keep in mind that these logs establish who knew what and when, so they may become legal documents at some point in the future. You have to balance the need for timely information with the potential for litigation. As unfortunate as it may be, sometimes too much documentation leaves the company open to lawsuits, even when the company has acted as best it could given the circumstances. We don't suggest you do anything illegal or unethical—quite the opposite—but you may want to talk with your legal counsel to understand what can and cannot become evidence in the event there is a lawsuit that stems from some sort of business disruption. If something can become a legal document or be used as evidence in some manner, you should be aware of that going in. Your legal counsel may have recommendations about how to record data to minimize the possibility of litigation while maintaining accurate, useful logs.

In the absence of specific legal advice on how to develop logs, the best general advice is to record only the relevant information and stick to the actual facts, not conjecture. Instead of "Barnett seemed confused by the request to review the equipment," you might simply say, "Barnett was contacted regarding reviewing the equipment at 11 P.M., 2/22/07" or "Barnett had numerous questions regarding the request to review equipment. Issue escalated to Barnett's boss, Martina." All these statements are true but the first statement contains conjecture—was Barnett confused or did you just assume he was because of the look on his face? If you state in your log that Barnett was confused, this might be the basis of a lawsuit claiming that appropriate action was not taken in a timely manner. Stating only the facts keeps everything moving forward and does not unnecessarily open the door to legal problems down the road.

On the other side of the legal coin, there may be legal or regulatory requirements to log certain events or make notifications within a certain timeline. Event logs can help you operate within these legal requirements as well. If you operate under these constraints, be sure to include these requirements in your BC/DR plan, perhaps with hard copy templates of the event logs, so that your team knows clearly what the logging or notification requirements are in the stressful aftermath of a business disruption.

Change Control

Change control is a necessary element in any project and BC/DR planning is no exception. There are two types of change control you'll need to develop. First, you need to devise a method of updating your BC/DR plan when change occurs in the organization that impacts your plan. Second, you need a method of monitoring changes to the BC/DR plan to ensure they don't inject additional uncertainty or risk into your plan. Let's look at both of these scenarios briefly.

As companies grow and expand, numerous changes occur to the organization's infrastructure. This can include departmental reorganization, the creation of new departments, the expansion to additional facilities, and more. It also comes with changes to the IT infrastructure including the location and duties of servers, the implementation of new applications and technologies, and the reorganization of existing infrastructure components. All these kinds of changes impact the existing BC/DR plan. These elements should be addressed in the plan maintenance activities, which we'll discuss in Chapter 9. You can't control the change that occurs in the organization, but you can put in place a system for assessing change and how it impacts your BC/DR plan. In most cases, this occurs during the periodic review of the plan and we'll remind you of that in Chapter 9.

A subset of change control is version control. Be sure to include a process for managing revision history for your BC/DR plan. Many people choose to simply put a small table at the beginning of the document outlining the changes in chronological order. Table 6.1 shows an example of a revision history table that might be used in your BC/DR plan.

Table 6.1 Revision History Table

Revision Number	Revision Date	Detail
1.0	02.22.07	Finalize first version of BC/DR plan
1.1	03.20.07	Modify network diagrams in Section 4.2
2.0	06.05.07	Revise plan to include acquisition of ABC Co.
2.1	11.05.07	Include new specifications and contract for alternate site.

You can define what constitutes a major and minor revision. Typically, going from 1.0 or 1.1 to 2.0 is considered a major revision (when the number to the left of the decimal point increases); going from 1.0 to 1.1 or 2.11 to 2.2 is considered a minor revision (when the number(s) to the right of the decimal point increase). Clearly, the numbering scheme is not quite as important as keeping track of revisions, unless you work in a company that has a very formal system for revision control in place. A quick note in the Detail section can help clue you in to the changes in the revision. Some people also like to document more extensive information about the changes and this can be done in the beginning of the document.

For example, you could create paragraphs labeled, "Changes in Revision 1.1," and note the key changes made to the document. This helps you see at a glance how the plan has changed without reading the entire document. There are numerous systems for managing revisions and you should select one that is consistent with the way your company operates. Don't make it into a huge production or it may be circumvented, but do use some system for tracking changes so you don't have to compare two documents side by side to figure out what changed between revisions.

Distribution

Although the plan is not yet complete, you should devise a strategy here for distributing and storing the final BC/DR plan. The revision history will help you and the team with version control, but you will still need a method of distributing the latest revision or notifying the team that a new version exists. In some cases, the plan may be stored in a software program that performs version control and revision notification. In that case, you're pretty well set other than adding team members to the notification list. If you're not using such a program, you can still maintain the plan on a shared, secured network location and provide team members or team leads with access to the folder. Keep in mind that this document is a very sensitive document and all precautions should be taken to ensure it does not fall into the wrong hands, is not leaked to competitors or to the media, or otherwise compromised. Use standard security and encryption where this document is concerned. Distribute the document in soft copy via e-mail only as needed. If possible, simply e-mail a notification that a new version is available while maintaining the document in the secure location. Remind people that the document is sensitive and should not be copied, distributed, or otherwise handed out. The document should only be distributed to those who have a defined need to know.

Finally, be sure you create a process or method for *printing* the updated plan so you have a hard copy version available if systems go down. The BC/DR team lead or leads should all have a paper copy in a secure location, both on-site and off-site. When new versions are available, old versions should be shredded or destroyed in a secure manner.

Appendices

Any information relevant to your plan that does not belong in the body of the plan should be attached or referenced as an appendix. There are no strict rules about what should or should not be included in the appendices, but it's usually detail required for successful implementation of the plan that may pertain to only one group or subset of BC/DR teams. For example, you might include the technical specifications of mission critical servers in an appendix. As servers are moved, updated, or decommissioned, you can easily update the related appendix without modifying the plan itself.

Contracts with external vendors should be kept as appendix items so that they are located in one central place for reference. Your finance and/or legal departments may want to retain originals of these contracts, which is fine, but be sure to include copies in your BC/DR plan. If you have to activate your plan, you don't want to have to run around looking for someone from finance or legal to determine how and when you can activate your external contracts.

Templates for event logs, communications, and other predefined processes can be included. In event log templates, be sure to include time, date, event, notification requirements, legal, or compliance issues and other requirements so they're easily accessible in the event of a business disruption or disaster.

Key contact information should be included in the plan, but you may choose to include it as an appendix, especially if it changes frequently. If you choose to do this, you should include key contacts within the body of the plan and use the appendix for additional contact information, as appropriate. The reason for including the key contact information within the body of the plan is twofold. First, key contacts are integral to the successful activation and implementation of the plan. As such, that information should be incorporated into the body of the plan. Second, if that information changes, it should trigger a BC/DR plan revision. Key personnel need to be trained, they need to understand their roles individually and as part of the BC/DR team, and they need to be given the tools, resources, contacts, and information needed to do so successfully. If a key member of the BC/DR team leaves, for any reason, the person replacing them needs to be brought up to speed. This should trigger a quick review of the plan. If the successor has been assigned by virtue of position (the Facilities manager resigns and a replacement is hired), the replacement needs to be trained in all aspects of their duties with regard to the BC/DR plan. If the successor is not assigned and needs to be found, looking through the roles and responsibilities of this position can help you select the right person to fill the gap.

Any other information that is related to the plan that needs to be updated, maintained, and correlated to the BC/DR plan itself should be included as an appendix. Don't throw everything you can think of into an appendix and think you're covered. More is not necessarily better in this case, but do be sure to include key information you'd want to have quick access to in the event of a natural disaster or other significant business disruption. To give you a few ideas about what else might be attached to your plan in an appendix, we've provided the following list. Not all of these elements are needed by every company, but you can pick and choose based on your unique situation.

- Critical work space equipment and resource information and related vendor data
- Critical IT hardware, software, equipment and configuration information, and related vendor data
- Critical manufacturing, production and warehousing information and related vendor data

- Critical data and vital records information, including storage and retrieval information
- Alternate IT or work site information
- Crisis management center resources and information
- Insurance information including all relevant policies, policy numbers, and insurance contact information
- Service level agreements (that you must provide to customers or that vendors must provide to you)
- Standards, guidelines, policies, and procedures
- Contracts related to BC/DR
- Forms
- BC/DR Plan distribution list
- Glossary

Every company and every BC/DR plan is different, so there is no hard-and-fast rule about where information belongs, as long as critical data is included in a logical manner. If writing a plan or organizing data is not your strong suit, be sure to recruit assistance to draft a plan that makes sense. It should follow a logical progression and match the way your company does business to the greatest extent possible.

Additional Resources

What other resources do you need to successfully implement and maintain your plan? In the next chapter, we'll discuss emergency and business recovery plans, so some of this may come up in that context. However, if there are communication tools, equipment, or resources you think of as you develop your plan, they should be noted in a section called Additional Resources (or other similar heading) and they should be added to your WBS to ensure someone takes ownership of gathering these needed resources.

What's Next

When you complete the work in this chapter, you should have a fairly robust BC/DR plan in the works. It will have gaps related to specific emergency and disaster recovery efforts (see Chapter 7), and in training, testing, auditing, and maintaining the plan (see Chapters 8 and 9), but other than that it should be well on its way to completion. If not, step back and review your data, your plan, and your company to determine what is missing and how you can address those gaps.

Summary

Putting your business continuity and disaster recovery plan together requires pulling together the data previously developed and adding a bit more detail. Understanding the phases of the BC/DR plan helps you develop strategies for managing activities if you have to implement your plan. The typical phases are activation, disaster recovery, business continuity, and resumption of normal activities. The plan must also be tested and maintained, regardless of whether it's ever implemented.

Potential disruptions need to be categorized and we discussed three levels: major, intermediate, and minor. By clearly defining these for your organization, you can ensure you understand what recovery steps should be implemented. This will define how and when you activate your BC/DR plan. After the plan is activated, a trigger should define when disaster recovery tasks begin. These recovery tasks should be well defined in your BC/DR plan and we'll cover these in detail in the next chapter. The transition from disaster recovery to business continuity should also be well defined so that you can begin to resume business activities, though things will not be back to business as usual at this juncture. This is also discussed in detail in the next chapter.

Developing your BC/DR teams is a vital part of your planning. There are numerous roles and responsibilities in each phase of your BC/DR work and defining these and populating your teams in advance is crucial to your success if the plan is ever activated. In addition, these teams will need to be trained in implementing the BC/DR activities. Training is discussed in detail in Chapter 8.

After you've created your teams, you can further develop your planning tasks and assign resources, timelines, and budgets. You can identify task dependencies, develop milestones, and create completion criteria for key tasks. Since each company's set of tasks will vary widely, we presented only a sampling of high-level tasks related to acquiring an alternate computing site and contracting with vendors.

Communications plans are part of the BC/DR process because if a business disruption occurs, many different groups of people will need status updates and information. This includes employees, management, shareholders, vendors, customers, and the community, among others. You'll need to decide who needs to know what and when they'll need to know it. Then, you'll need to develop distribution methods appropriate to those groups (and to the circumstances of the disruption).

Event logs can help you manage the business disruption from start to finish, but remember that these may become legal documents later on. You may wish to consult with your legal counsel regarding what should and should not be included in the event logs. For example, it's generally considered fine to include facts but not conjecture or opinion. Sticking to the facts helps keep the log clear and concise and can avoid misinterpretation of data. In addition, there may be legal or regulatory requirements for event logging or notification, so be sure to include this in your process and make a note of it in any log files you develop (whether soft or hard copy).

Keeping track of document revisions is a bit of a "housekeeping" task but an important one when it comes to your BC/DR plan. Use a simple, concise method of ensuring that the plan is updated and that everyone has the latest plan. Develop a method for distribution and storage of the plan so that it's accessible to key personnel in the event of a disruption. Finally, include additional data such as technical requirements, service level agreements, and vendor contracts as appendices to the main BC/DR plan. Keeping all relevant data with the plan can make plan implementation and maintenance much easier.

Solutions Fast Track

Phases of Business Continuity and Disaster Recovery

- ☑ The various phases of the BC/DR cycle include activation, disaster recovery, business continuity, maintenance/review. Plan maintenance and review occurs periodically, regardless of whether or not the plan has ever been activated.

- ☑ The activation phase occurs when a disaster or business disruption occurs and it is determined that the plan should be implemented. Clear directives on how and when to activate the plan should be included.

- ☑ The disaster recovery phase includes the tasks that must be undertaken to stop the impact of the event and to begin recovery efforts. This includes damage assessment, risk assessment, salvage operations, as well as the evaluation of appropriate alternatives and solutions.

- ☑ The business continuity phase entails the activities required to restore the company to business operations. This assumes disaster recovery has been completed and that the business is up and running in a limited mode. This is not yet business as usual, and may involve the use of temporary solutions and work-arounds.

- ☑ Maintenance and review are similar phases. Maintenance requires a review of the plan from time to time to ensure everything is still current and that changes to the company or its infrastructure are reflected in the plan.

- ☑ Review occurs after the plan has been activated and implemented. Gathering lessons learned and updating the plan with new information gleaned from the experience helps the organization avoid making the same mistake twice and helps the organization learn from the experience.

Defining BC/DR Teams and Key Personnel

- ☑ You should have already identified key personnel, positions, skills, and expertise needed for your BC/DR activities. In this phase, you should form your BC/DR teams based on those stated needs.

- ☑ There are many different types of teams you may need. In smaller companies, people may take on multiple roles. Be sure teams are large enough to accommodate the potential that one or more team members may be unavailable during or after a disaster (for a variety of reasons).

- ☑ Key personnel should also be identified at this time. This may include certain members of the executive or management team, people or vendors with specific skills needed, and the like.

Defining Tasks, Assigning Resources

- ☑ Tasks, owners, timelines, budgets, dependencies, and completion criteria are among the details that should be developed for BC/DR plan activities. Additional detail and checklists are provided in Chapter 7.

- ☑ All the tasks needed to activate, implement, manage, and monitor the BC/DR plan in action should be defined. You can create project plans for each subsection of work or develop detailed checklists.

- ☑ Be sure to note key internal and external dependencies for tasks. Milestones should be added to your project plan or as items on a checklist.

- ☑ Maximum downtime and other time-based objectives should be noted and addressed within the project plan or checklist.

- ☑ Contracts for alternate sites, equipment, and other products/services should be defined in the BC/DR plan as well. The finalized contracts can be added as appendices to the final BC/DR document. Include service level agreements, where applicable.

- ☑ Address MTD and other constraints along with legal or regulatory requirements that must be met in the aftermath of a disruption to ensure continued compliance.

Communications Plans

- ☑ There are many different kinds of communications needs during and after a major event, disruption, or disaster. You should develop plans for communicating with various stakeholders in these cases.

☑ Management and employees require communication regarding the current status of the business, where/when/how to report to work, who to contact for information, and so on. This often is handled by the HR team, who can be tasked with managing the employee communication plan.

☑ External communications are needed to contact key customers, vendors, suppliers, and contractors to notify them of the event and the company's next steps.

☑ Communications with local, national, and international media may be required. In these cases, it's best to have someone from the company who is trained in media relations handle these communications. PR firms often offer plan development, training, and guidance in the aftermath of an event.

☑ In some cases, the information communicated can become the basis of legal action in the future. Therefore, you should consult with your company's legal counsel or an expert in media relations to determine how, what, when, and where information should be communicated. This should be done before any emergency communication is needed.

Event Logs, Change Control

☑ Event logs, like emergency communication, can become the basis of legal action, so be sure to understand the requirements and constraints various kinds of emergency reporting may have on your company.

☑ Event logs help you keep track of what's going on, what's been done, and what needs to be done next. Keeping detailed logs in real time helps keep track of details that might later be lost.

☑ Your company may be required to meet certain legal or regulatory reporting requirements. Event logs can be helpful in ensuring you meet those requirements. Consult with legal counsel if necessary and include these requirements in soft or hard copies of your logs.

☑ Changes to the BC/DR plan should be tracked and noted so that team members can easily determine if they have the latest revision of the document as well as the general nature of those revisions. Be sure to develop a distribution system that notifies team members of new revisions, provides a method for accessing new documents, and reminds teams to print and store the documents in locations accessible both on- and off-site.

☑ BC/DR plans should be treated as confidential documents. They should be handled and stored in a secure manner and old copies should be destroyed appropriately.

Appendices

☑ Information that should not be included in the body of the plan but that is nonetheless vital to the plan should be included at the end as an appendix.

☑ Appendix data can include event log or other document templates, vendor contracts, technical specifications, service level agreements, customer contacts, or any other relevant data that would be useful to have along with the BC/DR plan if/when it's activated.

Frequently Asked Questions

The following Frequently Asked Questions, answered by the authors of this book, are designed to both measure your understanding of the concepts presented in this chapter and to assist you with real-life implementation of these concepts. To have your questions about this chapter answered by the author, browse to **www.syngress.com/solutions** and click on the **"Ask the Author"** form.

Q: Our IT department consists of three people and our company has one location with 50 employees. It seems all the information in this chapter is a bit of over kill. Any comments on that perspective?

A: Yes. There is a lot of information in this chapter and some of it may not apply to all companies, including small companies like yours. However, it's important that you review and consider all this information so that you can develop a comprehensive plan appropriate to your organization. When it's all said and done, you may not include the bulk of what's included here, but you will most likely feel confident you're not missing something major. Because each company is unique, there really is no one-size-fits-all solution. Instead, we have to raise as many potential points as we can and allow you to incorporate or exclude that data as you see fit. If you're in a small company, your plan may be very short. It might include plans for storing your backups off site, plans for buying or renting new computers after a disaster, and a contact list—if that's what you and your company feel is important. At the end of your planning process, you should have a plan that you're comfortable with—a plan that will enable you to recover from a minor or major business disruption in the time and at the cost you've designated.

Q: Our planning process has gotten bogged down in this very step. We managed to perform the various analyses, but now no one seems to have the time or energy to actually develop the plan—to actually define what we'll do, step by step. Any suggestions for how to move forward?

A: It can be difficult to maintain momentum, especially when the analysis activities take a fair amount of time and effort. Teams can lose participants or simply lose focus and motivation. As the team leader for the project, you have several options. First, you should understand the current environment in your company. Are there competing pressures for time and resources? Are these impacting your project? If so, it might be time to check in with your project sponsor and get his or her assistance in reigniting the interest in this project. Second, you should look at the project as a whole. In some cases, the project begins looking so large and overwhelming at this juncture that people just sort of melt down. If this appears to be the case, you can find ways to break these steps down into smaller chunks of work so that people can see progress. For example, you can develop plans for what you would do in light of a minor disaster or event. This is usually a manageable task to consider and can provide a sense of accomplishment to the team.

Once these steps or procedures are defined, you can move on to intermediate and then major events. Building upon prior work can help teams feel they are making progress and make each subsequent task seem a bit less daunting. You can also spend time breaking the project down into smaller subprojects. For example, have the IT team work on their response to minor, intermediate, and major events. Have the communications team and the other subteams do the same. Then, you can convene meetings to have these teams sync up or coordinate their plans. You may have other tried-and-true methods you use to get a stalled project back on track, but understanding the underlying cause is the first step. In the case of BC/DR planning, the usual suspects are lack of continued corporate commitment or focus and the sense of unending, daunting or overwhelming work for the team. Addressing these underlying causes often gets the project moving forward again.

Q: Someone in the job before me created a BC/DR plan but it doesn't track at all with what you've presented so far. I'm not sure the best course of action at this point. Any suggestions?

A: Yes. The key is to ensure that the needed data is included in the plan. In reading through your old plan, even if it doesn't track at all with the way we've presented the material, you should be able to get a sense of whether or not the plan contains the necessary data. In essence, ask yourself if you had to use this plan if you would know how to recover from a major disaster. If the answer is no or maybe, then you don't have all the information you need and the plan is incomplete. You can decide to scrap it and start from scratch or you can try to fill in the holes while updating it with the most current information. Sometimes it's easier to start from scratch but you can make this determination. A detailed review of the plan to be sure it contains the necessary elements and the most current information will probably suffice. There is no single "right" way to develop your plan. The bottom line is this: Will your plan provide you enough guidance during the stressful aftermath of an emergency, disaster, or disruption to get the business up and running again? If the answer is yes, you're in good shape. If the answer is no, you need to perform a gap analysis on your plan and find out what's missing and how to address those gaps.

Crisis Communications 101

Contributor Profile

Patty Hoenig, Marketing, Media, and Public Relations Specialist

Patty Hoenig has extensive experience in marketing, media, and public relations. She has held various positions in these fields throughout her career and has developed a number of crisis communications plans. Her experience working with small and large companies, including those in heavily regulated industries such as banking and finance, enable Ms. Hoenig to provide practical and actionable advice to her clients.

Ms. Hoenig holds a Master in Business Administration (MBA) degree from the University of Phoenix and a Bachelor of Art (BA) in Journalism degree from the University of Arizona.

Background

Waiting until a crisis strikes to develop your communication plan can only spell disaster, even for the smallest company. A good, effective plan for crisis communication cannot be created quickly in those first few hours following an accident or mishap or mistake. The stress of the situation causes people to respond differently and having a plan already in place for crisis management and communication will facilitate effective communication to all parties concerned.

Though it is a crucial part of a business continuity and disaster recovery plan, a crisis communication plan is an important part of business communications planning in general. Crisis communication is a formalized plan that ideally should be in place before you open your doors, created in tandem with the business plan. It evolves as the business evolves, it changes as the company players change, and *most important*, a crisis communication plan is practiced on a regular basis, just like a fire drill. This may surprise you and unfortunately, many (or most) companies do not do this. However, if you wait until you have a disaster to try out your crisis communication skills, you almost certainly will stumble or fall.

One of the most severe long-term impacts of a disaster is the loss of confidence by employees and their families, investors, customers, suppliers, regulators, and the community at large. In many cases, poor handling of the crisis communication piece results in an exacerbation of the existing problems and leads to further decline in sales and customer confidence. Hiding behind closed doors in the face of a crisis used to be the standard management technique, but with the 24-hour news cycle and the rise of the Internet as a news source, companies can ill afford to simply allow information to "manage itself." Saying nothing also allows others to fill in the blanks. Rumors, innuendoes, and outright lies about the company, the situation, the cause, the impact, and all other details need to be managed and controlled by the company, not the public at large.

Three Simple Rules for Crisis Communication

There are a lot of things your crisis communication plan should cover, but let's give you the bottom line first. There are three simple rules to effective crisis communication, which, if followed, should yield the best result possible within the constraints of the facts and circumstances. That means that if the company made a serious mistake, these three rules won't change that, but they will help to mitigate some of the damage.

1. Always tell the truth.
2. Appoint a single spokesperson to be the face and voice of the company with the media.
3. Provide information that addresses who, when, what, where, why, and how.

It's that simple. Now, let's talk about some of the implementation considerations of these three rules.

Rule #1: Always Tell the Truth

Lying almost always comes back to bite you, and in the case of crisis communications, it is a virtual certainty. You might say that even in this litigious society, the truth is your best option; but it's actually more accurate to say that because of this litigious society, the truth is your best option. Although it might feel like you're providing fuel to the litigation fire when providing a truthful accounting of events, circumstances, and actions, especially if you or your company is at fault, you're actually providing information about what occurred. This will likely come out one way or another. If your company made a massive or minor error that led to the event, the truth will eventually come out. Someone, somewhere will know the truth and before you know it, it shows up on blog after blog and suddenly it's on the 24-hour news channel. So, stick with the truth.

There are numerous case studies demonstrating the power of the truth. When companies own up to mistakes they've made and take action to prevent these mistakes from being repeated, they invariably come out on top. It can be a long, painful journey, but the organization is stronger as a result. The truth, as difficult as it may be, is always the best option when dealing with a crisis of any kind.

However, telling the truth doesn't mean spilling your guts and disclosing every last detail you know. Although you should stick with the truth, short simple communications are better than long, all-too-revealing communications. Also, when assessing Rule #2, keep in mind that some people tend to talk more when they're nervous. These people need to be thoroughly trained to keep their cool and practice the "less is more" philosophy of communication or they need to be in the background and not appointed as your corporate spokesperson.

TIP

There's a saying I like to share with my clients and it's this: "You don't need a good memory if you always tell the truth." Whether you remember the small details of an event or not, you will remember the highlights (or in most cases, the lowlights). If you're busy trying to cover up the truth, you'll have less time to actually deal with the situation. If you don't like the truth, you need to change the situation, not the story you tell about it.

Rule #2: Appoint a Single Spokesperson

The single spokesperson should be someone well-trained in public and media relations and specifically, in crisis communications. This person provides a single point of contact and a consistent message to all communication channels. Although you may have multiple communications teams working internally to address various needs such as employee or customer communications, you should have one and only one official corporate spokesperson during a crisis.

If the situation escalates, it's possible that your spokesperson will have to defer to the CEO or top executive at the firm. Sometimes the media wants information directly from the person charged with overseeing the company and no one else will do. However, in all other instances, assign one and only one person to talk with the media and to represent the company for the duration of the crisis.

TIP

Not everyone is suited to being a spokesperson. The person that represents the company has to stay calm under pressure and respond in a thoughtful but concise way. You do not want someone standing before the microphones airing the company's "dirty laundry"; you do want someone who understands how much to share of the truth and how to present it. Certain words are emotionally charged and will do nothing but inflame a bad situation. Using a different word can convey the same meaning but without the emotional impact. "Casualties" is less distressing than "dead bodies" for example. Both are correct and true, but the first is more emotionally neutral. These are the kinds of nuances your spokesperson should understand and be able to employ naturally.

Rule #3: Provide Formatted Information

If you've ever taken a Journalism 101 class, you know that the basic information every reporter needs to convey is who, when, what, where, why, and how. Who was involved, when did it happen, what happened, where did it happen, why did it happen, and how did it happen? If you fashion your communications to address these basic needs, you'll help the media do their job quickly and efficiently. Also, as you attempt to craft answers to these basic questions you know reporters will ask, you can spot potential pitfalls and find ways to carefully craft your answer. We all know there is a difference between speaking truthfully and fanning the smoldering flames. For example, certain words evoke certain images and emotions. Avoiding loaded words during a crisis can help deescalate the issue instead of fueling it. If you think about how to convey the information in a factual manner that helps quell fear and chaos, you're doing your company, its employees, and the general public a service. Think of the difference between "There were numerous casualties, we don't have an exact count or any other details at this time." and "It was a blood bath, there were bodies everywhere and we're still trying to pull dead bodies out of the rubble." Both statements may be true, but the second statement takes the emotion and drama up a notch.

Your crisis communications plan and training should include language that can be used to help neutralize the crisis by providing factual data without editorials, emotions, and other extraneous data. Also, before releasing specific data, ensure that it is legal, ethical, and appropriate. For instance, you should not release names of casualties or victims until families have been notified. You should not release information about the potential cause (how, why) of the event until it's been cleared with executives, investors, finance, or legal representatives. Providing templates with guidelines for crisis communication along with training and practice can help ensure crisis communications that meet everyone's needs without unnecessarily escalating the problem.

Directional Communications

Your crisis communications need to go in all directions almost simultaneously. You need to communicate out to the community, the media, and the public. You also need to communicate up the chain of command so corporate executives can be kept informed about the progress of events. You need to communicate laterally to others within the firm, including other communications teams and other BC/DR functional teams.

Also keep in mind that part of your training for crisis communications includes educating executives and employees about corporate policies related to public communication. Essentially, everyone needs to buy into the single spokesperson model. You want to avoid having random employees speaking to the media out of turn, providing inappropriate or incorrect information. Set clear guidelines about these kinds of situations so everyone is clear how it should work. Of course, you probably can't avoid having an employee go home and post something on a blog, but you can help educate employees as to why they should adhere to these kinds of policies. Providing information about how this policy helps employees (the *what's in it for me* method) can help avert some of these errant communications that can confuse or contradict ongoing efforts.

Practicing Your Plan

Practicing a plan allows people within the company to understand their role and what they must do as part of the team. When people understand what is expected of them, what their roles, responsibilities, and restrictions are, they are empowered and the tendency toward fear and panic in the aftermath of an emergency can be reduced. If a crisis occurs, company spokespeople can be (and should be) totally in control of the message. The person talking to the press is calm, cool, and in charge because he or she has practiced this many times. He or she will have the answers and information necessary at their fingertips and will know how to answer difficult questions with composure.

The people who have to act behind the scenes can reassure employees and keep any anxiety at bay immediately because they, too, have practiced these skills through repetition and drills. They, too, will be able to answer the employees' questions quickly and easily because they will have the written plan at their disposal. They'll understand how, when, and where to communicate key information and will avoid making serious gaffes that could have serious repercussions for employees, the company, its executives, and the community.

A well thought-out and practiced plan for crisis communication can be a corporation's strongest tool. It allows the crisis management team to gain quick and effective control over chaos that typically ensues from a disaster or serious business disruption. An effective plan defines who to speak to, what to say (or not to say), and how best to say it.

Chapter 7

Emergency Response and Recovery

Solutions in this chapter:

- **Emergency Management Overview**
- **Emergency Response Plans**
- **Crisis Management**
- **Disaster Recovery**
- **IT Recovery**
- **Business Continuity**

☑ **Summary**

☑ **Solutions Fast Track**

☑ **Frequently Asked Questions**

Introduction

The most basic rule about planning for emergencies is this: Keep it simple. The more complicated your emergency response plans are, the less likely they will be effective in a real emergency. It's sometimes easy to overengineer a plan in the relative calm of everyday business activities. When an emergency strikes, people are not likely to remember a lot of rules, procedures, and details. As you read in the preceding case study, "Crisis Communications 101," there are three basic rules to remember. It's pretty easy to remember three rules. So, when you create your emergency response and disaster recovery activities, you should strive to keep things simple. Once the emergency has subsided, you can begin to put more complex plans into place to begin restoring business operations.

We're not going to go into tremendous detail on emergency response, but we will provide a few pointers. If you want to create a detailed emergency response plan, you can work with local emergency responders who will be best able to provide details relevant to your community, its resources, and its geography. We'll also discuss computer incident response, disaster response, IT recovery, and business continuity.

In addition, we've provided several detailed checklists for emergency and disaster response and recovery in the appendix material of this book that you can use to develop your own detailed checklists. We chose to place them in the Appendix so you have easy access to these lists in one location rather than having to leaf back through the book looking for these checklists. We'll refer to the lists in this chapter and refer you to the Appendix to view related checklists.

Emergency Management Overview

Regardless of how your company is organized, managed, and run, your emergency management process should follow a very simple rule: assign clear roles. If no one knows who's in charge or who has the authority to make decisions, nothing gets done. On the other hand, if everyone believes they have the authority to make decisions, chaos will reign. The aftermath of Hurricane Katrina is testament to this problem—everyone assumed some other organization was in charge, no one knew to whom to turn for solutions. The Federal Emergency Management Agency (FEMA) was assumed to be in charge but was clearly late in gaining control of the situation. As a result, thousands of people were without food, water, ice, and shelter for an extended period of time. The lesson from this catastrophe that companies can learn is this: someone has to be clearly in charge and take immediate and effective action.

Throughout earlier chapters in this book, we've referred to the fact that emergency responders may not be able to get to your company for an extended period of time because they will prioritize your business lower than hospitals, schools, or nursing homes, to name a few. Therefore, your BC/DR plan should include some sort of internal emergency response capability in the event emergency responders are not available.

NOTE

Whenever possible, use your community's emergency responders to assist you in an emergency. Dial 911 or contact emergency services in your area as quickly as possible after a disaster or emergency occurs. At the same time, have your emergency response team respond to the incident. In many cases, first responders can help save lives by providing early care until trained professionals arrive. Administering CPR, for example, can help keep someone alive until paramedics arrive. Whenever possible, be sure to contact your emergency responders for assistance since most company's employees lack the training and experience to provide the same level of emergency support.

Emergency Response Plans

Emergency response plans stem from the risks you've identified for your company. Remember, though, the emergency response is the *immediate* response to the incident. If fire breaks out, the emergency response is evacuating the building and calling the fire department while perhaps having trained employees use fire extinguishers to try to control the blaze. These are the basics of a fire emergency response. However, there are other kinds of risks your company faces and these also require emergency response plans. Rather than creating a separate plan for every type of event that could occur, it's often advisable to create a basic emergency response checklist that can be used regardless of the emergency. The basics don't change—contact appropriate emergency personnel, get people out of harm's way, determine if there have been fatalities or injuries, determine if anyone is missing or unaccounted for, determine the source of the emergency, take measures to contain or halt the source of the problem if possible, and so on.

Develop an emergency response plan that meets the needs of your company without getting too complicated. A simple response plan that covers a variety of similar emergencies will help ensure things run more smoothly if an emergency does occur. For example, there might be several different reasons you would choose to evacuate your building—a fire, internal flooding (burst pipes, etc.), or a bomb scare. The threat sources are different, but the action is the same. Therefore, look through your risks and identify which emergency actions would be needed. Then, group them together so you can develop just three or four emergency responses, if possible.

The basic set of emergency response tasks are these:

- Protect personnel
- Contain incident

- Implement command and control (Emergency Response Team, Crisis Management Team step in)

- Emergency response and triage (medical, evacuation, search and rescue)

- Assess impact and effect

- Notification

- Next steps

The response procedures include protection of people first, containment of the emergency second, and assessment of the situation third. Regardless of the type of plan you create, these should be your priorities. Although it seems intuitive that you'd address the health and safety of people first, it's not always the first thing that comes to mind when an emergency strikes, so having a well-rehearsed set of procedures for emergency response that focuses on getting people to safety first then addressing the emergency will help form an appropriate response if something does occur.

Each plan should include:

- Roles and responsibilities

- Tools and equipment

- Resources

- Actions and procedures

Roles and responsibilities identify who's on the team and what they should do in an emergency. Tools and equipment for those emergency roles should be identified. This might include fire extinguishers, first aid kits, hard hats, haz-mat suits, walkie-talkies, shovels, and more. Any tools identified by the ERT should be purchased and stored in a suitable location. A list of these supplies should be maintained and someone on the ERT should be responsible for periodic inventory as well as testing and replenishing of supplies. For example, first aid kits have various medicines such as antibiotic creams and aspirin that expire and should be replaced periodically. Other resources the ERT might need should be acquired or identified. If specialized equipment such as a fire truck with an extension ladder would be needed to reach top stories of the building, that should be noted. The local fire department should be contacted to determine whether they have appropriate resources (such as a truck with an extension ladder). If equipment is not available, alternate plans should be created that address the specific needs. The company should also develop numerous evacuation scenarios and procedures that address the possibility of a fire in the upper floors of the building. Finally, actions and procedures should be developed that the ERT will initiate in the event of an emergency.

We've provided a detailed emergency response checklist for you and it's located in the appendix materials so you can easily refer to it later. If you take a moment now to mark this page, flip to the back of the book, review the list, and then come back, you'll see that there is extensive detail in the list. It provides a generic step-by-step process that you can tailor to your company's specific situation so that you have a solid emergency response plan in place. The plan must be executed by people, so let's take a moment to discuss the role of the emergency response team.

From the Trenches...

Powering Up after Katrina

Hurricane Katrina has become an icon for many people. The enormity of the storm and its impact took most disaster planners off-guard and few organizations responded effectively in the aftermath of the storm. For many, the most immediate need was for electrical power. Imagine trying to restore power in an area where power poles were torn down, transmission lines were shredded, employees' homes were destroyed, roads were blocked, and communications were nonexistent. That's the situation faced by Mississippi Power's CIO Aline Ward. For a fascinating recount of what Aline Ward did to restore power to the area, visit this link: http://media.techtarget.com/digitalguide/images/Misc/AwardMississippi.pdf. It's a real world view of the aftermath of a natural disaster of massive proportions and how one person managed to bring order from the chaos to get power back to the area in record time.

Emergency Response Teams

Your company should have an emergency response team with defined roles and responsibilities for team members. Each person should clearly know the bounds of their authority and to whom they should turn for help or for escalation of issues. In previous chapters, we've referred to a Crisis Management Team (CMT), which may or may not be the same as an Emergency Response Team (ERT). If you're in a small company, it may be the same set of people, but in many cases these are not the same people because the skills required are different.

The ERT leader is responsible for activating and coordinating the emergency response and for notifying civil authorities such as the police or fire department, contacting hospitals or paramedics, and so on. The ERT leader also should be a member of the Crisis Management Team and should coordinate closely with the CMT to ensure that the appropriate level of BC/DR activation occurs in a timely manner. Emergency response and disaster recovery activities can occur in parallel. Typically, only trained members of the ERT can address the actual emergency. Members of the CMT can begin assessing damage, evaluating options, and implementing the BC/DR plan as soon as possible.

The ERT is also responsible for ensuring that the proper communication equipment is available prior to an event, for activating and distributing that communication equipment in an event, and for communicating appropriately throughout the event.

Emergency response team members should receive training on the aspects of the job they'll be expected to perform in an emergency. If team members are expected to fight small fires by using fire extinguishers, they should be trained not only on the use of the fire extinguishers but on how to fight fires. This includes safety procedures for fire fighting as well as methods for fighting different types of fires. Training is critical to ensure team members' safety and effectiveness in an emergency.

Emergency response training may include:

- Relocation and evacuation safety and techniques

- Fire fighting equipment, safety, and techniques

- Search and rescue safety and techniques

- Hazardous material handling

- Chemical spills or leaks (liquid, airborne, etc.)

- CPR, first aid, and emergency medical skills

- Water safety, water rescue

- Cold weather survival

- Emergency shut off/shutdown procedures

- Damage assessment and control

Obviously, the type of training required depends largely on your company, the nature of its business, and its geographical location. Identify the types of emergency response training that would be helpful for your staff to have and develop training plans to ensure training occurs periodically. Skills should be tested, rehearsed, and refreshed from time to time. Also, develop some method for responding to the loss of ERT members through retirement, attrition, or transfer. Finally, be sure several people on the ERT have similar skills and training so your team does not have a single point of failure. If only one person knows how to shut off the main electrical breaker and he or she is injured in an explosion, you have a problem (well, several major problems, actually).

It's also helpful to assign ERT members roles and responsibilities outside of emergency situations related to continued preparedness. For example, the ERT might be responsible for staging emergency training sessions or simulations on an annual basis that the entire company participates in. It might also be tasked with periodically checking fire extinguishers (e.g., are they where they should be, are they well-marked, are they functional, have they been tested, have they expired?) or checking emergency lighting from time to time. This keeps the team in tact and functional during nonemergencies, which can help them work together as an effective team during an emergency. It also helps maintain safety measures for the company, which is another risk mitigation strategy. If no one is responsible for checking fire extinguishers, there's a good chance you'll run into a problem if a fire actually does flare up. Define roles and responsibilities for ERT members that help reduce your company's risks and liabilities.

Crisis Management Team

There are hundreds, if not thousands, of books on crisis management available and if this is an area of interest, you should do additional research to delve into the details of this topic. As you know from watching the news or reading blogs, there are all kinds of crises that companies have to manage, not all of them related to BC/DR. In this section, we'll cover the basics of crisis management with an eye specifically toward BC/DR activities.

When you declare an emergency, disaster, or crisis event that must be managed, you begin implementing your BC/DR plan. The crisis management team (CMT) is the team responsible for making the high-level decisions; for coordinating efforts of internal and external staff, vendors, and contractors; and for determining the most appropriate responses to situations as they occur. They should be well versed with the BC/DR plan and the various team leaders for BC/DR activities either should be part of the crisis management team or should report to them.

Emergency Response and Disaster Recovery

The CMT oversees the emergency response team and the disaster recovery team(s). Once an emergency occurs, the emergency response team leader should take charge of managing the emergency itself, and the leader of the crisis management team should begin coordinating efforts between ERT, civil emergency responders (if appropriate), and other initial activities related to the BC/DR plan. The ERT leader should be a member of the CMT and should report to the team periodically throughout the emergency response. The ERT should be quickly released back to emergency duties while someone from the CMT documents the information provided by the ERT. This is part of the event log that should be initiated and maintained throughout the event. In addition to coordinating the emergency response, the crisis management team also coordinates activities related to initiating the disaster recovery efforts. Once the ERT leader has notified the CMT that the actual emergency has ceased and that disaster recovery can begin, the CMT takes over coordinating all activities. Typically, once the disaster recovery efforts conclude and business continuity efforts begin, the crisis management team winds down and operations may resume through normal management channels. This is a decision each company must make based on its unique structure, but in general, the CMT leader should manage the situation until it makes sense to hand over control to the operations team.

One very important note. You should clearly define the point at which the CMT stands down and normal operations take over. If you fail to clearly identify this line of demarcation, you risk having turf wars, power struggles, and people working at cross-purposes. Create a clear set of criteria for when the CMT hands over operations so that there is no question in anyone's mind about how the transition should occur. This is usually not a major issue in companies where the members of the CMT are members of the senior management team. In some companies, however, there may be confusion over roles, responsibilities, and authority, so be sure to clearly delineate these in advance.

Alternate Facilities Review and Management

The CMT is responsible for overseeing the activities related to disaster recovery and business continuity at alternate sites. They should review the activities leading up to activating the alternate site and should be the ones with final authority over decisions that need to be made related to the alternate site, such as bringing in additional services, equipment or vendors if original arrangements do not meet current needs. They are responsible for resolving problems and issues that arise and should be the final decision makers for escalated issues.

Communications

Crisis communications covers a lot of territory and may involve numerous teams working in a coordinated fashion, but the messages being communicated should originate from or be approved by the CMT. In an emergency situation, you should avoid having multiple sources of communications going out since it can cause confusion, error, frustration, and worse. Though you don't want to create a bottleneck in your communication stream, in the early stages after a business disruption or emergency, strive to have the CMT clear any messages going out. This not only will ensure that the message is correct and consistent, it will keep the CMT in the loop as well. This establishes a two-way communication channel between the CMT and the teams working on disaster recovery activities and helps in the coordination of activities and teams. This is critical for disasters or disruptions that also disrupt communication lines.

Human Resources

Representatives from Human Resources should be included on the CMT so that they can specifically address the needs of employees and maintain a communication channel with employees through preplanned methods. They should track employees who may be injured from the event or not available for work due to leaves of absence, vacations, and so on. They should provide support for injured employees and their families including facilitating access to emergency or ongoing medical or psychological services. They can also assist employees with financial, legal, and insurance issues related to the injury or death of an employee or family member. They should prepare and update an employee head count to determine who is available for recovery operations and who may be available later for business continuity activities. If temporary staff or contractors are needed, they can help select, manage, oversee, and monitor temporary staff as well as manage timecards and other payments for such staff. Last but perhaps most important, they can determine the status of payroll and ensure employees get paid in a timely manner. This is one of the biggest concerns employees will have in the aftermath of disaster, and having someone actively manage and monitor this process can alleviate some of the stress of the situation. Pro-actively addressing these concerns will also reduce the number of calls, e-mails, and contacts related to questions about payroll, freeing up time to address other HR-related concerns.

Legal

Depending on the nature of the disaster or disruption, you may need to have the CMT contact legal counsel. The firm's lawyers or legal representatives may need to review or approve emergency contracts; review language in agreements with vendors, suppliers, or contractors; review documents related to injury, death, or property damage; or address regulatory and compliance issues. As soon as the CMT is activated, it should be someone's specific responsibility to contact legal counsel and notify them of the event so they can provide appropriate information, feedback, and guidance throughout the remainder of the event and during its aftermath.

Insurance

As we've discussed, insurance is a risk transference method and one used by many, if not all, businesses today. In some cases, your firm may be required to hold certain types of insurance; in other cases, it may be voluntary. Your BC/DR plan should have contact information for your insurance company representatives and they should be notified upon activation of the CMT. The CMT may also perform an initial damage assessment and document it for the insurance company. This might include taking photographs or video images as well as making detailed notes. Members of the CMT team should also begin gathering documents related to insurance claims and submit loss estimates to the insurance company. Finally, someone on the CMT should review the insurance documents to determine exclusions, limitations (financial, time, location, cause, etc.), or maximums on various policies. Any issues with insurance should be escalated to management and/or legal counsel for review and resolution.

Finance

The CMT should also have representatives from the financial department available to assess the status of the company. This might include assessing the cash availability of the company, the viability (or advisability) of processing employee payroll early, or to provide advances to employees. Financial representatives also need to assess the status of the accounts payable and receivable to ensure bills and invoices are issued in a relatively timely manner and that revenue and payments are received in a timely manner as well. A process for managing, tracking, and monitoring expenditures during the disaster or disruption should be implemented and managed by the financial representative(s) on the CMT. Estimates for repairs and other expenditures should be submitted to this team for review and approval. Upon resumption of business operations, the financial team should assess the status of the company's finances and report to executives or senior management.

Disaster Recovery

We discussed the different phases of business continuity and disaster recovery in Chapter 6, including activation, disaster recovery, business continuity recovery, and maintenance/review. In this section, we're going to discuss the disaster recovery activities in a bit more detail. This detail belongs in your BC/DR plan, but breaking it out into sections in this manner will help you process and manage the massive amount of detail required to address these activities properly. Once you've developed your emergency response, disaster recovery, and business continuity responses, you can (and should) include that information in your BC/DR plan. We've included various checklists in the Appendix that you can use as the basis for creating your own checklists or project plans. These can be included in the body of your BC/DR plan or as appendices at the end of your document for ease of use.

Activation Checklists

You may find it helpful to develop a variety of checklists, which can be extremely useful in making quick decisions for moving forward. Since you and your team may not have time to rehearse these plans frequently, checklists can help remind you of critical steps to take, regardless of the situation. Activation checklists should delineate all the activities and triggers that should take place prior to and during plan activation. This begins with some sort of disruptive event occurring, someone notifying the BC/DR team, and someone determining that the BC/DR plan should be activated as a result of the disruptive event. Remember, there may be some minor events that do not trigger the activation of the BC/DR plan, so deciding what criteria will be used to activate the plan in whole or in part should be part of the process. The checklist included in the Appendix provides a framework for you to develop your own activation process.

Recovery Checklists

The recovery phase also has specific tasks that should be undertaken. The specific steps to be taken should be defined in your BC/DR plan. If you've looked at the various risks and potential impacts of these risks, you should have numerous scenarios that require planning. By developing plans for various scenarios, you will have the steps you need in almost any type of disaster because even though the details of the disaster may vary, the steps you need to take will be the same in a major disaster or a minor disaster. As with the activation phase, there is a long list of items you can use for this stage of work. We've provided extensive recovery checklists in the Appendix. Remember, all these lists are intended solely to get you thinking about how you will manage your company's BC/DR efforts, so you will need to modify them accordingly.

IT Recovery Tasks

The tasks needed to recover IT systems are probably quite familiar to you, but they should be delineated within your BC/DR plan. Each subteam should have a clear set of guidelines and procedures for how and when they will perform their work. Be sure to note dependencies within the checklist so that teams don't work at cross-purposes. You can add items to the checklist as checkpoints for these purposes, much like milestones are used in project plans. IT recovery checklists are also included in the Appendix of this book.

Part of IT recovery involves responding to, stopping, and repairing problems caused by system failures, security breaches, or intentional data corruption or destruction. Depending on the nature or severity of the attack or incident, you may need to activate a computer incident response team (CIRT). Let's take a moment to discuss computer incident response and the team that performs these tasks.

From the Trenches...

Training Is Not Optional

When disaster strikes, most people resort to what they know best; they fall back on their training. The same is true of IT professionals. In the face of a major system outage or security breach, IT staff will do what they've been trained to do. Training is not an option for emergency preparedness, it is a *requirement*. Emergencies by their very nature are incredibly stressful and chaotic. People, by their very nature, feel most comfortable in any situation when they know what to expect and what to do. In an emergency, they won't necessarily know what to expect, but they will know what to do if they've been trained. Training is also important for CIRT teams because security incidents can be devastating to a company. CIRT members should know what to look for and exactly what actions to take in order to address a potential security breach or other serious incident. It doesn't help to shut down a server if the firewall has been breached; it doesn't help to shut down e-mail if the virus has infected a server. In addition to general IT skills, CIRT members should represent the various areas of expertise required in your IT department including servers, infrastructure, security, database administration, and applications, to name a few. CIRT members also should have

Continued

checklists or step-by-step instructions to follow for standard incident types such as security breach, firewall breach, virus outbreak, and so on. This helps reduce stress and ensures everyone follows standard procedures to halt the immediate impact of any computer-related incident.

Computer Incident Response

Most IT departments have some process in place for addressing and managing a computer incident. An incident is defined as any activity outside normal operations, whether intentional or not; whether man-made or not. For example, the theft in the middle of the night of a corporate server is an incident. A Web site hack or a network security breach is also an incident. A database corruption issue or a failed hard drive is also an incident, but for the purposes of this discussion, we're going to stick with the emergency kinds of incidents and leave the more routine incident handling to your existing IT operations procedures. For example, we'll assume you can handle a bad hard drive or a failed router through standard operating procedures and we won't cover that here. What we will cover are the incidents that require a swift and decisive action to stop the incident from continuing. This includes events such as a network security breach or a denial of service attack and events such as a fire in the server room or a flood in the building.

The first step in this process is to form a computer incident response team. You may already have a team in place that addresses computer incidents such as security breaches. If that's the case you have the foundation of a computer incident response team (CIRT) that can be used in the event of a more widespread disruption such as a fire, earthquake or flood. The members of the team, like the ERT, should have defined roles and responsibilities. As with the ERT, team members should also be trained in their roles. For example, if you have staff responsible for monitoring network security and they notice a potential breach through a particular port, they should also know how to shut down that port and have the network permissions that enable them to do so. If all they know how to do is monitor the log file or traffic, for example, and have no idea how to shut down a port or stop the problem, it could be hours before the problem is addressed. Therefore, members of your CIRT should have training and appropriate network permissions to address these problems.

CIRT Responsibilities

In order for the CIRT to be effective, its duties must be well defined. There are five major areas of responsibility for the CIRT team. These are:

- Monitor
- Alert and Mobilize
- Assess and Stabilize

- Resolve
- Review

Monitor

Every network must be monitored for a variety of events. Some of these are failure events that indicate a problem has occurred such as a hardware failure or the failure of a particular software service to start or stop appropriately. Other events are tracked in log files for later review or auditing. These might include failed login attempts or notification of a change to security settings, for example. Other incidents may include unusual increases in certain types of network traffic or excessive attempts to login to secure areas of the network. Whether the event stems from intentional or unintentional acts, the network needs to be monitored. The CIRT should be involved with helping to determine what should be monitored as well as assisting in monitoring the network. Not all events have significance and sometimes it's only through seeing recurring events that a pattern can be discerned. Therefore, having experienced team members monitor the network will help reduce the lag time between an unwanted event and a response.

While a serious security breach might not cause you to activate all or part of your BC/DR plan, suppose you had some very strange activity on four of your corporate servers and the CIRT member couldn't determine the source of the anomalies. Is this a disaster or not? If it's caused by fire in the server room, yes. If it's caused by an errant software update that was just applied, maybe not. The point is that your CIRT team should monitor the network activity and take appropriate action regardless of the source of the problem. In some cases, this will involve activation of the BC/DR plan, in other cases it won't.

Alert and Mobilize

Once an unusual, unwanted, or suspicious event has occurred, the CIRT member should alert appropriate team members and mobilize for action. This may involve shutting down servers, firewalls, e-mail, or other services. As part of a BC/DR plan, this can also include being alerted that the event or disaster disrupted network services, such as a data center fire or theft of a corporate server after a fire in another part of the building. Alerting and mobilizing should have the effect of stopping the immediate impact of the event.

Assess and Stabilize

After the immediate threat has been halted, the CIRT team assesses the situation and attempts to stabilize it. For example, if data has been stolen or databases have been corrupted, the nature and extent of the event must be assessed and steps must be taken to stabilize the situation. In many cases, this phase takes the longest because determining exactly what happened can be challenging. If you have members of your team that have been trained in computer forensics, they would head up this segment of work. If you do not have members of your team trained in this area, you should decide whether it would be advisable to provide this training to staff or hire an outside computer forensics expert. Outside consultants can be helpful in this case for the simple fact that they work in this arena day in and day out and are most likely more up-to-date and experienced in this area than staff that occasionally goes to training and rarely (if ever) puts that training to use. The decision is yours based on the skills, expertise, and budget of your company. Having in-house expertise can be a good first step and you can always hire an outside expert on an as-needed basis.

Keep in mind that you have defined maximum tolerable downtime and other recovery metrics. A review of these should be included as part of the assess and stabilize procedures so that plans and actions can accommodate these requirements.

Resolve

After determining the nature and extent of the incident, the CIRT can determine the best resolution and implement it. Resolution may involve restoring from backups, updating operating systems or applications, modifying permissions, or changing settings on servers, firewalls, or routers.

Review

Once the event has been resolved, the CIRT should convene a meeting to determine how the incident occurred, what lessons were learned, and what could be done to avoid such a problem in the future. Within the scope of a BC/DR plan, this might involve understanding how the recovery process worked and what could be done differently in the future to decrease downtime, decrease impact, and improve time to resolution.

From the Trenches…

Computer Emergency Response Team (CERT)

There are numerous terms and acronyms floating around regarding computer emergencies, computer incidents, and computer security. The grandfather of them all, however, is the concept of computer emergency response developed by the Software Engineering Institute (SEI) at Carnegie Mellon University. We mentioned this resource earlier in the book and thought this would be a good time to mention it again. The Web site has a vast array of information and resources you can access. When developing your BC/DR plan for the IT portion of your business, read up on the latest trends and knowledge on the Web site at www.cert.org. Head to this URL for details on creating a CERT team: http://www.cert.org/csirts/action_list.html. It's a great resource for IT professionals even outside the scope of BC/DR planning as well.

Computer incident response is an activity that spans disaster recovery, business continuity, and normal operations. It is likely the CIRT team will have day-to-day responsibilities as part of standard IT operations or that CIRT activities will be building into IT standard operating procedures. However, if an earthquake hits the area or a flood shuts down operations, the CIRT's expertise can be put into play immediately as part of the BC/DR response. Be sure to integrate CIRT responsibilities into your BC/DR plans.

The skills of CIRT members should be kept up to date so they are aware of and can respond to the latest threats, vulnerabilities, and issues on the IT realm. Although training is important for IT staff in general, CIRT members need to be aware of the constantly evolving threats and vulnerabilities. They need to have the tools and skills necessary to recognize and resolve problems in a timely and effective manner. This is accomplished in part through training. CIRT members must also take responsibility for staying up to date on the latest trends by reading technical journals, newsletters, Web sites, blogs, and other related materials.

Business Continuity

Business continuity begins when disaster recovery ends. As we've discussed, it's not a sharp cutover from one phase to the next. Though we've discussed this to some extent throughout the preceding chapters, we haven't really looked at what it takes to move from the disaster recovery phase to the business continuity phase specifically. As with the other topics in this chapter, we've included a business continuity checklist in the back of the book for your reference. We'll review the basics in this section.

The disaster recovery efforts include stopping the effect of the disaster and getting basic operations set up. For example, if your building was destroyed, disaster recovery would include salvaging anything from the building you could, activating an alternate work site, activating an alternate computing site (may be the same or different than the alternate work site), and setting up and restoring network components, servers, and systems. Now that disaster recovery, from an IT perspective, is complete, business continuity kicks in. These steps include managing business processes in work-around mode, if needed, and assessing the status of operations and beginning to normalize operations. For example, it's possible that some systems can be restored almost immediately, whereas other systems may take several days or a week to restore. The work-arounds in place may allow some operations to resume but others to remain dormant. Backlogs in some areas are created, data gets out of sync, and the state of the business is perhaps more chaotic now than it was during the disaster when it was clear that no business operations would take place. Therefore, having a plan for business continuity steps is critical to your eventual success.

Part of the challenge of the business continuity phase is determining what should be restored, what should be salvaged, and what should be replaced. There is certainly a time consideration that needs to be factored in along with the obvious financial considerations.

Repairing and replacing have their own sets of challenges and the options should be reviewed prior to making decisions to move forward. In order to process all the information needed, the various teams should work together to identify optimal solutions. Some of the factors to be considered include:

- Executive/administrative
- Business operations
- IT operations—infrastructure
- IT operations—end users
- Communications
- Facilities, security, and safety

As with the other emergency and disaster response activities listed in this chapter, we've also developed a business continuity checklist you can use as the basis for your business continuity planning activities. Since every business is different, the checklist you find in the Appendix is fairly generic. It lists major level activities you should consider including. Not all activities on the list will be appropriate for your organization. There may be areas *missing* from the checklist that you'll need to resume operations at your firm. However, if you start with these lists, there's a better chance you'll include what you need to successfully resume business at your company.

As you'll see in the checklist, the last two activities are reviewing what happened during the disruption or disaster and adding that knowledge to your BC/DR plan. Once your firm gets back to business as usual, no one will have the time to capture this data. It's vital that you capture lessons learned from the incident and build them into your BC/DR plan so that the mistakes made aren't repeated and the innovations or positive lessons learned can be incorporated. This is part of plan maintenance, discussed in detail in Chapter 9, but it also should be part of your BC/DR activities as well.

Summary

In this chapter, you learned about emergency plans and emergency responses that should be included in your BC/DR plan. Emergency response is the initial response to a disaster or disruption. The first response should be to get people out of harm's way and to determine if there are fatalities or injuries. Secondary efforts should be to stop the source of the problem whether that's through calling civil emergency responders (fire, bomb squad, police) or through attempting to address the problem with an emergency response team (fighting a fire, turning off gas or electric sources, containing hazardous spills, etc.). Emergency responders should be trained in appropriate skills such as safe building evacuation methods, CPR and first aid, fire fighting, hazardous material containment, and others. Emergency plans should be well conceived and well rehearsed because people will fall back on their training in an emergency.

The crisis management team may activate the emergency response or the emergency responders may notify the crisis management team of an event. In any case, the crisis management team coordinates emergency efforts and activates the BC/DR plan based on the specifics of the situation. The CMT is also responsible for coordinating recovery efforts and should manage these activities through the business continuity stage. Roles and responsibilities should be well defined to avoid confusion or working at cross-purposes. Activities the CMT typically manages can include the emergency and disaster response, activating alternate work sites and facilities, managing corporate communications, interfacing with insurance and legal representatives, and working with the finance department. You can define other appropriate activities for your CMT to reflect the specifics of your business.

Because disasters are by their very nature chaotic events, it helps to have checklists you and your team can use to manage activities in the aftermath of a major disaster or disruption. We've included numerous checklists in the Appendix of this book so you can easily refer to them and use them in your planning activities. Disaster recovery tasks fall into two major categories: activation and recovery. Activation includes all activities related to assessing a situation and determining what recovery plans should be implemented as well as taking initial steps toward that end.

Within disaster recovery, there are specific IT recovery tasks that should be performed as well. Separate IT recovery checklists should be created so that you have a clear plan about how to recover from various events. These checklists should include information regarding the maximum tolerable downtime (MTD) and other recovery metrics that have been established. The lists also should include timelines, milestones, and dependencies that need to be addressed. Some companies form computer incident response teams (CIRTS) or computer emergency response teams (CERTS) to respond quickly and effectively to computer-based incidents. The activities of the CIRT occur in the day-to-day operations of the company (outside the BC/DR domain) and are also part of BC/DR activities. Defining how the

CIRT should operate and interact with your BC/DR plan is vital to ensure an effective response.

Business continuity activities begin after recovery efforts have concluded, though there is usually some overlap. Business continuity activities include the limited resumption of business operations, typically in manual or work-around mode. These activities pose a unique set of challenges from an IT and operations perspective because data must be managed differently until IT systems are fully back online and normal operations can resume. The business continuity checklist should include steps needed to resume limited operations, it should identify requirements and dependencies, and it should include timelines, milestones, and checkpoints. The resumption of normal business operations typically occurs when the company either reoccupies its original facility and all equipment is back up and running, or when the company decides on a permanent business location (which may be the alternate site or newly acquired site). Criteria for determining the cutover to "normal operations" should be developed and the CMT should hand over operations to the management team toward the end of the business continuity phase. Clearly defining this cutover as well as roles and responsibilities will help prevent confusion during this last phase of activity.

Solutions Fast Track

Emergency Management Overview

- ☑ Emergencies are chaotic events that require a coordinated response.

- ☑ Lack of a coordinated response after Hurricane Katrina exacerbated the problems.

- ☑ Emergency responders should be contacted first but understand what their priorities will be in the aftermath of a serious event.

- ☑ Companies should be prepared to be somewhat self-sufficient in the immediate aftermath of an event.

Emergency Response Plans

- ☑ Emergency response plans deal with protecting people first, property second.

- ☑ Emergency responses should attempt to contain, control, or end the emergency. This includes evacuating buildings, fighting fires, turning off utilities, and other response activities.

- ☑ Emergency response teams (ERT) should have the skills required to address the specific needs of your company's operations.

☑ Training is imperative for ERT members. Training should be refreshed and tested periodically.

☑ Training for ERT members may include fire fighting, CPR, first aid, hazardous material containment, and other skills appropriate to the location and nature of your business.

☑ Emergency response checklists help keep people calm and focused on next steps. Develop emergency response checklists in conjunction with expertise from your ERT and local civil emergency responders (fire, police, hazmat, bomb squad, etc.).

Crisis Management Team

☑ The crisis management team (CMT) may activate the emergency response or it may be activated by the ERT.

☑ The CMT manages, directs, and oversees the disaster recovery efforts.

☑ CMT responsibilities include emergency and disaster response as well as coordinating efforts related to alternate facilities and work sites, communications, human resources, insurance, legal, and finance.

☑ CMT roles and responsibilities should be clearly delineated.

☑ Maximum tolerable downtime (MTD) and other recovery metrics should be well understood by the CMT and addressed by recovery plans.

Disaster Recovery

☑ Activation checklists can be used to determine if, how, and when to activate the BC/DR plan. In some cases, activation of part of the plan may be warranted.

☑ Clear activation checklists help responders understand what steps to take and help them make better decisions in the confusion that surrounds major disasters or disruptions.

☑ Disaster recovery checklists should include MTD and other recovery metrics so the CMT can make decisions appropriate to these requirements.

☑ Disaster recovery checklists should address the safety and well being of personnel first, then address physical facilities, buildings, equipment, and other business assets.

IT Recovery

- ☑ Every company should have a computer incident response team (CIRT) that responds to incidents related to IT equipment.

- ☑ Incidents may be unusual activity, intentional or unintentional breaches, failures, and so on.

- ☑ CIRT activities are both day-to-day and part of BC/DR activities.

- ☑ The responsibilities of the CIRT include monitoring, alerting, mobilizing, assessing, stabilizing, resolving, reviewing all IT-related incidents ("incidents" as defined by the team).

- ☑ CIRT skills should be kept up to date so they are aware of and can respond to the latest threats, vulnerabilities, and issues on the IT realm.

Business Continuity

- ☑ Business continuity activities typically involve the resumption of limited business operations.

- ☑ These activities typically involve manual and work-around systems while equipment and IT systems are being fully restored.

- ☑ The decision to move to a permanent facility, whether returning to the original location, staying at the alternate site, or acquiring a new location, typically triggers the final stage of business continuity and signals the resumption of normal operations.

- ☑ Business continuity checklists should be used to ensure that required systems are in place and functional. Checklists should also contain references to timelines, milestones, dependencies, and other business metrics.

- ☑ Once business continuity activities end and normal business resumes, the BC/DR teams should review lessons learned so they can be incorporated into the BC/DR plan.

Frequently Asked Questions

The following Frequently Asked Questions, answered by the authors of this book, are designed to both measure your understanding of the concepts presented in this chapter and to assist you with real-life implementation of these concepts. To have your questions about this chapter answered by the author, browse to **www. syngress.com/solutions** and click on the **"Ask the Author"** form.

Q: We have a small company of 60 people, our IT department has three full-time employees and one part-time employee. We don't have the ability to create all these different teams you mention. Any advice on how we should approach this?

A: There is no one-size-fits-all approach to business continuity and disaster recovery planning because every company is so unique. However, the intention of this chapter was to inform you about the various needs that occur in the aftermath of a disaster or major disruption. These things are needed whether your company has five or five thousand employees, the basics don't change. What will change is each company's response to these events. For your company, it might make sense to ensure that five to ten of your staff have CPR and first aid training. You could ask for volunteers so that those selected for training are those most interested in receiving it and administering it should it be needed. Many communities have organizations that provide low- or no-cost first aid and CPR training, so this should be first on your to-do list. Second, you should have basic emergency equipment including first aid kits and fire extinguishers. Staff should be trained in how to use the fire extinguishers. Again, you may be able to contact your local fire department for training or lectures on appropriate safety measures during a fire. You should have evacuation plans in place and should practice fire drills from time to time. Finally, you should have a checklist you can use to safely evacuate the building, contact emergency responders, contact management (if not present), and initiate disaster recovery plans. Your plans should match the size of your company and its staff's capabilities. Following these basic steps should help you develop a plan that meets your needs.

Q: I work in a company located in the downtown area of a large city. I have no idea how we would respond to a fire in the building if the fire was not on our floor or how we would know to evacuate the building if something happened at another company. Any suggestions?

A: Yes. The building manager should have an emergency plan in place for the building and you should inquire what those plans are. If they do not have a plan in place, you may recommend, suggest, or require that the building manager develop these plans in conjunction with other companies in the building. In some localities, these basic plans may be required for occupancy, but in many cases, no emergency plans other than evacuation routes, emergency lighting, smoke detectors, and fire alarms are required. That may be adequate based on the size and location of the building. However, if you believe it is not adequate, you should encourage (or require) your building manager to address the deficiencies. If you're in a situation where that's not possible, you may want to create an ad hoc safety committee with representatives from the various companies in the building to devise safety, notification, and evacuation plans for your building. This addresses two concerns. First, other companies may have useful resources or expertise that can be used in an emergency. Second, you'll meet the other tenants and learn about the nature and risks to their businesses. This may help you identify secondary risks you would not otherwise know about. The same holds true if your business is located in a complex where there are multiple buildings housing a variety of businesses. Knowing who your neighbors are, what they do, and what they can help with in an emergency is always a good idea.

Chapter 8

Training, Testing, and Auditing

Solutions in this chapter:

- **Training for Emergency Response, Disaster Recovery, and Business Continuity**

- **Testing Your Business Continuity and Disaster Recovery Plan**

- **Performing IT System Audits**

☑ **Summary**

☑ **Solutions Fast Track**

☑ **Frequently Asked Questions**

Introduction

At this point, you have your BC/DR plan pretty well defined and ready to go. The next step in the process, as shown in Figure 8.1, is training, testing, and auditing. Training includes training staff on their roles and responsibilities related to the BC/DR plan as well as training them in the specific skills they'll need to carry out their roles effectively. Testing is the process of testing the plan, and there are various methods for doing so that we'll discuss in this chapter. Finally, there is the process of auditing the IT systems that form the foundation of most BC/DR plans.

Figure 8.1 Business Continuity and Disaster Recovery Project Plan Progress

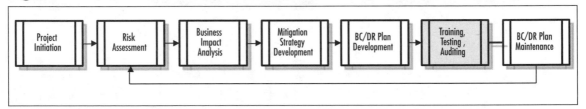

There's an interrelationship between testing, training, and auditing as shown in Figure 8.2. Performing one impacts the other two—when you test the plan, you're training and auditing to some extent.

Figure 8.2 Training, Testing, and Auditing Activities

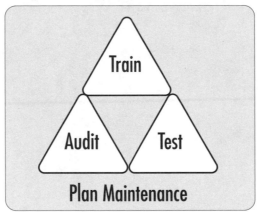

Training, testing, and plan maintenance are all bound together. Testing the plan trains staff and maintains the plan. Training staff tests and maintains the plan. As you train staff and test your plan, you will likely find areas that require modification. These modifications are

made through the change management process defined as part of the plan maintenance phase. The information you glean from training and testing can be extremely useful in honing your plan in advance of a disruptive event. Testing and training go hand in hand, so let's begin by discussing training. We'll discuss plan maintenance in Chapter 9.

Training for Disaster Recovery and Business Continuity

There are two distinct parts of disaster recovery and business continuity training. The first is the actual physical response to the disruption or emergency. That might involve evacuating a building if there's a fire, grabbing a fire extinguisher to douse a fire in the server room, or finding the water main if there's flooding inside the building. These actions all require some basic training so responders know what to do and how to do it safely. There's little point in a responder grabbing a fire extinguisher and subsequently being burned by the fire because he or she did not know how to properly use the equipment or properly extinguish a fire. That's one aspect of training. The second aspect of training has to do with ensuring that the various response teams know how to implement the BC/DR plan and that they have the skills needed to do so. For example, you might want to provide periodic training for your IT staff so they can stay up to date on the latest threats and security measures or training for alternate BC/DR staff on performing a system restore and verification routine.

Emergency Response

Your BC/DR team should have an emergency response team (ERT) identified and these team members should be trained in appropriate emergency response activities. Each company should identify the likely emergency responses needed and provide training in these activities. If your firm is located in an area prone to flooding, earthquakes, hurricanes, or tornados, you should provide training in emergency response related to these events. In addition, basic first aid and CPR training should be part of all emergency responders' training, and some companies find it useful to provide this training to all employees.

The specialized skills for the ERT might include fire fighting techniques or building evacuation procedures, for example. These specialized skills require training in order to protect the safety of the responders and to enable the responders to be effective. As mentioned in Chapter 7, your local fire or police department may provide this type of training or may be able to recommend firms that provide this type of training.

Your BC/DR plan should include the designation of an ERT as well as a list of required training/skills, certification requirements (if any), as well as periodic refresher courses. The ERT leader should be responsible for managing this. He or she should ensure team members have the training and/or certifications required and should arrange for the periodic testing and refreshing of these skills.

We discussed training needs for emergency responders in Chapter 7, so we mentioned it here briefly, primarily as a reminder to you to address and include emergency training in your plan. Let's focus now on disaster recovery and business continuity training.

Disaster Recovery and Business Continuity Training Overview

Disaster recovery is a crucial step that can mean the difference between the company's eventual recovery or failure. Training can help improve the chances for eventual success. Disaster recovery and business continuity training includes defining the scope and objectives for the training, performing a needs assessment (gap analysis), developing training, scheduling and delivering training, and monitoring/measuring training. In this section, we'll discuss disaster recovery and business continuity training as one since they are so closely related. However, as you develop your training plans, you may find it helpful to separate these two phases out so you can pinpoint distinct training needs. Remember, too, that you may choose to perform training while testing your plan. It depends largely on how you approach your testing. We'll discuss testing in detail later in this chapter, so you may revise your thinking on this once you're read through the entire chapter.

Training Scope, Objectives, Timelines, and Requirements

Ideally, you should develop a training project plan that ties in with the BC/DR project plan. The training plan should include a statement of scope (what *is* and *is not* included) as well as a list of high-level objectives. These objectives might be parsed out to include objectives for each of the implementer groups (emergency responders, crisis management team, damage assessment team, disaster recovery team, etc.). In addition, the timelines for training various teams should be developed. Keep in mind that some people may be members of more than one team, so training and training subjects should take that into consideration. Then, develop requirements for training. One of the easiest ways to make sure training meets its stated objectives is to clearly define the objectives, then list the requirements to meet those objectives. For example, suppose you want to provide training for your computer incident response team (CIRT). For simplicity's sake, we'll use a very limited set of objectives, but it will give you a good idea of how to approach this section of the project. The data is organized in Table 8.1 for your reference.

Table 8.1 Sample CIRT Training Outline

Topic	Details
Scope	Train all net admins on monitoring network traffic for security-related issues. Does not include training net admins on how to set up auditing or enabling log files for security monitoring.
Objectives	1. Develop awareness of current security threats. 2. Develop understanding of log files to monitor. 3. Understand what to look for in log files. 4. Understand how to investigate suspicious log file entries, data, or trends. 5. Understand how to respond to suspicious network activity.
Timeline	Initial training will be developed and delivered within 30 days. Training is a two-hour session. Refresher courses will be held quarterly for 30 minutes. Attendance by all net admins is required.
Requirements	1. Locate latest threat data and trend information. 2. Location of [specified] log files. 3. Ability to read and understand log entries. 4. Ability to understand and spot trends. 5. Ability to take [specified] action to address suspicious or malicious network activity.

This example is simply to demonstrate that you should develop scope, objective statement, a timeline, and a set of requirements for your training. It also shows you that you can do this relatively quickly and that it doesn't have to become a massive project itself. As you test your project plan, you'll also find areas that should be addressed by training, so you will likely need to revise these plans once or twice as you go through the training and testing phases.

Performing Training Needs Assessment

The needs assessment phase is essentially a gap analysis. You should review current skill sets against required expertise to carry out various functions and determine what sort of training would best fill the gap. In many cases, training needs become evident during the testing of the plan. Later in this chapter, we'll discuss specific steps you can take to test your plan. As you test your plan, you'll see areas where specialized or updated skills and knowledge will be required to successfully execute the plan. You can make note of these potential skill gaps during your plan testing and circle back to include these in your training plans. Remember, a training needs assessment should be performed on the same periodic basis as your plan testing schedule or on some other periodic basis. People leave the company, are promoted, or change jobs. You need to ensure that at any given moment, your organization has the skills it needs to implement your BC/DR plan successfully. In many cases, a company's routine training plans will cover many (if not all) of the essential skills, but any skills that would not normally be covered through routine training should be flagged for special consideration.

> ### Tip
>
> If you work in a small company, you may need to cross-train people to perform mission-critical functions if your BC/DR teams are not large enough to reduce the risk of small companies. Also, teams should be familiar with other teams' tasks, objectives, and requirements so that teams can cooperate in a seamless fashion in the chaotic aftermath of a serious disruption.

Developing Training

Many companies have limited time or funds available for training, much less for BC/DR training. However, many studies support the thought that companies that train their employees benefit not only from improved productivity but greater loyalty as well. Targeted training to maintain or improve skills, especially those related to mission–critical business functions, can be accomplished relatively quickly and often at a reasonable cost. As with other risk factors in BC/DR planning, the risk of having untrained personnel can easily be mitigated through training, and it may also help drive productivity within the organization. (*Hint*: That's the business case you use to get your BC/DR–specific training budget approved.)

Common Challenges...

The ROI of Training

Many companies have little time or money for training, especially if the company is under tight financial constraints. Many top-level managers look at the line items with an eye toward the bottom line and training is one of the items that gets slashed early in the budget-tightening process. However, experts agree that might not be the best long-term move.

"There is evidence that suggests that training can have productivity payoffs," says Robert D. Atkinson, vice president of the Progressive Policy Institute (PPI) and

Continued

director of the Technology & New Economy Project. "Training can have positive ROI (return on investment) because it can lead to productivity improvement.

"There's also a lot of evidence that when...companies introduce new technology..., the benefits of that technology are significantly enhanced if companies concurrently train their workers."

(Source: www.ppionline.org/ppi_ci.cfm?knlgAreaID=107&subsecID=175&contentID=253143)

Granted, training costs have to be aligned with organizational and financial constraints, but most companies can find creative ways to develop and deliver cost-effective training. The proliferation of online training courses along with local resources makes finding affordable training easier than ever.

When developing training, create clear, specific, measurable outcomes. A measurable outcome means that it either *was* or *was not* accomplished. Either Jill can restore the database from backups using the written procedures or she can't. Either Tony can safely shut power off to the manufacturing floor or he can't. Also keep in mind that not all training for your BC/DR plan will be extensive training. Some may be as simple as showing Tony where the power shut off is and how to perform a power shutdown for the manufacturing floor. Other training, such as how to restore various IT systems that are closely integrated or interconnected, may require training in several knowledge areas as well as hands-on experience (ideally in a similarly configured lab environment) performing the activities in the requisite order. When appropriate, problems should be designed into the training so students can also learn how to troubleshoot and think creatively when things don't go according to plan.

Training should provide some sort of materials (printed, soft copy, web-based, etc.) that capture and reinforce the skills and knowledge presented. The training should also be designed to use several elements such as written, classroom lecture, hands-on (lab), and field (exercises). The more ways you use to deliver training, the more likely it is students will absorb it. Finally, use a final quiz or exam to ensure students have grasped the key concepts and can apply them appropriately. The final test or exam should reflect the training outcomes identified.

In the next section of this chapter, we'll talk about training staff on the BC/DR plan. The outcomes and other deliverables for BC/DR training should be developed as with any other type of training.

Scheduling and Delivering Training

Scheduling and delivering training is a secondary challenge after getting the training budget approved. These days, you can often find various training programs online that people can attend on their own schedule. If you use a flexible online learning system (either your own or an external one), be sure to set timelines and test for knowledge along the way. For example, if you decide that some of your network admins should attend an online course provided by a third-party training provider, you should develop some method of assessing whether or not the net admins learned what they should have. Some online courses are better than others, some test knowledge better than others. Be sure to verify the quality of the training in advance and find ways to verify that students learned the required materials.

If training is developed and delivered in-house in a classroom or lab setting, it may be a bit more difficult to manage. If you develop training that moves quickly, is interesting, engaging, and relevant to the students, it's much more likely you'll be able to get students to attend your training sessions. If necessary, you may need to call upon the organizational clout of your project sponsor to help you get the training scheduled and delivered in a reasonable timeframe.

TIP

Since some of the training will be specific to BC/DR, you may find people saying, in essence, "it can wait, there's no rush." You'll have to find creative ways to counter that, but one way that might work is to say, "If you knew that the building would burn to the ground next week, would you want this training to have already happened?" In most cases, the answer is yes. Since fires are the most common business disaster and can occur completely without warning, you might be able to gain consensus on a reasonable training timetable. If you can tie your BC/DR training into other business objectives, you may have an even greater chance of success.

Monitoring and Measuring Training

The first step in monitoring and measuring training is the development of clear objectives and outcomes for the training. If you don't know what should be accomplished in training, you won't be able to determine if the training was effective.

Exams and hands-on demonstrations of skills can be extremely effective in testing and verifying knowledge. For physical skills such as using a fire extinguisher or performing CPR, both a test of knowledge and a demonstration of skills is best. The same is true for some types of "logical" skills such as restoring a server or verifying user permissions. In some cases, the best you'll be able to do is verify that the training occurred and that several basic concepts were retained by students. An example of this might be restoring an enterprise resource planning (ERP) system that cannot be easily recreated in a lab setting.

Monitoring also involves ensuring key personnel have actually attended required training and have not somehow accidentally fallen through the cracks. If staff emebers leave or move into different positions, replacements need to be trained, so you need to develop some method of periodically checking your key BC/DR staff positions and ensure individuals are still in place and ready to perform their assigned BC/DR duties. These vary widely from one company to the next. You may be able to work with your HR group if they have an established system for tracking employee training and certification in place.

Training and Testing for Your Business Continuity and Disaster Recovery Plan

There are four basic ways to train staff regarding the BC/DR plan, and these also simultaneously test the plan. These are paper walk-throughs (or tabletop exercises), functional exercises, field exercises, and full interruptions. Regardless of how you implement it, you need to cover specific elements in your training. Team leaders, in particular, need to know how and when to activate the plan as well as how to notify, assemble, and manage their teams. Specifically, they need to know how to:

- Use the plan effectively.

- Understand their individual and team roles and responsibilities.

- Notify, assemble, and manage their team members.

- Operate as a cross functional team member.

- Communicate effectively across organizational boundaries in a stressful situation, often without the aid of common communication tools such as phones, e-mail, or other devices.

The most basic part of the training is understanding the plan and how to utilize it. That includes understanding how and when to activate it and how to implement the steps defined. If your BC/DR plan ends up being 50 pages long, you can be sure that no one will take time to read it if the building is on fire. The role of training is both to familiarize people with the plan elements and processes and to reinforce the basic knowledge of the plan. In an ideal scenario, the plan document is accessible immediately upon notification of a disruptive event and someone starts managing the plan. However, in the real world, there's a small probability that things will progress in an ideal manner. Therefore, having a team well versed in the initial steps of the plan will provide an effective, early response. Be sure that your training objectives reflect the specific knowledge you need students to gain such as how to use the plan, what the boundaries of their assigned roles are, and so on. Clear, specific, and measurable outcomes for BC/DR plan training are as important as for any other type of training.

Everyone involved with the BC/DR implementation needs to understand their specific roles and responsibilities once a plan is activated. Training should address both the BC/DR process itself as well as the specific skills needed by team members to be effective in their designated roles. For example, a database administrator may be part of the IT damage assessment team. She may be an outstanding DBA but may not have the specific skills to know how to approach the IT damage assessment process. She should be trained in the process of performing the IT damage assessment as well as in the overall BC/DR process. That way, she will understand how and when the IT damage assessment is performed, how it impacts other BC/DR activities, and how to perform the duties of that role. Another example is an administrative assistant who is also tasked with being the crisis team coordinator. He might have the skills to manage multiple tasks at once, communicate and update people effectively, and so on, but he needs to understand the specific roles and responsibilities of the coordinator role. If there are tasks within that role he doesn't understand or know how to do, appropriate training needs to be provided. He may not know, for example, how to use emergency communication equipment such as walkie talkies. A simple thing to learn, perhaps, but not something you want to take time to teach him in the midst of a serious emergency.

Team leaders head up their individual teams (be sure to assign alternates or backups for key roles) and they must also be able to work effectively as part of the ERT or CMT. That means there has to be a leader assigned or selected for the crisis management team. Without such a designation, it's likely there will be confusion or perhaps a bit of jockeying for position. Leaders like to lead. Leaders can be extremely effective members of a team if they are confident the team has a competent leader. Otherwise, they'll naturally try to step in to fill the gap. That's fine if only one person steps up, but it's a problem if four or five (or more) people step up. Therefore, understanding roles and responsibilities is a key part of the initial training.

Many companies will implement a crisis management team comprised of leaders of other teams. This structure means that departments that have little interaction during normal business operations may have to work closely together during an emergency. It may even mean that someone higher in the organizational hierarchy is reporting to someone lower in the hierarchy during the emergency. Think of this scenario. Perhaps you have an area director on your CMT because she understands operations. Suppose the person designated to head up the CMT is the Facilities manager because he has experience in CMT as well as several related certifications. In the early stages of a disruptive event, the Facilities manager, as head of the CMT, is directing all activities, including those of the director of operations. This may be an appropriate situation but clearly, everyone has to be comfortable with this structure. The director has to be comfortable taking orders, temporarily, from the Facilities manager. There are numerous scenarios you can construct in which various levels of the organization have to work together seamlessly without anyone pulling rank inappropriately. Your training should address these cross-functional needs, define lines of authority and decision-making, and ensure that all team members are comfortable with the decision-making structure of the BC/DR process.

Finally, training should address the communication needs across the organization. As we've discussed previously, there are numerous communication needs throughout the life-cycle of a disaster and the team should understand this. The training should address the various communication groups (groups to whom the CMT should communicate), the appropriate frequency and content of the communication, and the appropriate distribution mechanism. Remember that during a disruption, your teams may not have access to standard communications equipment so communications plans and training should address various contingencies.

Now that we've looked at basic training elements, let's look at four commonly used methods of training, to which we referred earlier. These are the paper walk-through, the functional and field exercises, and full interruption. Figure 8.3 depicts the relative accuracy and organizational disruption each type of test generates. The least disruptive type of test is the paper walk-through, and it's the one most organizations do. The results from a paper walk-through are obviously going to be less accurate than functional or field tests. However, paper walk-throughs, if done well, can still yield extremely helpful results that can be incorporated into the plan to incrementally improve it.

Figure 8.3 Relative Disruption and Accuracy of BC/DR Plan Test Methods

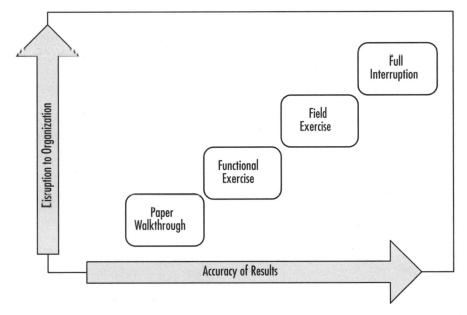

Paper Walk-through

In most companies, if you can manage to schedule a paper walk-through of your BC/DR plan once a year, you've scored a major victory. As gloomy a prediction as that is, it reflects the reality in today's organizations. However, if you've managed to get approval to put together your BC/DR plan, you can make a pretty strong case that without a walk-through, you'll never know if it works or not. It's like carrying a spare tire that's flat—it's of absolutely no consequence until you need it. You want to know if your BC/DR plan will work if needed, and the only way to determine that is to test it out. A paper walk-through will take time to step through but it's time well spent. There are eight discrete steps you can take to run an effective paper walk-through. These steps also apply to the other types of training (functional, field, etc.).

Develop Realistic Scenarios

The first step is to develop realistic scenarios for your walk-through. You should develop scenarios based on those risks determined by your assessment to be the highest risk, highest likelihood, and highest impact. Although it may be interesting or fun to walk through some oddball scenario (space aliens land and their magnetic field erases only the zeros on all disk drives), it's not particularly useful. Focus on the things most likely to occur. Start with a fire in the building, since statistically speaking, that's the disaster most likely to strike businesses. Also create scenarios that involve your highest risk/impacts. Remember, you will likely need to perform several walk-throughs based on various threats. However, it is possible after you've run through several scenarios that your team is familiar enough with the process that future walk-throughs can use a single scenario that covers the all the bases. Ideally, you'll perform a paper walk-through for each of your major risks. Given the time and budget constraints most of you are facing, that's probably not realistic, but it at least can be held as the ideal.

Develop Evaluation Criteria

The key to any successful test of your plan, whether it's a paper walk-through or a full interruption, is to have criteria by which you'll evaluate the success of that training. We'll discuss test criteria in a bit more detail later in the chapter as well. For your paper walk-through, you might develop criteria that include:

- How well participants were able to follow and utilize the plan
- How well participants were able to communicate across team lines
- How well the checklists or defined steps worked to achieve the stated objectives
- How confident participants felt with their implementation of the plan
- How confident participants feel about implementing the plan in the future

Provide Copies of the Plan

Members of the crisis management team should be given the latest copies of the plan in advance of the walk-through. The hope (but usually not the reality) is that they'll look through the plan prior to the walk-through. However, the likelihood is they will not, so your training and testing need to work on the assumption that prior reading or familiarization will not occur (despite what people might claim). In addition, individual team members that might be participating, such as members of a damage assessment team or an emergency response team should be provided their section of the plan. If helpful, you may want to create a flowchart of your plan's processes in order to help individual team members visually see and understand how things should proceed. This often helps individuals understand their roles within the larger plan and operate more effectively as part of the larger team. Figure 8.4 shows a portion of a sample flowchart. The adage "A picture's worth a thousand words" is very true in this case. Checklists and simple flowcharts can be helpful in an emergency if staff are familiar with how they work and how to utilize them.

Figure 8.4 Sample Flowchart of BC/DR Plan (Partial)

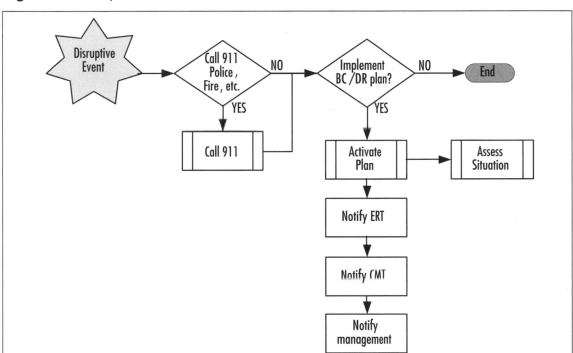

Divide Participants by Team

If your walk-through includes members of different teams, having them sit together can help the flow of the walk-through. If they need to confer or make notes among themselves, they can do so more effectively by being in close proximity to one another. It also helps reduce cross-talk and interruptions. Be sure to have alternates attend the training and work along-side their counterparts. If you have vendors you've designated as team members, they should also be included in the training.

Use Checklists

If you have checklists for your key processes (such as those shown in the appendix materials), be sure to provide copies of these checklists and ensure the team uses these checklists. If they find steps that are out of order, missing, or redundant, they can correct the checklists quickly. Like flowcharts, using checklists will also help maintain direction and forward progress during the walk-through.

Take Notes

Someone should be tasked with keeping notes about the process, major issues that arise, and the like. If you run the walk-through with various teams, each team should be responsible for keeping notes on their process and their section of the plan as well.

Identify Training Needs

As you train staff in the use and implementation of the plan, you should specifically keep an eye open for additional training needs. Be sure to ask training participants to make a note of any skills they believe they need in order to effectively carry out the BC/DR plan. Those closest to the job are in the best position to identify skills gaps and you can develop a list of training needs from these run-throughs. Of course, you might end up with a long wish-list, so you'll have to prioritize and sort through the training requests to determine what is high priority and what can wait (or is not needed).

Develop Summary and Lessons Learned

After the walk-through, you should compile and summarize the notes collected. You should summarize the lessons learned from the exercise and schedule a follow up meeting. This follow-up meeting should be held a day or two after the walk-through (i.e., not immediately following the walk-through, but not four weeks later) so that participants have a chance to think about the walk-through and bring their thoughts, suggestions, and feedback to the follow-up meeting. You can use the data collected from this process to modify future walk-through sessions and to modify the BC/DR plan as needed.

An annual walkthrough of the plan is often used as a combination of plan familiarization, training, and testing. In some cases, that may be adequate, but this type of exercise is really the bare minimum. Also, be sure to flag your team members in some manner so that if someone leaves or is promoted, for example, you can either notify the alternate or designate and train a replacement. One of the biggest risks you face to your plan is people not being familiar enough with it to implement it properly when needed. This happens for a variety of reasons and a regular training schedule can help reduce that risk.

Functional Exercises

Functional exercises are used to actually test some of the plan's functionality. In most cases, if you want to test all the functionality, you'll plan field or full-scale interruptions, discussed in the next sections. However, it's often helpful and adequate to perform a paper walk-through along with functional exercises. Functional exercises train staff in critical procedures or functions needed to respond to and address the disruption.

Typically, functional exercises make use of scenario-based scripts and run for two to three hours. The team is divided into two groups Alternates make an excellent second group for training purposes. A script starts off the sequence of events, which typically takes about 15 to 20 minutes. The ERT and CMT teams have to respond to the scripted events using their training and BC/DR plan. The second group, we'll call them the alternates here, act as nonteam members. For example, if the scenario includes evacuating the building, alternates may behave as employees might—panicking, not following instructions, and so on. If the scenario involves assessing injuries, alternates can have scripted injuries the ERT team has to deal with. The alternates use a menu of responses or events based on the specific scripted scenario to prompt the team members. The goal of this type of functional training is to get members to work as a team, to help members understand their roles and responsibilities, and to communicate effectively under stressful conditions.

TIP

If confusion crops up during the functional exercises, be sure to stop the exercise and clarify. Although it may break from the realistic scenario you've created, the primary purpose is training and testing the plan. If gaps or errors become obvious or there massive confusion during the exercise, stop and address it immediately. It might mean spending a couple of minutes clarifying roles and expectations or making notes about areas of the plan that need modification. The few minutes you spend clarifying can make all the difference in the confidence and competence of staff if they are ever called upon to put their training to the test. The same holds true for field exercises, discussed later.

As with any other type of training, you should have clear objectives and outcomes identified for functional exercises. For example, if you're going to teach staff how to restore a database from three weeks' worth of incremental backup tapes pulled across the Internet from a remote data vault, you should list the key knowledge you expect staff to gain. This might include:

- How to determine that the database needs to be restored (i.e., is the local copy destroyed, corrupted, offline, etc.)

- How to access the data vault backups (location, login credentials, accessing data, etc.)

- How to restore the data (what order, what locations, what settings, etc.)

- How to verify the restore (verification of file names, sizes, locations; sample test scripts, etc.)

A functional test of the BC/DR plan follows the same path. If you want to test some of the functions of your plan, develop step-by-step instructions and have participants use those steps to test the function. As we've mentioned, testing the plan is a great training tool and functional exercises go a long way toward both ends.

Field Exercises

Field exercises involve fairly realistic exercises based on likely scenarios. You've undoubtedly seen stories on these kinds of exercises on your local news stations. From time to time, local emergency responders exercise their skills by practicing scenarios. If you would like to practice your emergency and disaster recovery response using full-scale field exercises, you may be able to coordinate such exercises with your local emergency responders. They may welcome the opportunity to test their skills and to help train your staff in the process. If so, you have an excellent resource at your disposal that will not only test and hone your skills but provide valuable input into your disaster planning.

Most companies barely have the time or resources to do an annual paper walk-through of their plan, so it's not likely you'll be able to run through a real-world scenario. That said, if your company works in a dangerous industry (hazardous chemicals, explosives, power, etc.), you may want (or be required by law) to perform field exercises to assess and improve readiness. It's not until a situation is unfolding, even in a simulated manner, that some problems with a plan come to light. As useful as paper walk-throughs and functional exercises can be, they may still leave knowledge gaps or plan problems that you just won't know about until a real situation presents itself. Field exercises can reduce the risk of plan gaps but at a much greater expense of time and resources. For some companies, this investment makes sense.

Full Interruption Test

Like a field exercise, a full interruption test can be for the organization or just for specific systems within the organization. It activates all components of the plan and interrupts all mission-critical functions. The full interruption test will also activate the alternate work sites or facilities and off-site storage facilities, and the plan is actually implemented in whole. This type of full interruption test can be announced or unannounced. Clearly, an unannounced test simulates a real disruption or disaster more accurately than an announced test, but is also more disruptive.

Most companies are unlikely to be willing to disrupt their operations long enough to perform a full interruption test. However, there may be instances when a full disruption of a single business unit (rather than the whole company) is an acceptable trade-off for the knowledge and readiness that can be achieved through this type of realistic simulation.

Training Plan Implementers

If you have specific personnel designated as plan implementers, you may want to develop a specific training session for these staff members. They should understand exactly what their roles are and how to implement the plan, should it be necessary. Because the situation in the aftermath of a serious disruption or disaster is extremely stressful and chaotic, plan implementers should rehearse implementing the initial steps of the plan frequently enough that they're very comfortable with it. They should know by rote what to do, who to contact, and what steps they need to take in several likely scenarios. This memorization and practice of key first steps will help them if they are called upon to implement the plan. As we mentioned earlier, people fall back on their training in an emergency, so the plan implementers should be extremely comfortable with their responsibilities in this regard to prevent a total breakdown of the plan in the aftermath of a serious event.

Testing the BC/DR Plan

There are numerous reasons for testing the plan. The obvious reason is to make sure that the plan will work in the event of a real disruption or disaster. However, the underlying reasons that testing helps the plan work more effectively is that testing serves these purposes:

- Checks for understanding of processes, procedures, and steps by those who must implement the plan
- Validates the integration of tasks across the various business units and management functions
- Confirms the steps developed for each phase of the plan's implementation
- Determines whether the right resources have been identified
- Familiarizes all involved parties with the overall process and flow of information
- Identifies gaps or weaknesses in the plan
- Determines cost and feasibility

As you read through the training section of this chapter, it probably became clear to you that training and testing a BC/DR plan are closely integrated. One way of training staff on the implementation of the BC/DR plan is to test the plan (and training will test the plan). If you choose to test your plan through BC/DR training, be sure to include the items listed here as objectives or deliverables for your training.

TIP

If you have designated an alternate site, off-site data storage, or backups, these should be tested for failover capabilities periodically. Don't wait for disaster to strike to test these important capabilities.

Understanding of Processes

The processes, procedures, and steps taken by the various team members once the plan is activated (including how and when to activate the plan) should be the primary outcome of the testing phase. This phase should uncover any missing processes. It should also identify and verify processes and their interdependencies. Mission-critical functions should be restored first and the plan processes should address these priorities effectively.

In addition, the linear progression of the plan itself (first do this, then this) should be understood by participants. By walking through these processes, participants both learn the processes, and can verify that they make sense. Often, a BC/DR plan is created by a specialized team of subject matter experts, but it's not until the people who may be called upon to implement that plan (who may not be the same SMEs) that flaws are found. Any problems found with the plan through this phase should be noted and the change management process should be used to modify the plan appropriately. We'll discuss change management in Chapter 9.

Understanding the processes also includes understanding the work-arounds and manual processes that should be implemented during BC/DR activities. If you've identified moving to manual systems or work-arounds if certain systems fail, these processes and procedures should be identified, tested, and verified. In addition, they should be looked at from the perspective of how manual processes might interact with automated systems. In some scenarios, you might have one or two key systems down and other systems still up and running. How will the manual and automated systems interact, how will manual processes be tracked and managed, how will work-arounds impact systems still running? These are the kinds of questions that must be addressed when examining the processes of the BC/DR plan. All work-arounds, manual processes, and associate forms and paperwork should be included in this test phase.

Validation of Task Integration

Any walk-through or test of the plan should involve key personnel from mission-critical business functions as well as members of the BC/DR team. During the validation of task integration, these business subject matter experts will be best able to identify if the tasks are listed in the right order, with the right dependencies, with the right requirements, or resources, and such. The integration of tasks is often where plans fail in implementation due to the complexity of most businesses today. This is particularly true when looking at IT systems, which are at the heart of most recovery efforts. If tasks are not properly identified and sequenced, it can take hours, days, or weeks to uncover the source of the problem. The time and place to do this is in the plan testing phase, not during an emergency.

Confirm Steps

In addition to testing the tasks and their integration, the testing should confirm each of the steps delineated in the plan. This confirms that all necessary steps are listed and in the correct order. It's often when you're walking through the plan step-by-step that you discover errors or omissions. If you're fortunate enough to have captured the correct data the first time around, this step will confirm that your plan is as complete as possible.

Confirm Resources

At each step during the testing, you should ask and answer, "What resources are required to perform this step?" When you're thinking through scenarios, it's easier to identify needed resources. These might include people, skills, equipment, and supplies. It doesn't do much good to teach employees how to administer first aid if there are no first aid kits in the building. This step of the testing should look at needed resources for each step. For example, you need to be sure that the resources are not simultaneously required by two different teams or in two different places, just as you would in any other type of project resource management plan. If you do not have those resources at the time of the test, you should flag these steps as incomplete or in need of resources and create an action item to obtain these resources as soon as possible.

Familiarize with Information Flow

Communications are extremely important during a business disruption or disaster and are very difficult to maintain in those circumstances. This section of the test identifies who needs to know what and when. It identifies where information must flow and how it will flow. It identifies information needs for the mission-critical business functions as well as for the ERT and the CMT groups. As staff become familiar with the flow of information through the BC/DR plan, they are more likely to have a heightened awareness of this flow during an actual event. Some communications will inevitably break down during a disaster, but the training you provide here by testing the information flow of the plan will help reduce the likelihood of a serious communication and information flow breakdown. In addition, the heightened awareness of information flow here helps build awareness of information flow through the organization on a normal day-to-day basis. This can help bridge communication gaps that currently are impacting operations and productivity.

The other type of information flow you need to address is the flow of data through IT systems and the organization. As you test your plan, you can identify how data flows through systems and determine whether your disaster recovery and business continuity plans addresses this appropriately. In large companies, there are numerous data and IT systems interdependencies that have to be identified. Testing your plan can help you look at data flow in light of BC/DR activities and make necessary adjustments.

Identify Gaps or Weaknesses

As you test the plan using checklists, paper walk-throughs, and simulations, the plan's gaps and weaknesses, if any, should become evident. It's usually not until we put something into a realistic scenario that we can see whether or not there are any problems. If you identify gaps or weaknesses, these can be addressed through modifications to the BC/DR plan. Omissions are often spotted as well—What *is* the number to call to replace your servers? Who *is* the contact person to report injured staff who can't report to work? Other details can be missed during the creation of the plan, such as where licensing information is stored or whether a particular backup will run on a new server or CPU type. The technology issues can be massive and overwhelming and though you probably can't test every scenario, you can test the most likely ones.

Determines Cost and Feasibility

It's difficult to completely understand potential costs of implementing the plan when you're creating it. You can create realistic scenarios and estimate potential costs, but as you test the plan, you're likely to understand more fully the potential costs for implementing, managing, and maintaining the plan. This information can be helpful in finalizing your plan or in revising your plan to meet your company's budgetary constraints. In addition, the overall feasibility of the plan will be tested. Again, it's relatively easy to think that someone could perform steps in a particular order or achieve certain milestones in a particular time frame when you're developing the plan. When people actually put parts of the plan to the test, it's likely that some aspects are simply impossible to implement or manage as expected. The feasibility of the various steps, processes, and work-arounds is tested and can be revised to reflect the reality of the situation rather than the perfection of the situation on paper.

By testing the plan through training and through looking at these specific issues, you will have the best possible plan in place short of actually putting it to the test. Of course, the irony of the situation is that despite your best efforts on the BC/DR plan, you hope you never need to find out how good that plan is. Having trained and tested using the methods described in this chapter, along with any training and testing methods appropriate for the unique needs of your organization, you increase your odds of successfully pulling your company through a major (or minor) disruption or disaster. We all know that few things in life mirror the perfect world of planning. The following conversation with a seasoned BC/DR executive sheds light on the difficulties of implementing and managing BC/DR planning in the real world.

From the Trenches...

How It Works in the Real World

Debbie Earnest, an experienced IT professional, has a background in manufacturing, infrastructure, and software. She managed a DR group for about 18 months, so she's been on the front lines of BC/DR planning. She's currently working for a major B2B software services company based in the United States. She was kind enough to take time out of her busy schedule to sit down and talk with us about her experience with business continuity and disaster recovery planning and implementation.

What are the significant challenges in BC/DR planning?

Many companies are in a bit of denial. Everyone says they need an ERP system with data flowing seamlessly through the company. That's great on paper, but how do you create that? Any hiccup upstream affects multiple systems downstream. The complexity of systems integration and interoperability is difficult in normal day-to-day operations, never mind in the midst of a crisis.

This complexity becomes a significant challenge in BC/DR planning that can quickly become overwhelming. When you begin looking just at mission-critical functions, there are so many interdependencies and so many points of failure that it becomes difficult to figure out where to start or how to cover them all. You can spend X dollars on disaster recovery planning and systems and you may still not be able to recover. That expenditure is then seen as a waste of money and, of course, that comes with its own set of problems. Even though your plan may not have addressed just one of a thousand failure points, if the one you missed is the one that comes into play, your plan may fail despite your best efforts.

BC/DR should really not be driven by IT, but in reality, it is. Companies cannot function without IT systems and IT staff is used to thinking about what could go wrong and how they should plan to avoid it. They are pretty good about risk management because they know that even a corrupt database can cause an outage for a period of time, or a fire in the UPS in the data center can cause a disruption. Many companies are so large or so complex, you just can't do business on paper anymore. Even trying to set up viable work-arounds or manual methods is next to impossible.

Continued

Another significant impediment to BC/DR planning is corporate mergers. It's difficult enough to develop a sound BC/DR plan for one company, but when you have the day-to-day IT tasks coupled with the tasks of integrating two disparate technology systems, BC/DR planning is almost impossible. There's no real clear solution to this problem other than to continue to build BC/DR systems into your plans as you move forward.

Perhaps the biggest problem is the budget. Unfortunately, when companies are looking for places to tighten their belts, they usually start cutting the BC/DR staff and activities. I worked for a company that was doing about $1B annually and when things got tight, they began eyeing the dedicated BC/DR jobs for elimination. Those people saw the writing on the wall and all eventually left the organization. This exemplifies the problem. Companies may think BC/DR is a great idea, but when it comes to including it in the budget and defending it at budget meetings, it doesn't happen.

What advice do you have for those trying to develop BC/DR plans?

Companies must figure out what is absolutely critical. Business leaders often can't identify critical components—to them every system they work with is critical. Therefore, the best approach is to say, "OK, this system goes down. What happens on Day 1? What happens on Day 2? Day 3?" By asking them these kinds of scenario-based questions, you can discern the relative priorities of the various systems. In some cases, the best you can do is develop a BC/DR plan that addresses the top three business functions and even then, it still may not be adequate. Again, it goes back to the number of failure points that exist in the organization. In some companies, it's manageable. In large corporations, it's pretty much impossible to cover everything.

The goal, from my perspective, is to never get into a disaster situation—build in safety valves along the way. The first line of coverage is to build it into your systems. IT people are used to thinking about downtime and outages and have developed almost an instinct for providing the first line of DR defense—primarily stemming from the rigorous availability demands from companies today.

In BC/DR planning, you can try dividing it into two basic levels: Level 1 = Data Loss, Level 2 = Building Loss. Since IT staff is pretty good about figuring out how to avoid data loss, the Level 1 defense can be the best line of protection. If you have sound methods in place for avoiding data loss, you are reducing your risk significantly. Identify what the Level 1 systems are and how they provide protection.

You also have to look at alternate sites in advance, if that's going to be part of your Level 2 plan. To try to get an alternate site after a disaster will take you twice as long and cost twice as much, so planning in advance is definitely the preferred approach.

Finally, you have to test and practice the plan. In some companies I've worked in, we simply did a paper walk-through of the plan once per year and revised it accordingly. Although that may not be ideal, it's about as much as you can squeeze out of some organizations for BC/DR purposes.

From your perspective, what's the bottom line?

The real issue is time and money—there's never enough of either. There is no perfect BC/DR plan, but having one is better than not having one. You have to ask, based on your company, what is the bare minimum? At the very least, plan for that. Then set expectations so no one is looking for perfection, it just doesn't exist.

Test Evaluation Criteria

Before embarking on the testing phase, you should develop clear evaluation criteria for your tests. In many cases, the easiest way to create test criteria is to go through the various check-lists or steps in your BC/DR plan and create corresponding questions. Let's look at an example involving the notification step in the activation of the plan.

1. Was the primary team member able to begin the notification process successfully?

2. How many team members were contacted?

3. How long did it take to notify team members?

4. Were there any missing or incorrect phone numbers?

5. How many team members were contacted via their primary methods vs. alternate methods?

6. How many team members were not on the notification list?

7. Were there any names on the notification list that should not have been?

8. Would this have worked if phone systems were out?

You can create a set of questions for each phase of the plan and use these to evaluate the test results. You can then measure the performance against the ability to complete each step, the thoroughness of each step, the effectiveness of each step and the accuracy and validity of each step.

Recommendations

Based on test results, you should develop recommendations. These recommendations may result in modifications to the BC/DR plan, but they may result in modifications to other areas as well. For example, you might find areas in which staff need additional training. You might find through these tests that there are areas of the business not included in the plan that should be or that there are operational changes that are recommended based on the test results. Recommendations for each team as well as for each phase of the plan should be developed. Be sure to include a process for incorporating recommendations so they are actually utilized and not overlooked.

Performing IT Systems and Security Audits

By definition, an audit is the systematic examination against defined criteria. If your company is required to comply with laws or regulations, you have no doubt been through rigorous audits. The audits you perform to conform to these regulations may help in your BC/DR planning and may need to be included in your plan. For example, if you must comply with HIPAA (Health Insurance Portability and Accountability Act) standards, your

BC/DR plans must address these issues and your audit of the plan has to include these parameters as well. Your audit should include both business continuity and systems audits.

IT Systems and Security Audits

Auditing IT systems involves a set of tasks that help reduce the risk of an intrusion or attack. Audits are concerned primarily with ensuring the company maintains data confidentiality, integrity, and availability, because these are the areas that typically come under attack. In some cases, this can disable a company's critical business functions; in more extreme cases, it disables the company's entire operations and creates a significant legal or financial liability for the firm as well (see the case study that follows Chapter 1 for more on the legal implications of data security).

An IT systems audit typically focuses on conducting a systematic evaluation of the security of various IT systems by measuring how well it conforms to established criteria or requirements. It includes an assessment or review of the network and systems' physical configuration and environment, the configuration of the software, the handling (storage, transport, access, etc.) of data, sensitive data in particular, and user access. Security audits are often performed in conjunction with compliance efforts, though even companies not subject to compliance regulations should undertake periodic IT audits.

Hardening systems is a risk mitigation strategy that is employed by virtually every company using IT systems today. Hardening systems, as you're probably aware, consists of taking actions to minimize the attack footprint of a system or network. This includes actions such as removing network protocols not in use, disabling ports or services not being used, removing unused user accounts, reducing permissions to the least possible, and automating the updating of anti-virus and anti-spyware data files, to name just a few examples.

With respect to BC/DR planning, systems auditing should include several key elements. These include:

- Ensuring IT risk mitigation strategies are in place and properly implemented/configured.

- Ensuring systems identified by the BC/DR plan are still in place and functioning.

- Identifying areas where new technology has been implemented and may not be incorporated into the BC/DR plan.

- Identifying areas where technology has been retired or modified, resulting in the need to revise the BC/DR plan.

- Reviewing the processes identified in the BC/DR plan with respect to IT systems to ensure the steps and processes are still correct, complete, and relevant.

- Verifying that the IT incident response team (CIRT, CERT or whatever term you use) is in tact and has a clear understanding of roles, responsibilities and how to implement the IT-specific segments of the BC/DR plan.

- Reviewing data regarding various systems to ensure they are still compliant with the BC/DR plans. These systems include operating systems, networking and telecommunications equipment, database and applications, systems backups, security controls, integration, and testing. Any of these areas is subject to frequent change. An audit can help assure the BC/DR plan will still work if implemented.

This is not an exhaustive list, but it provides examples of what types of data an IT audit within the scope of a BC/DR plan might include. The key is to identify how IT systems have changed (or remained the same) and assess how and where that impacts the BC/DR plan. Most IT systems are not static and even gradual changes over time can end up creating a significant change to the way a BC/DR plan must be implemented. Referring to the interview with IT professional Debbie Earnest earlier in the chapter, you can see that with the complexity of systems, the proliferation of corporate mergers and acquisitions, and the ever-changing technological landscape, your best bet for keeping your BC/DR plan up to date will be through the IT audit process. Periodic auditing is an excellent operational practice for IT and it doesn't take too much more effort to include a check of key elements from the BC/DR plan during these audits. We'll discuss maintaining the BC/DR plan in more detail in the next chapter, but keep in mind that IT audits are the easiest way to maintain the BC/DR plan from an IT perspective primarily because they involve a periodic review that is likely already part of your standard operating procedures. Adding a few extra steps to your audit plan is easier than trying to perform a fully separate BC/DR audit every quarter or every year. It's also easier to address gradual changes as they occur than to try to assess how much change has occurred since the prior year's review of the plan.

Summary

Training and testing your BC/DR plan are tightly integrated activities. Training staff for the specific roles, responsibilities, and actions they take during the implementation of the BC/DR plan also tests the plan. On the flip side, testing the plan trains staff in the implementation and management of the plan. Therefore, these two activities should be viewed as a whole and plans for training and testing should be complementary rather than redundant.

Training activities should be defined for emergency responders. These skills often are taught by community organizations such as the local fire department or other local organizations. Skills include building evacuation, fire fighting, and first aid. Training should include safety procedures as well as instruction on the use of specialized equipment. These skills should be reviewed and refreshed periodically through exercise, drills and simulations, if possible.

Training for business continuity and disaster recovery is a slightly more difficult undertaking. Training can take any of several forms and the training activities are also plan testing activities, so there's a great deal of overlap. Training for BC/DR should include training team members on their specific roles and responsibilities during the implementation of the BC/DR plan. It should also include training on specific skills needed to effectively implement and manage the plan. Cross-functional teamwork and communications should also be part of the BC/DR training.

In order for the training to be effective, you should develop clear, specific, and measurable outcomes for your training. This should include scope of training, requirements for training, and learning outcomes expected. You may need to perform a training needs assessment before developing the training requirements. As you test the plan, you'll also identify areas that may require additional staff training, and these can be added to your training requirements. Developing the training can be done in conjunction with developing the testing plan for your BC/DR plan in order to achieve some efficiency in your efforts. Finding time to schedule and deliver training is a challenge in most organizations, so if you can find a way to tie these efforts or outcomes into larger business objectives, you might have greater success. The results of training activities should be monitored and measured to ensure the training achieved its objectives and that revisions to the training based on input and feedback can be incorporated in the next iteration.

Testing the plan helps train team members on the use of the plan, on their specific roles and responsibilities, and on communicating across the organization. Testing the plan will also help you identify processes, procedures, steps, or checklists that are incorrect, have gaps, or require revision for some reason.

There are four primary ways plans are trained and tested, though there are an infinite number of variations. A paper walk-through is the easiest and least disruptive way to test your plan, but it yields the least accurate results as compared to other methods. However, because it is least disruptive, it *is* the easiest for most organizations to implement and the results can help improve the quality of the BC/DR plan significantly. Functional exercises test subsections of the plan and the functionality of various components. An example of a functional exercise is having IT test the steps in the BC/DR plan related to restoring a server from remote backups. These types of tests can help uncover problems that would otherwise go unnoticed, but they take more time and resources to perform than paper tests. Field exercises and full interruptions certainly provide the most realistic simulations, but most companies will be reluctant to plan and pay for this type of training. In some types of industries, this type of exercise is a requirement either for health and safety reasons or due to legal or regulatory requirements. Regardless of which type of training and testing you undertake, you should pay special attention to the skills and training needs of the plan implementers. They should be well versed in how to activate and implement the plan so that they can do so relatively easily if a disruption or disaster occurs.

Testing the plan checks for understanding of the processes, procedures, and steps defined. It validates the integration and dependencies of tasks across various business and functional units. It also helps determine if the right resources have been identified for the various steps. Ultimately, it familiarizes the implementers with the entire process and uncovers potential gaps, errors, or omissions. Finally, the cost and feasibility of implementing a plan can be better assessed through testing.

IT systems and security audits are typically part of company's standard IT operating procedures and they may be required by law or regulation (HIPAA, etc.). In addition, BC/DR-specific IT audit tasks can be included in standard auditing procedures to reduce the amount of additional work that might be required to test the BC/DR plan. Some of the elements you might choose to audit in this manner include ensuring the IT risk mitigation strategies have been implemented per the BC/DR plan, ensuring the processes and procedures for IT work-arounds are feasible and meet requirements and identifying changes to technology that impact (or are impacted by) the BC/DR plan.

Solutions Fast Track

Training for Emergency Response, Disaster Recovery, and Business Continuity

☑ Training, testing, and auditing are tightly integrated activities used to validate the BC/DR plan.

☑ Emergency responders should be trained on the specific skills needed to respond to emergencies. This includes safety procedures and well as the use of specialized equipment.

☑ Disaster recovery and business continuity plan training should begin with a definition of the scope, objectives, and requirements for the training.

☑ A needs assessment should be performed and results from plan testing should also be used to identify training needs.

☑ Training should be monitored and results should be analyzed so any changes to future training or to the BC/DR plan itself can be made through the change management process.

☑ Training also includes how to activate, implement, and utilize the BC/DR plan effectively. It should help team members understand their specific roles and responsibilities within the constraints of the plan. It should help them with cross-functional teamwork and communications as well.

☑ There are number ways to train and test a plan. The four most common methods are paper walk-through, functional exercise, field exercise, and full interruption. These methods train participants while simultaneously testing the plan.

Testing Your Business Continuity and Disaster Recovery Plan

☑ Testing a BC/DR plan often is done in conjunction with training exercises such as the paper walk-through, functional or field exercises, or a full interruption. Testing helps uncover areas of the plan that require revision.

☑ The plan test should assess the processes defined to determine if they will work as specified and if they are feasible.

☑ The test helps identify task integration and dependencies to determine if there are gaps, errors, or omissions in the steps identified in the plan.

☑ Testing helps train participants in the information flow that is required throughout each step or phase of the BC/DR plan.

☑ A good BC/DR test plan should have well-defined scope, requirements, outcomes, or objectives.

☑ One of the objectives of plan testing should be an assessment of the cost and feasibility of the overall plan. Any weaknesses in this area will require a revision to the plan to better align with the constraints of the organization.

☑ BC/DR planning is a difficult task that may not cover every single point of failure. Level 1 plans address data loss and from an IT perspective, these are the easiest to address.

☑ Don't aim for perfection; try to cover the basic needs of your mission-critical functions.

Performing IT Systems Audits

☑ IT audits are performed as part of most company's standard IT operating procedures and may be required by law or regulation.

☑ IT systems audits focus on conducting a systematic evaluation of various IT systems and this information can be incorporated into the BC/DR plan.

☑ Changes to IT technologies should be incorporated into the BC/DR plan.

☑ Auditing with an eye toward BC/DR requirements can be incorporated into standard operating procedures to reduce the impact of BC/DR work on the IT department.

Frequently Asked Questions

The following Frequently Asked Questions, answered by the authors of this book, are designed to both measure your understanding of the concepts presented in this chapter and to assist you with real-life implementation of these concepts. To have your questions about this chapter answered by the author, browse to **www.syngress.com/solutions** and click on the **"Ask the Author"** form.

Q: We've managed to create a BC/DR plan, but for the life of me, I can't get anyone to test the plan. Since I am the project manager for this project, I'm a bit concerned that if the plan has to be implemented that it might fail and that I will be stuck "holding the bag." Any advice?

A: Unfortunately, you're facing a fairly common problem. As you read in the interview with Debbie Earnest, many companies do not want to put much time or effort into BC/DR activities, especially when the budget is tight. Sometimes the difficulty in testing the plan is that people are afraid it will become a big, all-consuming activity. You might be able to reduce resistance by actually developing an outline of a test plan for different segments of the plan. For example, you might put together an outline for activating the plan. Write up one or two likely scenarios, based on your company's specific situation. Then, take the steps from your BC/DR plan (this is where checklists can be extremely helpful—be sure to see the Appendix for checklists you can use) and create a document that you can use for a walk-through. Bring this document to one or two of the key people and say, in essence, "Look, I need one hour of your time to go through this. It will be extremely helpful for the company in terms of testing this part of the BC/DR plan and I anticipate you might learn a few things that might be useful to you in your daily operations as well. When can we schedule an hour with [name required participants]?" Often, if you take a segment of the plan, you create a testing outline with scenario, you define the time requirement, and you help participants understand what's in it for them, you may get the buy in you need. Remember, too, that scenarios really help bring the plan to life, so although your plan should be realistic and based on your company's most likely situations (highest risk, highest vulnerability, highest impact), you can also have a bit of fun with them. Make them vivid—if you're not a good writer, find someone who is. Make the opening of the scenario read like a good crime novel and you might just hook your participants into enjoying the process. Even if your scenarios aren't scintillating, you may be able to get buy into test your plan by 1) segmenting the testing, 2) creating a test outline, and 3) defining a specific timeframe for the test activities.

Q: Our company has five divisions, and each division has multiple geographic locations. Our data is distributed across the enterprise and crosses a lot of organizational boundaries. We've managed to develop a preliminary BC/DR plan, but we can't agree on how to test it. We don't have the time or organizational tolerance to perform any sort of disruptive testing, but without that, some of us aren't sure the plan will work. Any suggestions on how to get everyone on the same page?

A: It's interesting because small companies and large companies have many similar challenges in BC/DR planning, but this particular challenge is really unique to medium- and large-scale companies. The problem, as you know, is that each business unit runs on its own timetable and has its own unique constraints and challenges. So, it's not likely you can schedule a functional exercise on a particular day and time that will work for multiple business units, for example. The best solution, which is not optimal, may be to ensure that subsections of the plan are tested using a paper walk-through. As with the previous question, you may be able to get buy in if you define a scenario, invite only key players, and set a time limit for the meeting. That means you (or your designee) will have to manage the meeting effectively—develop an agenda, distribute the agenda in advance along with required preparation, start the meeting on time, move it forward productively, achieve your stated objectives, gather feedback, and end on time. If you do a paper walk-through based on a single scenario, you might be able to get some participation. For example, you might be able to have the people in Division A run through a scenario that their Internet connections go down at three of their four international locations. What happens? Another scenario might be that a key system, say your ERP or CRM system, crashes. What happens at Division B? What happens at each office of Division B? Ultimately, you may have to be somewhat creative in getting people to test the plan. In some organizations, certain words are hot buttons, so you may find simply using different language to describe what you need might work. Instead of saying you need to "test the BC/DR plan" you might say you want to schedule time to "run through a few scenarios" or "take a look at our readiness" or "take a look at our key systems." Sometimes it's what you say, or don't say, that achieves your objectives. When it's all said and done, you have to do what you can to push the organization to test the plan in ways that are feasible within your unique constraints. Sometimes your project sponsor can help, sometimes an e-mail to one of the corporate executives asking for assistance in moving it forward can help. It's not an easy task in your situation, but with a bit of creativity, you may be able to find ways to test the key parts of the plan, which is better than not testing the plan at all. The upside is that any testing of the plan also trains participants.

Q: We've tried doing walk-throughs a couple of times and we've run into a weird problem. People get so into it that they go off on all sorts of tangents and before you know it, we're talking about what color the new chairs in the reception area should be. We seem to get bogged down in the same place each time and never progress beyond about the first third of the plan. What are we doing wrong and how can we move beyond this sticking point?

A: The good news is that you have people who want to participate and who engage with the process. The bad news is that the process is a bit chaotic. There are three suggestions on how to address this. First, you could hold the next walk-through as if it's an actual situation. That means that your CMT members are in charge and the leader of the CMT should step up and manage the process. If you were in a real emergency, the CMT leader should be taking charge of the situation, so the same should happen in a walk-through. It's possible the person designated as the CMT leader is not the right person for the job—he or she may not be good at leading in a chaotic situation. If so, you (and the team) will have to find the best way to address this—whether that's training for the CMT leader or replacement of the leader with someone who has more abilities in this area. Second, you should review your walk-through process. If it's not well defined, it will devolve into a discussion of the furniture or the drapes or whatever else comes to mind. In addition, the process can be broken down into distinct segments so that clear deliverables are met. This helps the leader manage the group by stopping at key places, recapping events, activities, and lessons learned, and then moving forward. Finally, someone can be assigned the role of "meeting monitor." Ideally, this person is not involved with the paper walk-through itself but can monitor when discussion begins veering off. This person's role is to remain somewhat neutral and to simply monitor time and progress against objectives. If discussion starts moving off course, the meeting monitor interrupts, helps the group decide what to do, then redirects activities back to the plan. Options about what to do can include creating a side list (sometimes referred to as a "parking lot" where issues are temporarily parked) for later follow up, it can mean simply stopping the conversation and getting back on track, or it can mean the group makes a conscious decision to continue the discussion because it is vital to the successful outcome of the training or testing. This "meeting monitor" role can help augment the leader who may be trying both to manage the meeting and familiarize himself or herself with the role of BC/DR leader.

Chapter 9

BC/DR Plan Maintenance

Solutions in this chapter:

- BC/DR Change Management
- Strategies for Managing Change
- BC/DR Plan Audit
- Plan Maintenance Activities
- Project Close Out

☑ Summary

☑ Solutions Fast Track

☑ Frequently Asked Questions

Introduction

Maintaining the plan you've developed may end up being the biggest challenge you face in the entire business continuity and disaster recovery plan process. If you found lack of enthusiasm or outright resistance to the BC/DR process, you may find that support for maintaining the plan simply vanishes. However, there is some good news amidst this gloomy outlook. First, you actually *have* a plan to maintain. People within the organization have participated in evaluating the business, developing mitigation strategies, and perhaps even testing the plan. The other good news is that, as we've discussed throughout this book, there are many areas in which you can incorporate BC/DR strategies and activities in your standard operating procedures. For example, many of the IT strategies implemented to provide continuous (or very high) availability are strategies that are also BC/DR risk mitigation strategies. We've pointed out that it's extremely helpful to incorporate BC/DR strategies in your operational plans whenever possible to reduce the outright resistance you may face to BC/DR planning.

In this chapter, we'll discuss various considerations for maintaining your BC/DR plan, especially in the face of indifference or resistance. As you can see from Figure 9.1, we're in the last phase of our BC/DR planning project.

Figure 9.1 Business Continuity and Disaster Recovery Project Plan Progress

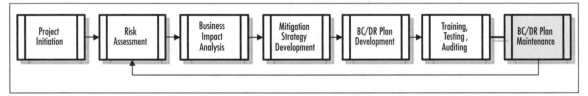

BC/DR Plan Change Management

Change is constant in organizations—change in operations, change in technology, change in personnel, change in regulations—the list goes on. You might be wondering how you can possibly reflect these changes in your BC/DR plan without having a full-time dedicated BC/DR team. It is challenging, but there are a few strategies you can use to reduce the complexity and enormity of the task. Change management has several discrete steps, as depicted in Figure 9.2. As you can see, the first step is to monitor changes. There are numerous sources of change that we'll discuss shortly. The next step is to decide how the changes impact your BC/DR plan. Not all changes have an impact on your plan, but you need to assess change before you can make that determination. If a change impacts your plan, the next step is to determine how to address the change in your plan. This typically involves cycling back and performing a modified version of your risk assessment, business impact analysis, and mitigation strategy development. This iterative process can be accomplished relatively quickly in many cases, but your assessments will have to specifically look at the suggested change and the impact on the entire plan.

Figure 9.2 Change Management Process

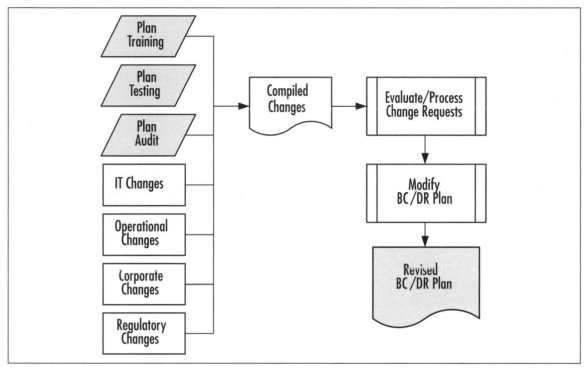

We've discussed plan training, testing, and auditing in Chapter 8, so let's continue now by looking at how these inject change into the BC/DR plan and then we'll examine other sources of change that impact a BC/DR plan. You'll find a checklist for change management tasks in the Appendix as well.

Training, Testing, and Auditing

In Chapter 8, we discussed the activities related to developing, delivering, and evaluating training. You learned that training often involves testing the plan, and that testing the plan trains staff on how to implement the plan and carry out the tasks assigned. Changes will naturally come out of these processes, and that's part of the purpose of training and testing. It's difficult, if not impossible, to develop a perfect plan the first time through. It's not until you try putting the plan to work that you discover steps out of order, errors, omissions, or redundancies. As you deliver your training and perform your testing, you should capture a list of changes that need to be made to the BC/DR plan. These changes should be submitted for review. Not all requested changes should be made for a variety of reasons. We'll discuss the change review process in the section entitled "Strategies for Managing Change" later in this chapter.

Changes in Information Technologies

The IT audit discussed in Chapter 8 is one of the ways you can keep track of changes to IT, but clearly this area is the one most subject to change and risk. You and your IT team are more than likely extremely familiar with reviewing and assessing change—from the location and duties of various servers to the implementation of new applications to the reorganization of existing infrastructure. As you know, even the most innocuous changes can suddenly inject all kinds of problems into your network systems. As you continue to manage your day-to-day IT operations, you should consider including an additional step in some of your processes that remind you to assess the process against BC/DR. For example, you likely have a process for evaluating the implementation of new technology. Consider adding one step in the process that says, "Assess impact of this new technology on BC/DR plans." It is a deceptively simple step and admittedly, it can open up a whole set of problems or questions you'd like to avoid. The flip side is that if two or more technologies are being considered, there may be one that contributes to BC/DR more than the others. In that case, you might be able to build in additional BC/DR capabilities with little effort.

As systems are upgraded, swapped out, modified, or retired, be sure to include a line item task to consider the impact on BC/DR plans. In some cases, this will be simply an item to be checked off. In other cases, you might discover that changing a system will have a large impact on your BC/DR plans. In those cases, you'll have to balance the potential change (for better or worse) against other alternatives. Regardless of your final decision, be sure to flag these changes so the BC/DR plan can be updated as a result.

From the Trenches...

Incorporating IT Changes

IT systems are at the heart of most BC/DR plans. As a result, changes to IT systems often have the biggest impact on the BC/DR plan. For most organizations, the easiest way to make sure IT changes are reflected in the BC/DR plan is to keep a list of IT changes and a brief assessment of the impact on the BC/DR plan. As the periodic review or audit of the BC/DR plan occurs, this list of IT changes can be incorporated. It might be as simple as keeping the list of changes in a spreadsheet or document in the same folder as the BC/DR plan. Since IT changes so frequently, the first questions to be asked should be, "How will this change enhance or degrade our ability to recover from a significant outage?" If the change will enhance it, you're set (though you still need to take an

Continued

integrated or holistic look at the change and how it impacts other parts of the organization and the BC/DR plan). If the change will degrade your ability to recover from a significant outage, you need to assess whether this is a wise move or whether there are other options available that will meet your needs and be neutral or positive as it applies to BC/DR.

Changes in Operations

During your risk assessment, you determined the mission-critical business functions that needed to be addressed in your BC/DR plan. Clearly, operations are not static and changes over time to operations may impact the BC/DR plan. Reorganization, expansion, new departments, new facilities, and new management structures can all impact operations in a variety of ways. In some cases, changes in operation happen slowly over time and these changes may go unnoticed as it relates to the BC/DR plan. The BC/DR plan audit (discussed later in this chapter) can be an effective method of reviewing operations against the BC/DR plan. If the business's mission-critical operations have changed over time or if the processes used to accomplish these functions have changed, the BC/DR plan is at significant risk of failure and should be revised. For example, if your company has slowly moved from bricks-and-mortar retail to e-commerce, many key processes may have changed. If the mix has shifted slowly over time, you might not notice it until you test the plan or perform a BC/DR audit. Obviously, the key is to be sure your BC/DR plan addresses your mission-critical business functions and if those shift over time, you plan needs to be updated. Changes to operational processes should be implemented as needed, but it would help if your operations staff understood that any changes to their key processes should be flagged so the BC/DR team can review the impact of those changes on the plan and revise as needed.

Corporate Changes

Corporate mergers, acquisitions, spin-offs, restructuring, and other types of corporate changes can have a major impact on the BC/DR plan. As these changes are considered and discussed, the BC/DR team should assess the potential impact to the plan. Of course, in many cases, these activities are not publicly announced until the deal is sealed, so your team may be caught off guard. As you read in the interview with Debbie Earnest (see Chapter 8), these kinds of changes are among the most challenging to deal with from a BC/DR perspective. IT staff will have a big enough challenge figuring out how to incorporate the required IT changes for daily operations much less trying to figure out how all this impacts BC/DR activities. Per Earnest's suggestion, the best you can do most of the time is to continually look to incorporate BC/DR activities into your normal operations and planning activities and to continually look to protecting data first. Sometimes, the BC/DR elements can be addressed through standard IT planning processes with an additional line item task.

Assessing how the plans impact BC/DR may help the team choose from among several viable alternatives or it might point out a path that optimizes immediate and BC/DR capabilities.

Legal, Regulatory, or Compliance Changes

Changes to the legal, regulatory, or compliance landscape will certainly trigger required changes to your BC/DR plan. For example, if laws change regarding data security, you will have to review your BC/DR plan to determine whether your existing plan meets these new requirements or whether you'll need to implement additional tools, technologies, or processes. As with other changes, it's sometimes as simple as looking at the current BC/DR plan and determining that no change is needed. Other times, a major change may require you to cycle through all phases of the BC/DR project planning stages and create a plan for implementing required changes. In most cases, changes in this arena will impact operations or IT and the impact to the BC/DR plan will be addressed through those channels.

Strategies for Managing Change

Two key strategies for managing change are having a process for monitoring and a process for evaluating change requests. It's usually easier to monitor change and respond to it as needed over time rather than sitting down once a year and trying to remember (or determine) what's changed since your last review of the plan. The easiest way to monitor change throughout the organization, as it relates to BC/DR plans, is to include an additional step or two in standard operating procedures. These steps can be as simple as "Determine impact, if any, on BC/DR plan. If impact exists, submit BC/DR change request to [insert position responsible for managing BC/DR change requests]."

Common Challenges

Removing Roadblocks

Most IT professionals are accustomed to dealing with change requests. In the BC/DR process, there are really two separate and distinct activities: change notifications and change requests. *Change notifications* are those changes that are being made, regardless of impact to the BC/DR plan. These include changes to the organization, personnel, operations, or the larger regulatory environment. These changes need to be addressed by the BC/DR plan. *Change requests* are elements that trigger changes to the BC/DR plan. In some cases, a change notification triggers a change request. Other times, a change notification does not trigger a change request. You may choose to

Continued

take these two functions out and ask your operations staff to incorporate a *change notification* process. In this way, they can notify the BC/DR team of change. The BC/DR team can then evaluate that change and generate a change request for the BC/DR plan, if appropriate. The rationale behind dividing this process in two in this manner is that operations staff might resist the use of a "change request" because, from their perspective, the change is not being requested, it is being implemented. Therefore, the use of the term change notification may help reduce resistance within the organization to the process of keeping the BC/DR team up to date. A simple change of terminology may melt resistance to the process of keeping your team up to speed on organizational changes that may impact the BC/DR plan.

Monitor Change

Implementing processes for monitoring change can make your job of maintaining the BC/DR plan much easier. Develop processes that can be incorporated into everyday processes so that as changes occur, they can quickly be assessed for their potential impact on the BC/DR plan. If the change has no impact, it can be ignored (from a BC/DR perspective). If the change will have an impact, a change request should be submitted to the BC/DR team. (*Note*: We'll use the term *change request* to keep it simple, but it will refer to either a change request or notification.) Remember, a change request may be a simple matter of noting that the leader of the Emergency Response Team has changed. This change request should trigger the appropriate revision to the BC/DR plan including contact names, phone numbers, and team rosters. The same is true of other types of changes. If the IT group is implementing a new server, there may be a change request generated to note the new technical specs of the server in the event of an outage, or it may trigger a quick review of the specs for servers at the alternate computing site to ensure the alternate site still meets BC/DR needs.

People

People leave organizations, they get promoted, or they move into different jobs. A periodic review of changes to the organization can help you determine if there have been personnel changes that impact your plan. This can be part of the BC/DR plan audit, discussed later in this chapter.

Process

Changes to processes should be monitored as well. Subject matter experts or members of the BC/DR team can be tasked with monitoring changes to key processes and flagging changes for BC/DR review. Many corporate processes remain fairly unchanged over time; however, some companies that are in high growth mode or that are streamlining operations, for example, might have significant changes to their daily operations. Changes to mission-critical functions should be reviewed with the highest priority since these changes could potentially cause the BC/DR plan to fail if implemented without these changes.

Technology

Changes to IT have been discussed, but some technology may fall outside the scope of IT management. If your company works with scientific equipment, manufacturing equipment, or other specialized technology, changes in this arena should be monitored and assessed to determine whether the BC/DR plan requires modification. Often changes in technology create changes in processes, so the trigger for review and modification may come from either area. However, a process for triggering a review should be included in your technology implementation plans to make BC/DR plan maintenance as low maintenance as possible.

Evaluate and Incorporate Change

The change review process should be a well-defined process for your BC/DR plan and someone should be responsible for processing change requests. In some cases, this is the BC/DR project manager, in other cases, it may be a role assigned to a team member or it may be managed through some other existing process.

Most project managers use a change management process for managing their projects and the same types of processes are useful here. As you know, not all changes requested can or should be implemented into a plan. Additionally, even if a change should be made to a plan, there are numerous considerations before incorporating the change into the plan. If you have a standardized change management process that you've worked with successfully in the past, you may want to use it here. Be sure to review your process to ensure it's appropriate to change management in the BC/DR process.

Not all changes should be incorporated, and reviewing changes individually and then as a group will help you and the team determine which changes should or must be implemented. Any changes that are required by law, regulation, or compliance clearly must be made, though there may be several different approaches to incorporating the change that will meet the requirements. In that case, further analysis may be required. Ultimately, for each change you consider, you need to determine the impact on the other elements of the BC/DR plan. For example, if your company moves to a new location, you need to assess the threats again. There may be a chemical processing plant in the new neighborhood that you have to consider. It might be that your business moved in order to get away from that chemical processing plant and to reduce certain risks. Some changes increase your risks, other changes reduce your risks. Some changes may have a strong effect on the business impact analysis outcomes, others may have no effect at all. For each change, you'll need to cycle briefly through all your assessment steps to see how a potential change impacts (and is impacted by) each area. After the assessment, if you and the team decide the change should be incorporated, it can be implemented and the BC/DR plan can be revised accordingly. This also means that the change should be incorporated in the training, testing, and auditing processes and procedures.

Here are some points to consider:

1. Compile all change requests and prioritize based on potential risk, vulnerability, impact (if applicable).

2. Determine if any change requests are required for legal, regulatory, or compliance reasons. If so, flag these as required changes.

3. Review compiled change requests, review for redundancy, relevancy, etc. Revise compiled list as appropriate.

4. Prioritize compiled list. For each item, determine how the change impacts (or is impacted by):

   a. Selected risks and threats

   b. Threat vulnerability

   c. Business impact analysis

   d. Risk mitigation strategies

5. Assess potential cost, risk profile (does it inject or reduce risk?), desirability, feasibility, and interaction with other elements of the plan.

6. Determine if change request should be incorporated, delayed, rejected, or closed.

7. For each change request incorporated, document impact to BC/DR plan in detail. Advise change requestor of change acceptance, if appropriate.

8. For each change *incorporated*, determine need for additional training or testing activities. Trigger notification for training, testing, or auditing if appropriate.

9. For each change *delayed*, document reason for delay and how change will be processed later. Communicate decision to change requestor, if appropriate.

10. For each change *rejected or closed*, document reason for denying change. Communicate with change requestor, if appropriate.

11. For all approved changes, make revisions to BC/DR plan, note change in plan, and notify plan stakeholders of plan revision, if appropriate.

BC/DR Plan Audit

You might wonder why you would audit your BC/DR plan if you're also performing training and testing. The plan audit is a process in which you review the BC/DR plan against specific requirements. For example, you may review it against the organization's business practices, objectives, strategies, or changing financial situation. You may also review the plan against external constraints such as legal or regulatory requirements.

The audit does not test the plan. From an audit perspective, there is no assurance that the steps and processes included in the plan will work. The audit does not train people in the use of the plan or in the skills needed to implement and execute the plan. The audit is a more impartial review of the plan to assess whether it meets the company's overall needs. An audit should be performed as a standard project and an audit plan should be created. This plan should include, at minimum:

- Audit scope, timeline, requirements, and constraints
- Review of corporate risks and risk management strategies including BC/DR
- Review of business impact
- Review of BC/DR plan development activities
- Review of BC/DR plan test plans and activities
- Review of BC/DR plan training plans and activities
- Review of BC/DR change management and plan maintenance processes

This assists in maintaining the plan because gaps or weaknesses in any of these processes or activities can be spotted and addressed. Reviewing these elements may result in the generation of change requests that should be processed by the BC/DR team.

Plan Maintenance Activities

There are a number of activities beyond change management that can help you keep your plan up to date and ready to go. We've listed a number of these activities here and of course, if you think of others that will help your team, be sure to add them to the list.

1. If the plan is revised, the BC/DR team members (or those who should have the latest copy of the plan) should be notified in a timely manner.

2. The plan should use a revision numbering system so team members know whether they have the latest version of the plan.

3. Review, update, and revise key contact information regularly. This includes staff, vendors, contractors, key customers, alternate sites and facilities, among others.

4. Create a BC/DR plan distribution list that is limited to authorized personnel but that includes all relevant parties. This distribution list should include off-site and remote facilities that may be used in the event of BC/DR plan activation.

5. Be sure there are up-to-date copies of the BC/DR plan off-site in the event the building is inaccessible.

6. Be sure there are up-to-date paper copies of the BC/DR plan on-site in the event IT systems go down.

7. Implement a process whereby all old versions of the plan are destroyed or archived and new versions replace them. This helps avoid a scenario where team members are working from different versions of the plan.

8. Always check soft copy and remote storage copies of your plan when changes are made to the plan. If you store copies off-site or at your alternate work site, these versions should be updated any time the plan is modified.

9. Whenever significant changes are requested or implemented, test the plan. This will ensure there are no new areas of concern and will help train staff on the changes.

10. Integrate BC/DR considerations into operational processes to reduce plan maintenance efforts in the future.

11. Assign responsibility for managing BC/DR change notification and requests to someone on the BC/DR team. The project management adage that a task without an owner won't get done is especially true here.

12. Document plan maintenance procedures and follow these procedures to avoid introducing additional risk into the project.

13. Incorporate training into the change process so changes to people, process, technology that are incorporated into the BC/DR plan also trigger changes to training plans.

14. Be sure to include BC/DR plan testing, training, auditing, and maintenance activities in your IT or corporate budget for future activities related to BC/DR.

Project Close Out

At this juncture, you should be ready to close out your BC/DR project. If you've been working through each step as you've read through these chapters, you should have a fairly comprehensive and rigorous BC/DR project plan in place. If you decided to read the material, create the project plan, and then initiate project work, you now have a clear roadmap for how to proceed. In either case, the result should be a clear, comprehensive, and reasonable business continuity and disaster recovery plan that should address the major threats to your company and mitigate risks to the most critical business functions. You should have developed procedures related to monitoring change, implementing change, and maintaining the BC/DR plan that can be folded into standard corporate operations to reduce the BC/DR effort going forward.

Now that you've completed work on your plan, you may be ready to launch into a training or testing activity or you may be ready to put the whole project away until the next review period. Regardless of what you decide your next steps are, you should take time to do several project close out activities.

1. Be sure all documentation is complete and finalized.

2. Be sure the BC/DR plan is distributed to appropriate personnel.

3. Announce plan completion to project sponsor and other project stakeholders.

4. Announce plan completion to company to increase awareness and celebrate success.

5. Announce training or testing plans, if appropriate.

6. Hold a project review session to discuss lessons learned and incorporate into process. This should not be held at the same time as a project close out or celebration. This should be a working meeting to capture best practices and lessons learned.

7. Hold project close out meeting to celebrate completion and recognize individual efforts, as appropriate.

8. Complete any staff reviews related to project work.

9. Submit summary or close out report to project sponsor, executive team, or other stakeholders, as appropriate.

10. Update legal or compliance documentation to reflect BC/DR readiness, as appropriate.

11. Set date for next BC/DR audit, review, testing, or training.

Your BC/DR plan will never be perfect and there may be times when it seems it is never complete. However, if you have taken the time and expended the effort to work through the suggestions throughout this book, you should have a solid BC/DR plan that provides a clear roadmap for staff so they know how to keep your business running even when disaster strikes. Along the way, you and your team may have learned a lot more about your company, how it operates, and what contributes to its success. It is our hope that you will never need to find out just how good your plan is and that your efforts will help improve your business operations outside of the realm of disaster readiness.

Summary

Once your BC/DR plan is developed, you need to implement methods for managing change. This includes monitoring changes to the organization that may impact, or be impacted by, the BC/DR plan. Change to the BC/DR plan comes from a variety of sources, including the training, testing, and auditing activities discussed in Chapter 8. Change in IT infrastructure, systems, and processes is the most common organizational change and one that potentially has the biggest impact on the BC/DR plan. Changes in operations can also dramatically impact BC/DR plans. Some operational changes happen slowly over time and may go unnoticed until an audit or plan test. Other changes may be more obvious. In either case, changes to operations should trigger change notifications or change requests. Corporate mergers, acquisitions, spin-offs, and restructuring activities can also have a significant impact on BC/DR plans. In many cases, these changes cannot be anticipated and the BC/DR team may simply have to respond to changes as they occur. Changes in the legal, regulatory, or compliance arenas may trigger mandatory changes to the organization or to the BC/DR plan. These changes should be flagged as required and their impact on the BC/DR plan should be assessed.

Since organizations are always changing, you may find you have more cooperation through creating a change notification process. Since operations staff will implement changes to their processes as they see fit, you may be able to get them to notify the BC/DR team of changes so the team can assess the impact and generate a change request to the BC/DR plan, if appropriate. Another strategy for monitoring change can be to include an additional step in standard operating procedures that includes a quick assessment of the potential impact of activities on the BC/DR plan. People, processes, and technology are ever-changing in organizations and developing easy-to-use processes for monitoring change and the potential impact on the BC/DR plan can assist in plan maintenance.

When change requests are generated, the BC/DR team should have a clear, consistent methodology for evaluating and incorporating change. Not all change requests should be implemented for a variety of reasons. Using established criteria to evaluate change requests will help reduce the risk that changing the plan injects into the process. Factors such as cost, feasibility, desirability, interaction with existing processes, and risk impact should be assessed before changes are accepted. If a change is accepted, it should be incorporated into the plan, the plan should be revised, and plan stakeholders should be notified the plan has been revised. Updated copies of the plan should be distributed appropriately and old versions should be destroyed or archived. If requested changes are delayed or declined, the change requestor should be notified of the decision and the rationale for the decision.

The BC/DR plan should be audited periodically to review it from a business perspective. This audit typically does not evaluate the process, as does a test nor does it necessarily help in training. Its purpose is to review whether the plan meets a stated set of criteria such as business practices, legal, or compliance requirements. Along with testing and training, auditing the plan helps maintain the plan by identifying areas the plan is diverging from business practices or requirements. Should problems be found, change requests should be generated, evaluated, and incorporated as appropriate.

There are numerous plan maintenance activities that can be incorporated into standard operating procedures throughout the organization. In addition, there are various steps you can include in your process to help keep the plan up to date. These include triggers for updating staff rosters, contact information, and vendor lists. Creating a method for notifying the team regarding the availability of a revised plan and processes for updating plans at remote sites, off-site storage locations, among others will help ensure that your most recent version of the plan is available in hard and soft copy, both on- and off-site, to the people who are responsible for implementing the plan, if needed.

Finally, the project should be closed out as you would close out any other project. This might include providing a summary document to your project sponsor and to corporate executives, performing staff evaluations and reviews, notifying the company as a whole that the project was successfully completed, holding a project review session to gather lessons learned, holding a project celebration to recognize individual and team efforts, and most importantly, setting a date for the next BC/DR update, review, audit, or test.

After all your hard work and diligent effort, the best scenario will be that your plan is never implemented. Even though you may not see your plan in action, you may find that the process of creating this plan has improved your knowledge and understanding of your company and perhaps improved some of your company's business processes along the way.

Solutions Fast Track

BC/DR Plan Change Management

☑ Training, testing, IT auditing are three primary ways the BC/DR plan is updated and maintained. Each of these activities may generate change requests that help modify the plan in ways designed to improve the effectiveness of the plan.

☑ Changes in IT are constant and incorporating methods of assessing impact to the BC/DR plan in standard operating procedures will help reduce maintenance efforts.

☑ Changes in operations may happen slowly over time and be almost imperceptible, or they may happen quickly and in obvious ways. Developing a process for change notification within standard operating procedures can help reduce resistance to plan maintenance.

☑ Corporate changes include mergers, acquisitions, and downsizing. Corporate changes often are planned behind closed doors and are then announced. The

BC/DR team must respond to these changes by evaluating the potential impact to the plan.

☑ In many cases, the best approach to plan maintenance is to incorporate an additional step or two in procedures so that the potential impact to BC/DR plans can be evaluated.

☑ Legal regulatory compliance may trigger required changes to the BC/DR plan. These should be flagged for special handling to ensure they are incorporated per requirements.

Strategies for Managing Change

☑ Monitoring change in the organization can be a challenging task. Changes to personnel, processes, and technology create constant flux in organizations.

☑ Developing a change notification process separate from a change request process may reduce resistance to BC/DR plan maintenance activities.

☑ You should develop a methodology for evaluating and incorporating change into the plan. The process should include evaluation criteria and steps for prioritizing, assessing, and incorporating change.

☑ Changes that are incorporated should trigger a plan revision and team notification that a new plan version is available. Changes may also trigger the need for additional testing or training. If so, this should be flagged and appropriate activities should be scheduled.

☑ Changes that are delayed or rejected should be noted and the change requestor should be notified of the decision and the rationale for the decision.

BC/DR Plan Audit

☑ A BC/DR plan audit reviews the plan against business practices, objectives, and strategies.

☑ A BC/DR plan audit does not necessarily test or train the plan. Together, training, testing, and auditing are the three fundamental plan maintenance activities.

☑ Change in financial, legal, or regulatory environment can be spotted during a plan audit and should be addressed appropriately and immediately.

☑ Generate change requests for all changes resulting from audit activities. Determine whether these changes require training or testing.

Plan Maintenance Activities

☑ There are numerous activities beyond change management that can help keep the plan up to date. These include simple steps such as a periodic review of team rosters and contact information, to more complex activities such as developing a change management leader for the BC/DR team who processes change notification and change requests.

☑ Any time the plan is updated, old copies should be destroyed or archived. This includes hard and soft copies both on-site and off-site, as well as copies at alternate work sites and facilities.

☑ Team members should be notified any time there is a revision to the BC/DR plan.

☑ Incorporating various plan maintenance activities into the company's standard operating procedures can assist in keeping the plan up to date.

Project Close Out

☑ All project documentation should be finalized and stored during project close out. This may include reports to the project sponsor, corporate executives, or other stakeholders.

☑ If the plan addresses legal, regulatory, or compliance issues, appropriate steps should be taken to notify stakeholders.

☑ A meeting should be held to review the project and capture lessons learned. This should be held as a separate meeting, not as part of a project celebration.

☑ A project close out or celebration can be held to recognize the efforts of individuals and teams for their work on the project.

☑ The project work should not be closed out until a date is set for reviewing, auditing, training, or testing the plan is scheduled.

Frequently Asked Questions

The following Frequently Asked Questions, answered by the authors of this book, are designed to both measure your understanding of the concepts presented in this chapter and to assist you with real-life implementation of these concepts. To have your questions about this chapter answered by the author, browse to **www. syngress.com/solutions** and click on the **"Ask the Author"** form.

Q: Our plan was developed by a small group of people a couple of years ago. I was recently hired and found a version of the plan. However, I have no idea if the plan is up to date or not and no one seems to know if other version exist. Many of the original team members are gone to other divisions or have left the company. Any suggestions on how to deal with this?

A: The good news is that a plan exists. The bad news is you don't know if you have the latest one or not. First, you need to determine if you have the latest revision. If you've looked around for paper copies, checked various network shares for soft copies, e-mailed previous team members (if possible), and generally asked around, you probably aren't going to find out anything new. If you haven't used these avenues, do so first. Then, if your BC/DR plan calls for an alternate work site or facility, you may check to see if a copy of the plan exists there and compare it to the copy you have. Once you determine that you have the latest revision (or the latest one you can find), you can assemble a BC/DR audit team. Audit the plan and compare it to current business practices, constraints, and so on. This initial audit can help you determine if the plan is mostly up to date or mostly out of date. Then, based on your findings, you can determine the best next steps. If the plan is woefully out of date, it might be easier to start from scratch. If the plan is mostly up to date, you may find through the audit that you can update the existing plan. Once the plan is updated, be sure to schedule a plan test and training.

Q: We spent two years creating this massive BC/DR plan. We even tested it with a paper walk-through last year. There were numerous problems uncovered in that walk-through that show we have a problem not so much with our plan as with our operations. Our operations folks refuse to make necessary adjustments and we've hit an impasse. Can you recommend a course of action for dealing with this problem?

A: The situation you face is all too common in business today. Everyone is busy and very few people recognize the value of a solid BC/DR plan. It's unlikely that you, single-handedly, can sway your operations folks. If you're sure the problem is with operations and not with the BC/DR plan, your best option may be to develop several likely BC/DR scenarios related to the operational changes you believe are needed. Schedule a meeting with the person or people who can authorize these operational changes (assuming you'll be able to convince them) and essentially ask them what would happen to operations if your scenarios played out. If you approach them with the attitude that there is a disconnect and you want to figure out how you might be able to modify the BC/DR plan, rather than the attitude that they're doing something wrong and have to change, you might have greater success. This may seem somewhat disingenuous, but if you consider the fact that these operations folks probably know what they're doing, you may find that the problem actually is with the BC/DR plan and not with operations. Through the course of the meeting, if it becomes apparent that the operations are the problem, you may have success in getting the required changes implemented. Scenarios can be nonthreatening ways of examining problems that help everyone see the problem and potential solutions more clearly. Remain open-minded and understand that you may have to compromise to find a solution that works for the BC/DR plan and for operations. If a solution exists, your best bet in finding it is to work together with the operations folks to find it.

Q: I can't get anyone to agree on next steps. No one will agree to a plan review, an audit, a test—nothing. We've just finished creating the BC/DR project plan and I'm afraid it's just going to start gathering dust on the shelf after all our hard work. Any ideas?

A: It can be difficult to get people to agree to even *more* work on the BC/DR plan after an extended, extensive, or intensive effort. There are several strategies you might use. First, you might hold a project close out meeting to celebrate the success you've had. People can get motivated by recognition for past efforts and this may be all it takes. Another strategy might be to give the project team a rest and put a reminder in your calendar two or three weeks out to gather the team to discuss the next steps. You could let them know at your project close out meeting that you'll be contacting them for a brief follow up meeting in a couple of weeks. If you're concerned the team will scatter to the wind and you won't be able to schedule any of these types of follow up activities, you may need to work with your project sponsor or corporate executives to get a bit of organizational "muscle" behind your effort. Finally, if you've tested your plan and you're confident it's pretty solid, you may choose to set up a review, audit, or test in six months and fight the battle at that time. It's a difficult task to keep people engaged in this process, especially when they all have other jobs and competing demands on their time. Try to determine what's the least possible action needed and then see if you can get a buy-in.

Risk Management Checklist

Risk management includes the three elements of the risk assessment: threat assessment, vulnerability assessment, and impact analysis. This information is the input to the risk mitigation phase that concludes the risk assessment portion of the business continuity and disaster recovery project work.

Risk Assessment

The first step in business continuity and disaster recovery planning is the risk assessment, covered in Chapter 3. The business impact analysis is covered in Chapter 4. Included here are top-level items that should be included. You can modify this list to suit your specific needs. Refer to the specific chapters for detailed information on these topics.

Threat and Vulnerability Assessment

1. Identify all natural threats.
2. Identify all man-made threats.
3. Identify all IT and technology-based threats.
4. Identify all environmental/infrastructure threats.
5. For each threat, identify threat sources.
6. For each threat source, identify the likelihood of occurrence.
7. Based on likelihood of occurrence, assess company's vulnerability to each threat source.
8. Based on likelihood and vulnerability, prioritize list of threats to company.

Business Impact Analysis

1. Based on prioritized list of threats, assess impact of each threat on business operations.
2. Based on threats, perform upstream and downstream loss analysis.
3. Prioritize business functions into mission-critical, important, minor (you can customize categories to suit your needs).
4. For each mission-critical business function, assess the impact of the loss of this function.
5. For each mission-critical business function, assess the impact of various threats to this function.
6. Develop a prioritized list of mission-critical business functions with the highest business impact.
7. For the highest priority functions, identify the recovery time requirements including maximum tolerable downtime (MTD).

8. For business systems, business functions, and IT systems, identify the following: business process criticality, financial impact, operational impact, recovery objectives, dependencies, and work-arounds.

Mitigation Strategies

Risk mitigation strategies are developed after the risk assessment phase is complete. Strategies should be developed based on the mission-critical business functions and the risks to the company. Cost, capability, and recovery times are among the aspects to be considered. IT systems can be included in the risk mitigation strategies or can be addressed as a separate set of strategies. See Chapter 5 for details.

1. For each mission-critical function, identify risk mitigation strategies for consideration including risk acceptance, avoidance, transference, and limitation.

2. For each mission-critical function, identify the recovery requirements and potential recovery options.

3. For each recovery option considered, identify the time, cost/capability, feasibility, service level requirements, and existing controls in place.

4. For each mission-critical option, select the optimal risk mitigation strategy.

5. For IT systems, identify mission-critical IT systems, equipment, and data.

6. For each mission-critical IT component, identify risk mitigation strategies.

7. For each risk mitigation strategy selected, develop implementation plan.

Crisis Communications Checklist

It's likely you'll need more than one type of communication plan. This checklist provides the generic elements to consider and you can modify, as appropriate, for each type of communication plan you need to develop. Therefore, this list refers to a single communication plan but should be used for all communications plans you need to develop.

Remember the three rules of crisis communication:

1. Don't lie.
2. Appoint a single spokesperson.
3. Provide who, what, when, where, why, and how.

Communication Checklist

1. Define communications needs.
2. Develop communication plan objectives based on target audience (employee, customer, media, etc.).
3. Identify and detail triggers for activating the communications plan.
4. Delineate all assumptions related to the need, objectives, and triggers for the plan.
5. Develop distribution list and methodology based on likely communications scenarios (i.e., if e-mail or phones are down, how will information be communicated?). Develop list of distribution alternatives.
6. Develop list of all contacts needed for distribution of this plan.
7. List all legal or regulatory constraints that may impact message or timing of message.
8. Develop communication template to assist in crisis communication situations.
9. Develop message content (see next).
10. Identify message and distribution authorization or escalation channels.
11. Establish distribution channels.
12. Identify frequency of communication.
13. Keep communication log.

Message Content

The template for the message can include specific information that *should* always be conveyed such as corporate commitments, policies, or other data related to the incident. The template also should include areas in which caution is recommended. This might mean not disclosing employee names or home addresses, not releasing names of victims or casualty counts, and so on. Include specific language that can be used as well as a reminder to provide the who, what, when, where, why, and how of the situation.

1. Disaster declaration statement to be communicated to BC/DR team, employees, investors, shareholders, customers, vendors, contractors, as well as community and media contacts.

2. General disaster information including:

   a. Notification and clarification of event

   b. Impact of event

   c. Current status and condition of people, facilities, and equipment

   d. Frequency of updates, estimated time of next update

3. Specific information and instructions for various stakeholders and groups including:

   a. Employees

   b. Vendors, suppliers, contractors

   c. Customers

   d. Business partners

   e. Community and media

   f. Legal and regulatory notification requirements

4. Contact information for additional information (corporate spokesperson or communication team leads, as appropriate).

Business Continuity and Disaster Recovery Response Checklist

This is a basic checklist you can use to identify the primary steps in your response to any serious business disruption or disaster. Modify this checklist to include details pertinent to your company's BC/DR plan. This checklist can be used as a high-level response list and can be used as the basis for developing an action flowchart for response activities. You may choose to refer to additional checklists here to point the teams to more detailed lists in each of the response areas.

Disruptive or disaster event occurs.

1. Initial response.
2. Notification.
3. Problem assessment.
4. Escalation.
5. Disaster declaration.
6. Plan activation.
7. Plan implementation activities and logistics.
8. Disaster recovery phase implementation.
9. Business continuity phase implementation.
10. Resumption and normalization of business activities.
11. Review of event, revision of BC/DR plan based on lessons learned.

Emergency and Recovery Response Checklist

> **!** **WARNING**
>
> Nothing in this book, including information in this appendix, should be construed as legal, medical, or emergency advice. The data provided is for your information only and you should seek appropriate expert advice on these matters.

We discussed the different phases of business continuity and disaster recovery in Chapter 6, including activation, disaster recovery, business continuity recovery, and maintenance/review. In this appendix, we've provided numerous checklists to help you sort through the details. You can use these checklists to help develop your plan and also as appendices to your own BC/DR plan to provide people with step-by-step roadmaps for emergency and recovery responses.

This detail belongs in your BC/DR plan, but breaking it out into sections in this manner will help you process and manage the massive amount of detail required to address these activities properly. Once you've developed your emergency response and business continuity response data, you can (and should) include it in your BC/DR plan.

Activation Checklists

You may find it helpful to develop a variety of checklists, which can be extremely useful in making quick decisions for moving forward. Since you and your team may not have time to rehearse these plans frequently, checklists can help remind you of critical steps to take, regardless of the situation. We've included three short checklists in this section; you can expand upon them as desired.

Initial Response

1. Receive initial notification of possible, impending, or in-progress disruption or disaster.

2. Alert appropriate emergency response organizations (fire, police, etc.), if needed.

3. Access BC/DR plan.

4. Notify and mobilize damage assessment team and the crisis management team.

5. Assess damage, determine appropriate BC/DR activation steps.

6. Notify appropriate BC/DR team members.

7. Prepare preliminary event report or log. Communicate with appropriate parties.

Damage and Situation Assessment

1. Receive initial notification of possible, impending, or in-progress disruption or disaster.

2. Review preliminary event report or log.

3. Assess structural damage, health and safety impact and risks.

4. Determine extent and severity of disruption to operations.

5. Assess potential financial loss.

6. Determine severity based on predefined categories (see categories described earlier in this section)

7. If impact is minor, take no further action and continue to monitor situation.

8. Prepare final assessment and report, notify BC/DR teams of findings.

9. If impact is intermediate or major, declare disaster and update event report or log, communicate with appropriate parties.

Disaster Declaration and Notification

1. Review disaster level assessment, impacts, and other data gathered during initial response phases.

2. Activate BC/DR teams if they have not already been activated.

3. Review recovery options based on disaster assessment.

4. Select best recovery options for the situation, begin plan to implement recovery options (see next phase).

5. Notify management and crisis communications teams.

6. Prepare a disaster declaration statement that can be communicated to employees, BC/DR team and community contacts (see the case study that follows this chapter for more on disaster declaration statements and their dissemination).

7. Monitor progress.

8. Document results in event log, communicate with appropriate parties.

Emergency Response Checklists

There are numerous emergency responses required in the aftermath of an event. This list is not meant to be comprehensive nor should you assume items that may not be on the list are unimportant. In developing your emergency response plans, be sure to utilize local experts including fire, police, and search and rescue teams to provide input on what measures you and your company's employees can reasonably take and which measures should be left to trained experts.

Emergency Checklist One—General Emergency Response

1. Determine the nature and extent of the emergency.
2. Identify whether anyone has been killed or injured.
3. If injuries have occurred, dial 911 to report the emergency or dispatch emergency medical personnel, as appropriate.
4. Determine if any danger still exists. If so, take appropriate precautions or measures to prevent further death, injury, or damage.
5. Notify crisis management team.
6. Dispatch appropriate trained medical personnel to assist with triage or to manage the situation until emergency responders arrive.
7. Notify civil authorities regarding the nature and extent of the emergency.
   - Police
   - Fire
   - Search and rescue
   - Hazardous Materials Team
8. Notify corporate executives.

Emergency Checklist Two— Evacuation or Shelter-in-Place Response

1. Install, identify, and/or test alarms and emergency signals.
2. Identify parameters that would trigger building or facility evacuation procedures.
3. Identify parameters that would trigger shelter-in-place procedures.
4. Identify evacuation/shelter leaders.
5. Identify evacuation routes and assembly points.
6. Identify building search-and-rescue procedures.
7. Identify procedures for securing and shutting down facility.
8. Identify shelter-in-place procedures and internal assembly points (safe areas).
9. Identify method of ascertaining if anyone is missing or unaccounted for.

10. Identify communication methods and frequency following an evacuation or shelter-in-place.

11. Identify provisions needed for shelter-in-place (food, medical supplies, communications equipment, etc.).

Emergency Checklist Three— Specific Emergency Responses

Develop specific step-by-step emergency response checklists for highest risk threats. These will utilize many of the same steps as other responses but should be tailored to these events to provide consistent and fast response procedures for staff.

1. Fire—internal or external.

2. Flood.

3. Earthquake.

4. Hazardous materials spill control.

Emergency Checklist Four— Emergency Response Contact List, Maps, Floor Plans

1. External emergency contact numbers:

    • Police, sheriff

    • Fire

    • Hospital

    • Ambulance

    • Other

2. Emergency response team contact numbers:

    • Emergency response team leader

    • Medical staff

    • Evacuation or shelter-in-place leaders

    • Search and rescue staff

    • Crisis team manager and/or corporate executive contact

3. Maps:

    • Evacuation routes and assembly areas

- Shelter-in-place assembly areas
- Escape routes from site—primary and secondary (may need several options depending on disaster scenario)
- Floor plans
- Location of fire doors, fire extinguishers
- Location of utility closets, circuit breaker panels, power lines
- Location of gas, electric, water lines
- Location and nature of hazardous materials

Emergency Checklist Five— Emergency Supplies and Equipment

Depending on the size of your company, the location of the facilities, and the nature of the business, you may need other supplies than those listed. Be sure to develop a list of supplies and equipment needed, a schedule for testing needed equipment on a periodic basis, a procedure for performing periodic maintenance on equipment, and a process for performing a periodic inventory count of supplies.

1. First aid supplies (portable kits, additional supplies).
2. CPR training and equipment.
3. Fire suppression equipment (fire extinguishers, etc.).
4. Hazardous materials safety equipment.
5. Hazardous materials containment and clean up equipment/supplies.
6. Water, water purification tablets, shelf-stable food supplies (for shelter-in-place).
7. Clothing, blankets, and other materials (injuries, cold climates, shelter-in-place).
8. Emergency communications equipment (walkie talkies, batteries, etc.).

Recovery Checklists

The recovery checklists are broken out into numerous separate lists. Modify these lists to suit your organization's individualized needs.

Recovery Checklist One—General

1. Perform a quick assessment to determine which members of the BC/DR team are available to assist with recovery activities.

2. Identify any travel needs for BC/DR team members (if some are coming from other sites or locations). Be sure to consider the need for local transportation and lodging as well.

3. Identify who will be working at the original site and who will be working at the alternate site (if applicable).

4. Identify resources required including computer equipment, communication links (Internet, dial up, etc.), communications equipment (walkie talkies, cell phones, land lines, etc.), office equipment, office supplies, BC/DR plans, contact lists, and inventory lists.

5. If needed, arrange for access to site or alternate site for vendors, contractors, or employees traveling in from other locations.

6. Notify and activate alternate work site and/or crisis communication command center. Distribute contact information including location, personnel, and phone numbers to key personnel including management, BC/DR team, crisis management team, and HR as appropriate.

7. Provide local contact information and chain of command information (who should people contact for various recovery needs?).

8. Order replacement computer hardware, software, data and voice communications equipment.

9. Locate configuration information and most current backups.

10. Order faxes, printers, routers, cabling, copiers, tapes, tape backups, disk drives.

11. Order forms used in normal course of business. Develop forms needed for recovery operations if they do not already exist.

12. Ship key documents to alternate site.

13. Order stationery, business cards, other business-specific printed matter, if applicable.

14. Prepare process for receiving, tracking, and dispensing equipment and supplies.

15. Prepare process for receiving and tracking data backups and critical records.

16. Finalize preparations for restoring site or activating alternate site.

17. Document results in event log, communicate with appropriate parties.

Recovery Checklist Two— Inspection, Assessment, and Salvage

1. Provide damage assessment team with inventory or list of critical resources at damaged site.

2. Ensure all team members have proper safety equipment and have been trained in or reminded of their proper use.

3. Ensure all team members are aware of proper safety procedures and guidelines.

4. Provide team members with forms or process for assessing and reporting damage.

5. Inspect building, utilities (gas, electric, water).

6. Inspect for hazardous materials, chemicals, or hazardous conditions.

7. Inspect resources and vital records for damage including water, fire, water, dust, ice, or physical damage (crushed, tipped over, etc.).

8. Determine potential for further damage or hazard.

9. Determine potential for salvage and restoration.

10. Determine any timelines that may be relevant (equipment sitting in water, operating in extreme heat or cold, etc.) to the salvage operation and to prevent further damage and deterioration.

11. Record assessments in event log.

12. Acquire salvage and restoration equipment, as needed.

13. Remove hazardous materials, as appropriate.

14. Relocate equipment, records, and other salvaged resources, as appropriate.

15. Perform restoration, as appropriate.

16. Document results in event log, communicate with appropriate parties.

WARNING

Any items on the list may be performed by outside contractors, including those specially trained and certified in handling hazardous materials, chemical spills, and so on. Listing these items here does not imply that your team should perform these tasks, simply that they should be performed by appropriately trained personnel

Business Continuity Checklist

The business continuity phase follows the disaster recovery phase and is focused on resuming business operations. Operations are not normalized or fully restored during this phase but initial business operations, including those deemed mission-critical, are initiated during this phase. At the end of the business continuity phase, normal business operations should resume, which signals the transition out of the BC/DR plan and back into normal operations.

Resuming Work

Work may be resumed on a limited basis in the original building or work location if it can be occupied after the disruption or disaster. If not, an alternate work site (AWS) should have been set up during the disaster recovery phase if this is part of your BC/DR plan. Once set up, the AWS should be brought online so that employees can begin work. Work activities typically resume on a limited basis. The restoration of the original facilities or a decision to permanently use alternate facilities will trigger the move to normalization of operations.

Resuming Work

1. Receive notification that work site is fully set up (disaster recovery phase end point).

2. Ensure all employees are aware of work location (original site or alternate site).

3. Ensure all employees have equipment, tools, supplies, and resources needed to begin limited resumption of work.

4. Check that computer networks, user computers, and other IT resources are installed, configured, tested, and ready for users.

5. Test communications equipment including phone, Internet access, wireless connectivity, and the like.

6. Provide employees with appropriate site access.

7. Review BC/DR plan to understand which mission-critical functions should begin, in what order tasks should be started, and what dependencies exist.

8. Review BC/DR plan to review maximum tolerable downtime (MTD) and other key recovery metrics.

9. Develop plan for resuming operations based on outcome of review.

10. Identify areas where manual or work-around methods will be implemented.

11. Identify methods for tracking and managing all manual or work-around procedures that are not part of the standard operating procedures.

12. Identify backlogs that may be created as a result of partial resumption of services. Determine if these backlogs are acceptable and if so, how they will be managed once normalization begins.

13. If backlogs are not acceptable, determine what other systems, processes, or procedures must be put into place to avoid backlogs.

14. Determine the status of elements required to avoid backlog, develop plan to put needed elements in place before resuming activities.

15. Resume limited operations.

16. Monitor results.

17. Begin backup procedures to protect new data or new work product(s).

18. Develop status report for crisis management team.

Human Resources

1. Ensure Human Resources (HR) has accounted for all employees. Appropriate actions should be underway if any employees were killed or injured in the event.

2. Take appropriate measures to ensure HR data is available and has been updated.

3. Begin reviewing personnel issues to resolve problems that stem from the disaster or disruptive event including medical or counseling services, insurance issues, and financial issues.

4. Work with department heads to determine if positions need to be filled.

5. Work with department heads to determine if contractors or temporary workers are needed to assist in the restoration or resumption of work activities.

6. Review and implement payroll process. Distribute updated payroll information to all employees.

7. Develop status report for crisis management team.

Insurance and Legal

1. Review insurance and notify insurance carrier.

2. Conduct internal assessment of damage and potential insurance coverage.

3. Identify potential insurance gaps or language that may limit claim.

4. If necessary, contact legal counsel for advice and guidance regarding the disaster event (insurance, other liability, regulatory issues, etc.).

5. Provide copies of event logs, damage assessments, and other pertinent documentation to insurance carrier.

6. Submit appropriate paperwork regarding loss.

7. Prepare appropriate documentation for legal review or regulatory compliance.

8. Develop status report for crisis management team.

Manufacturing, Warehouse, Production, and Operations

Whether or not you move into an alternate work site, you will need to address issues with manufacturing, warehouse, production, or other operations that normally took place at your original site. If you've made arrangements for alternate sites for these functions, you should modify the list presented earlier ("Resuming Work") to reflect the specific needs of your manufacturing, warehouse, production, or operations in addition to the tasks listed here.

1. Inspect work site (original or alternate) to determine suitability for resuming operations on a limited basis.

2. Review equipment inventory list to ensure all needed equipment is present and operational.

3. Review materials inventory list to ensure all needed materials are available in sufficient quantities to resume operations on a limited basis.

4. Review manual or work-around methods for managing and tracking inventory, production output, and other data if needed IT systems are not back online.

5. Determine the method of tracking and managing all manual or work-around methods until systems come back online. Determine how backlog of data will be addressed once systems come back online.

6. Obtain backup data to determine status of previous, pending, and new orders in the system. Review inventory and shipping data to determine the status of all orders.

7. For all open orders, determine priority and status of each. Develop prioritized order list to determine manufacturing, production, or operational priorities with the understanding that work will resume on a limited basis.

8. Set up and test operations on a trial basis to determine if quality, quantity, and specifications for production are acceptable. If not, take corrective action.

9. If trial run is successful, notify key customers or clients of updated status and timeline for delivery.

10. If trial run is successful, begin operations on a gradually increasing basis. Create checkpoints at all critical areas to ensure production meets requirements at each step. Take corrective action as needed.

11. Identify transportation and shipping options at original or alternate site.

12. Begin shipping, receiving, and managing inventory as appropriate.

13. Develop status report for crisis management team.

Resuming Normal Operations

1. Review assessment from HR regarding status of all employees.

2. Review assessments from damage assessment team, crisis management team, and other teams related to the current status of the facility.

3. Review assessments from CIRT or IT team regarding current status of IT systems.

4. Review assessments from manufacturing, warehouse, production, and operations regarding current status.

5. Review building damage assessment reports, determine feasibility and desirability of returning to original facility.

6. If returning to facility, develop project plan for repairing damage. Develop scope, budget, and timeline for return.

7. If not returning to facility, determine options to locate and occupy new facility. Develop costs and alternatives.

Existing Facility

1. If staying in existing facility, get bids for repairs from contractors.

2. Select contractor, initiate and supervise repairs.

3. Obtain appropriate permits for occupancy.

4. Notify insurance company regarding facility. Update policy, as appropriate.

5. Develop list of equipment, supplies, furniture, and other resources needed for resumption of business in original facility.

6. Purchase and obtain necessary equipment, supplies, furniture.

7. Install IT infrastructure components including LAN cables, servers, routers, firewalls, and such.

8. Install communications equipment (phone lines, Internet access).

9. Install furniture.

10. Install and test computers, workstations, printers, faxes, copiers, and other office equipment.

11. Install and test building access and security measures.

12. Distribute necessary supplies (paper, pens, business cards, etc.).

New Facility

1. If staying in alternate facility, review existing contracts for suitability, contact appropriate representatives to negotiate new contract/arrangement for permanent occupancy.

2. Contact legal representative to review any new, modified, or updated contracts.

3. Obtain appropriate permits for occupancy.

4. If locating new facility, work with real estate professional (if desired) to locate suitable facility.

5. Negotiate for facility including tenant improvements and other improvements or modifications, as needed. Sign lease or contract after appropriate legal review.

6. Notify insurance company regarding facility. Update policy, as appropriate.

7. Develop list of equipment, supplies, furniture, and other resources needed for resumption of business in facility if such resources are not already in place and available.

8. Purchase and obtain necessary equipment, supplies, furniture.

9. Install any needed IT infrastructure components including LAN cables, servers, routers, firewalls, and such that are not already in place at AWS.

10. Install any additional communications equipment (phone lines, Internet access), as needed.

11. Install furniture as needed.

12. Install and test any additional computers, workstations, printers, faxes, copiers, and other office equipment, as needed.

13. Modify and test building access and security measures, as needed.

14. Distribute necessary supplies (paper, pens, business cards, etc.), as needed.

Transition to Normalized Activities

1. Determine appropriate timeline to transition to normalized activities.

2. Notify all BC/DR team members of schedule and tasks for transition.

3. Notify all department heads of timeline for transition.

4. Identify all operational concerns or constraints regarding transition.

5. Freeze production environment at alternate locations.

6. Perform full data backups of all critical data and vital records.

7. Ship all backups and critical records to original or new location ("location" from hereon).

8. Transfer all needed personnel, equipment, machinery, equipment, and supplies to location.

9. Restore and test systems at location.

10. Verify all business systems including IT, manufacturing, production, communications, and such are installed and functional at location.

11. Redirect network traffic, communications traffic (phone lines, voicemail) to location.

12. Provide appropriate physical (building) and logical (IT, network) access to employees.

13. Resume normal activities.

14. Initiate normal data and vital records backup routines.

15. Clean up and close down alternate sites according to contractual agreements.

16. Perform post-disaster review to compile and discuss lessons learned, mistakes made, and improvements found during the event.

17. Modify BC/DR plan based on outcome of post-disaster review.

IT Recovery Checklists

The tasks needed to recover IT systems are probably quite familiar to you, but they should be delineated within your BC/DR plan. Each subteam should have a clear set of guidelines and procedures for how and when they will perform their work. Be sure to note dependencies within the checklist so that teams don't work at cross-purposes. You can add items to the checklist as checkpoints for these purposes, much like milestones are used in project plans.

We've included items related to recovering office work space and business operations in this section because they are intertwined with IT recovery efforts. You can reorganize these checklists to suit your approach to BC/DR.

IT Recovery Checklist One—Infrastructure

1. Review BC/DR team member assignments, ensure all team members are present or accounted for.

2. Convene brief planning meeting to ensure all team members understand the situation, the recovery options selected, the requirements and other constraints.

3. Provide all team members with updated contact information (if appropriate) and chain of command for problem notification and escalation.

4. Ensure all team members have inventory lists, equipment purchase order or shipment information, and that they understand recovery procedures moving forward.

5. Review equipment at alternate site (if used) or at main facility (if used). Ensure all equipment needed for selected recovery option is appropriate and meets requirements.

6. Review procedures for receiving, tracking, and testing IT equipment.

7. Receive backups from storage facility or confirm online availability of backups.

8. Inspect and test backup media, if appropriate.

9. Review or develop floor plans for replacement equipment including IT systems, communications equipment, and infrastructure components.

10. Review network diagram to verify location and connectivity of infrastructure components such as routers, switches, hubs, and gateways.

11. Review network addressing scheme, system configuration data, and security configuration data.

12. Connect all IT components to network.

13. Run procedures to configure infrastructure components.

14. Configure or restore security settings and security devices including firewalls, gateways, and routers.

15. Restore network servers and other critical equipment via backups.

16. Redirect data and voice traffic to alternate location, if appropriate. If you are restoring at the original site, ensure data and voice traffic are properly routed and working.

17. Provide network access to designated employees at alternate site.

18. Test and verify all network connectivity and security settings.

19. Document results in event log, communicate with appropriate parties.

Recovery Checklist Two—Applications

1. Review recovery procedures for critical applications. Verify needed servers are restored and online, as appropriate.

2. Review mission-critical data to determine which applications should be restored first.

3. Review internal and external data or application dependencies. Take action, as appropriate, to ensure all dependencies are addressed in the correct order and timing.

4. Review security settings—acquire passwords or reset passwords, as needed.

5. Restore, configure, and verify operating systems if not already performed.

6. Restore, configure, and verify applications.

7. Restore application data from backups, as appropriate. Ensure data is the most current available.

8. Verify integrity of data and functionality of applications.

9. Notify key users of application availability. Inform users of procedures to address data backlogs, if appropriate.

10. Document results in event log, communicate with appropriate parties.

Recovery Checklist Three— Office Area and End-User Recovery

1. Review teams and check to ensure team members are present or accounted for.

2. Review MTD and other constraints to ensure compliance with recovery requirements.

3. Verify that team members have necessary alternate work space inventory lists.

4. Review equipment at alternate location, determine if it meets recovery requirements. Note any discrepancies or gaps.

5. Review or revise material receiving, inventory management, and distribution procedures so when new equipment and supplies arrive, they can be properly managed.

6. Review floor layout for alternate work space, determine location of office furniture and equipment including copiers, file cabinets, bookcases, and printer stands.

7. Review network diagram and connectivity. Ensure office layout accommodates existing network, communication, and power connection points. Modify as needed.

8. Receive and set up office furniture per plan. Assign work areas to team members.

9. Receive and set up computers, workstations, printers, and other IT-related user equipment.

10. Set up copiers, faxes, network printers, and telephones at designated locations.

11. Provide office supplies to team members as needed.

12. Provide documents, manuals, and other materials that may have been stored at and retrieved from an off-site storage facility.

13. Reroute voice and data communications to alternate work location. Notify key personnel of current location, contact information, and status.

14. Ensure connectivity to key servers, applications, and data.

15. Set up help desk or customer service function at alternate location.

16. Document results in event log, communicate with appropriate parties.

Recovery Checklist Four—Business Process Recovery

1. Verify user workstations, desktop, and laptop computers are restored and have access to necessary network resources.

2. Ensure all key personnel or designated users have usernames and passwords for alternate site access.

3. Complete workstation, desktop, or laptop restoration, as needed.

4. Retrieve critical records and forms from storage, if applicable.

5. Receive and process new transactions manually until transactions can be handled electronically.

6. Verify integrity of data on restored systems. When tests are completed satisfactorily, transition to processing transactions electronically.

7. Identify work backlog and implement processes to address backlog to enter data into systems.

8. Begin using restored systems for new transactions.

9. Begin data backup procedures to protect new data being entered into recovered systems.

10. Document results in event log, communicate with appropriate parties.

Recovery Checklist Five—
Manufacturing, Production, and Operations Recovery

1. Review maximum downtime and other constraints.

2. Assemble manufacturing, production, or operations recovery team (called "operations" from hereon).

3. Tour alternate operational areas to assess status or tour original operational areas to assess current damage and status. Review safety requirements against current status.

4. Review environmental conditions including heating/cooling, humidity, or dust levels, air filtration status (dust, odors, airborne contaminants, etc.). Determine if current condition and status meet operating requirements.

5. Inspect any stored hazardous materials or chemicals for safety.

6. Inspect and test, as appropriate, all safety devices including fire extinguishers, smoke detectors/alarms, emergency lighting, among others.

7. Verify sufficient electrical (or other power) exists to run machinery and equipment.

8. Verify teams have alternate facility operating procedures and inventory lists.

9. Review equipment against inventory lists and operating requirements. Address any gaps or discrepancies.

10. Review and revise, as needed, equipment receiving, inventory management, and equipment distribution procedures. Ensure that equipment and inventory arriving at the alternate (or damaged original) location is tracked and monitored.

11. Receive any critical equipment, parts, supplies, or materials from off-site storage, vendor shipment, or salvage from original location.

12. Receive and inspect any salvageable or existing inventory. Assess status, dispatch inventory as appropriate (destroy, store, repackage, reuse, etc.).

13. Review floor layout for manufacturing, production, or operational activities. Ensure proper connections including power, data, or network exist in the proper locations.

14. Place equipment in locations and install, connect, and test.

15. Install and test auxiliary equipment including printers, copiers, telephones, walkie talkies, radios, and other equipment needed for operations.

16. Provide operational and configuration documentation to team leaders or equipment operators, as appropriate.

17. Install and configure any IT-related equipment including interfaces, workstations, desktops, and such.

18. Set up connectivity between operations and IT systems at alternate locations, as needed.

19. Test equipment, machinery, and configurations.

20. Test output of operations for quality, quantity, and other required attributes.

21. Test voice and data network to ensure connectivity.

22. Ensure operators have information needed to begin production including logins, passwords, keys, or other necessary tools.

23. Review and implement any manual work-arounds for production or inventory management needed.

24. Review and implement any electronic production or inventory management procedures, as needed.

25. Begin manufacturing, production, or operations on limited basis.

26. Test and verify output. Expand or increase production as warranted.

27. Document results in event log, communicate with appropriate parties.

Training, Testing, and Auditing Checklists

Business continuity and disaster recovery training can be accomplished through testing the plan. Testing the plan results in training participants, therefore they are referred to as one activity here. Refer to Chapter 8 for details on training and testing activities.

Training and Testing

1. Identify scope, timeline, and requirements for training.

2. Determine training needs for each participant group (ERT, CMT, damage assessment, IT, etc.).

3. Develop training approach (may use testing methods for training, see item 6).

4. Develop training objectives.

5. Develop training or testing duration and cost estimates.

6. Develop training or test scenarios.

7. Develop training or testing method (paper walk-through, functional, or field exercises, full interruption).

8. Develop training or testing evaluation criteria.

9. Identify training or testing participants.

10. Identify training or testing resources needed.

11. Deliver training or conduct testing.

12. Evaluate training or testing based on evaluation criteria.

13. Collect and analyze lessons learned.

14. Revise training, testing, or BC/DR plan, as appropriate.

IT Auditing

1. Identify IT risk mitigation strategies selected.

2. Audit IT risk mitigation strategies to ensure they have been properly implemented and configured.

3. Audit IT systems to ensure systems identified in BC/DR plan are still in place and functioning.

4. Identify new technology implementations (planned or in progress) and assess against BC/DR objectives. Recommend revisions to technology plans or BC/DR plans as appropriate.

5. Identify technology to be replaced or decommissioned (planned or in progress) to assess the impact on BC/DR plans. Recommend revisions to technology plans or BC/DR plans as appropriate.

6. Audit all processes in BC/DR plan related to IT systems to ensure steps, processes, requirements, tools, supplies, and resources identified are still accurate, current, relevant, and complete.

7. Audit IT response team to ensure team is intact, ready to respond, has clear understanding of roles/responsibilities, has tools/resources to implement plan.

8. Audit existing systems to ensure compliance with current BC/DR plans including:

   - Operating systems
   - Networking and telecommunications equipment
   - Database and applications
   - Systems backups
   - Security controls
   - Integration and testing
   - Other (define)

BC/DR Plan Maintenance Checklist

Training, testing, and auditing are three activities that generate useful information about the BC/DR plan and therefore contribute to plan maintenance. Change management and BC/DR plan audits also contribute to keeping the plan up to date. Modify the following list to meet the requirements of your organization.

Change Management

1. Review contact list. Update and revise as needed.

2. Review vendor list. Update and revise as needed.

3. Review vendor contracts. Update, extend, revise as needed.

4. Review team membership (ERT, CMT, CIRT, etc.). Update and revise as needed.

5. Review team membership changes. Assess training needs.

6. Develop, document, and implement formal BC/DR plan change management processes:

    a. Monitoring changes that impact or are impacted by BC/DR plan.

    b. Evaluating change notifications and requests.

    c. Implementing appropriate changes to BC/DR plan.

    d. Testing, training and auditing revised plan.

    e. Notifying stakeholders of changes incorporated, delayed, or denied.

    f. Revising BC/DR plan appropriately.

    g. Distributing updated copies of the BC/DR plan to appropriate parties.

7. Review lessons learned from training, testing, and auditing. Assess impact to BC/DR plan, revise plan as needed.

8. Review changes to IT systems and processes. Assess impact to BC/DR plan. Make changes as needed.

9. Review changes to operations, including mission-critical business processes and functions. Assess impact to BC/DR plan, revise plan as needed.

10. Review changes to corporation including mergers, acquisitions, spin-offs, downsizing, and so on. Assess impact to BC/DR plan, revise plan as needed.

11. Review and revise risk assessment. Perform subsequent planning steps (impact analysis, risk mitigation, training, testing) to update BC/DR plan.

12. Update flowcharts and checklists, as needed.

13. Distribute revised plans to distribution list. Notify appropriate parties of the revised plan as well as how to obtain it and how to dispose of the outdated copies of the plan.

14. Destroy or archive old copies of the plan including hard and soft copies, on- and off-site copies, and copies that may be stored with trusted vendors, partners, or at alternate work sites or facilities.

15. Perform periodic audit BC/DR plan, incorporate recommendations and changes.

16. Perform periodic test of BC/DR plan, incorporate recommendations and changes.

17. Perform periodic training of BC/DR plan, incorporate recommendations and changes.

Index